Springer Series in

OPTICAL SCIENCES 126

founded by H.K.V. Lotsch

Springer Series in
OPTICAL SCIENCES

The Springer Series in Optical Sciences, under the leadership of Editor-in-Chief *William T. Rhodes*, Georgia Institute of Technology, USA, provides an expanding selection of research monographs in all major areas of optics: lasers and quantum optics, ultrafast phenomena, optical spectroscopy techniques, optoelectronics, quantum information, information optics, applied laser technology, industrial applications, and other topics of contemporary interest.
With this broad coverage of topics, the series is of use to all research scientists and engineers who need up-to-date reference books.

The editors encourage prospective authors to correspond with them in advance of submitting a manuscript. Submission of manuscripts should be made to the Editor-in-Chief or one of the Editors. See also www.springeronline.com/series/624

H. Schneider H.C. Liu

Quantum Well Infrared Photodetectors

Physics and Applications

With 153 Figures and 8 Tables

Springer

Dr. Harald Schneider
Head, Semiconductor Spectroscopy Division
Institute of Ion-Beam Physics and Materials Research
Forschungszentrum Rossendorf
PO Box 510119, 01314 Dresden, Germany
E-mail: h.schneider@fz-rossendorf.de

Fraunhofer-Institute for Applied Solid State Physics
Tullastr. 72, 79108
Freiburg, Germany

Dr. Hui Chun Liu
Quantum Devices Group Leader
Institute for Microstructural Sciences
National Research Council
Ottawa K1A 0R6, Canada
E-mail: h.c.liu@nrc.ca

ISSN 0342-4111

ISBN-10 3-642-44780-5 Springer Berlin Heidelberg New York

ISBN-13 978-3-642-44780-8 Springer Berlin Heidelberg New York

Springer is a part of Springer Science+Business Media.

springer.com

Softcover re-print of the Hardcover 1st edition 2007

Typesetting by the authors and SPi using a Springer LaTeX macro package
Cover concept by eStudio Calamar Steinen using a background picture from The Optics Project. Courtesy of John T. Foley, Professor, Department of Physics and Astronomy, Mississippi State University, USA.
Cover production: *design & production* GmbH, Heidelberg

Printed on acid-free paper SPIN: 11767657 57/3100/SPi 5 4 3 2 1 0

Dedication: To Carmen, Maria, and Alexander,
who make our lives most enjoyable

Preface

This book discusses the physics and applications of quantum well infrared photodetectors (QWIPs). The presentation is intended for both students as a learning text and scientists/engineers as a reference. We have assumed that the reader has a basic background of quantum mechanics, solid-state physics, and semiconductor devices. To make this book as widely accessible as possible, we have attempted to treat and present the materials in a simple and straightforward fashion. We chose the topics for the book by the following criteria: it must be well established and understood, and it should have been or potentially will be used in practical applications. We have addressed most aspects relevant for the field but have omitted at the same time detailed discussions of specialized topics such as the valence band quantum wells.

Our book is organized in two parts. Physics (Part I) begins with the basics of infrared detection (Chap. 2), valid for any photon detector (of course, the reader who is familiar with the general concepts of photon detectors may skip this chapter). Chapter 3 covers a broad theoretical background on semiconductor quantum wells and intersubband transitions, which will be referenced in the following chapters (the reader may skip Chap. 3 at the beginning). Chapter 4 on photoconductive QWIPs, continued by its further development, the photovoltaic QWIP (Chap. 5), is considered as the central part of the book – relevant for all the following chapters. Chapter 6 is devoted to optical coupling, which is crucial for practical implementation. The chapters on miscellaneous effects (Chap. 7) and related devices (Chap. 8) are intended for scientists interested in further developments, which the reader may choose to skip (without affecting the rest of the book).

"Applications" (Part II) starts with thermal imaging (Chap. 9), the main application field of QWIPs. After some general considerations on thermal imaging using QWIPs, we showcase typical fabrication steps of QWIP arrays, performance data of some QWIP thermal imagers, and their practical use in fields like medicine, environmental sciences, and quality control. Finally, Chap. 10 covers the second application field, i.e., ultrafast and heterodyne detection. This field is still mostly in the stage of active research, but might

Fig. 0.1. The authors in Freiburg (QWIP thermal image)

soon enter the application phase since demands have recently emerged for ultrafast mid-infrared and terahertz detectors.

We thank our colleagues and coworkers for their contributions, especially E. Diwo, S. Ehret, J. Fleissner, H. Güllich, N. Herres, L. Kirste, P. Koidl, T. Maier, R. Rehm, K. Schwarz, C. Schönbein, S. Steinkogler, M. Walther, and G. Weimann of Fraunhofer Institute for Applied Solid State Physics (Freiburg, Germany), W. Cabanski, J. Ziegler, and H. Korf of AIM Infrarot Module GmbH (Heilbronn, Germany), and G.C. Aers, B. Aslan, M. Buchanan, M. Byloos, R. Dudek, E. Dupont, P.D. Grant, H. Luo, P.J. Poole, A.J. SpringThorpe, C.Y. Song, and Z.R. Wasilewski of National Research Council Canada (Ottawa, Candada). We also appreciate the careful reading of the manuscript by M. Buchanan and A. SpringThorpe of National Research Council Canada and by M. Helm of Forschungszentrum Rossendorf (Dresden, Germany). HCL thanks the Alexander von Humboldt Foundation for the Friedrich Wilhelm Bessel Award which provided the opportunity of an extended stay in Freiburg (see Fig. 0.1) where most of the book was written.

Freiburg and Dresden, Germany, *Harald Schneider*
Freiburg, Germany, and Ottawa, Canada, *H.C. Liu*

Contents

Symbols

A	normalization and detector area
C	device capacitance
c	speed of light
$D^\star$	detectivity (subscripts SL, BL (or blip), s, gr, and det for signal limited, background limited, shot noise, generation-recombination, and detector limited, respectively)
D_L	lens diameter
E	photon energy or electron total energy
E_act	thermal activation energy
E_ex	exchange energy
E_f	Fermi energy
E_g	bandgap energy
E_n	quantum well eigenenergy ($n = 1$ – ground, $n = 2$ – first excited state)
ΔE	broadening half width
ΔE_C	conduction band offset
e	elementary charge
F	electric field
$F_\#$	optical system f-number
f	oscillator strength
f_B	Bose–Einstein function
f_het	heterodyne frequency
f_IR	frequency of infrared light
f_L	lens focal length
f_max	lifetime-limited maximum frequency
$f_{\mu\mathrm{wave}}$	microwave frequency
Δf	measurement bandwidth
g	gain (g_photo – photoconductive, g_noise – noise gain)
$\mathcal{H}_{\nu,\Omega}$	radiance – power per unit area and steradian
h	Planck constant
$\hbar$	reduced Planck constant

I	current
I'	first derivative of I–V
I''	second derivative of I–V
I''_0	second derivative of I–V in low frequency limit
I_{dark}	dark current
I_{photo}	photocurrent
I_{rect}	rectified current
I_S	signal current
$\mathcal{I}_\nu$	irradiance – total power per unit surface area
i_{em}	emission current ($i^{(1)}_{\text{em}}$ emission current from one well)
i_{het}	heterodyne current
i_{mix}	mixing current
$i_{\text{n,dark}}$	dark current noise spectral density
$i_{\text{n,J}}$	Johnson noise current spectral density
i_{p1}, i_{p2}, i_{p3}	photocurrents from individual one-color QWIP in a multicolor stacked structure
$i^{(1)}_{\text{photo}}$	direct photoemission current from one well
J_{dark}	dark current density
j_{em}	emission current density from well
j_{3D}	three-dimensional current density in barrier
k_{B}	Boltzmann constant
$\boldsymbol{k}_{xy}$	in-plane wavevector
l_{c}	coherent mean free path
L	inductance
$\mathcal{L}$	length
L_{b}	barrier thickness
L_{p}	length of one period
L_{w}	well width
$L_{\text{B}}(\lambda)$	background photon irradiance
m	effective mass in well
m_{b}	effective mass in barrier
m_{e}	electron rest mass
m^*	reduced effective mass in well
N	number of wells
N_{D}	two-dimensional doping density
N_{N}	noise electron number
N_{S}	signal electron number
N_{2D}	total escaped electron density
NEP	noise-equivalent power (subscripts SL, BL, s, and gr for signal, background, shot noise, and generation-recombination noise limited, respectively)
NETD	noise-equivalent temperature difference
n, n'	wavefunction indices
n_{r}	refractive index

n_{ex}	number of photoexcited electrons in upper subband
n_{2D}	two-dimensional electron density in well
n_{3D}	three-dimensional electron density in barrier
$\overline{n_{\mathrm{det}}}$	average number of absorbed or detected photons
$\overline{n_{\mathrm{x}}}$	average number of electrons collected in external circuit
P_{B}	background power
P_{E}	power
P_{het}	heterodyne signal power
P_{IR}	infrared power
P_{mix}	mixing signal power
P_{out}	microwave output power
P_{S}	signal power
$P_{\mu\mathrm{wave}}$	microwave power
$P_{\mathrm{v,lens}}$	radiation power incident onto detector
p_{c}	trapping or capture probability
p_{e}	escape probability
Q_{A}	Equivalent noise charge of analog-to-digital converter
Q_{B}	accumulated charge due to background
R	device dynamic resistance
R_{L}	transmission line impedance
R_{s}	series load resistor
$\mathcal{R}$	responsivity ($\mathcal{R}_i$ – current responsivity)
r_1, r_2, r_3	dynamic resistances of individual one-color QWIP in a multicolor stacked structure
S_i	spectral density of the mean square noise current
T	temperature
T_{blip}	temperature for background limited infrared performance
$\mathcal{T}$	transmission probability
V	effective barrier height
V_{b}	barrier height at zero bias
V_{μ}	amplitude of microwave voltage
$V_{\mu 0}$	amplitude of microwave voltage in low frequency limit
V_1, V_2	applied voltages
v	drift velocity
v_{het}	heterodyne signal voltage
v_{sat}	saturated drift velocity
v_1	velocity in first excited state
$v_{\mu\mathrm{wave}}$	microwave voltage
$\mathcal{V}$	normalization and detector volume
x	Aluminum mole fraction
z_{c}	classical turning point
α	absorption constant
α_{n}	source of noise
$\beta_1(\omega)$, $\beta_2(\omega)$	intrinsic and circuit frequency roll-off factors
γ	half-width at half maximum

Γ	full width at half maximum
ϵ_0	vacuum permittivity
ϵ_r	relative (ϵ_S – static, ϵ_∞ – high-frequency) dielectric constant
ε	emissivity
η	absorption quantum efficiency
$\eta^{(1)}$	absorption quantum efficiency for one well
λ, λ_1, λ_2	wavelengths
λ_c	cutoff wavelength
λ_p	peak detection wavelength
$\Delta\lambda$	full width
μ	mobility
μ_h	hole mobility
ν	photon frequency
ω	photon angular frequency
Φ	photon number per unit time incident on detector
$\phi_{B,ph}$	integrated background photon number flux density incident on detector
φ	azimuthal angle
ψ_n	envelope wavefunction
θ	internal angle of incidence
ϑ	polar angle and field of view half angle
τ_c	capture time
τ_{esc}	escape time
τ_{int}	integration or measurement time
τ_{life}	lifetime from continuum to well
τ_p	duration of signal pulse for one absorbed photon
τ_{relax}	intersubband relaxation time
τ_{scatt}	scattering time from subband to continuum
τ_{trans}	transit time across one well
$\tau_{trans,tot}$	total transit time
τ_{tunnel}	tunneling time

1

Introduction

Semiconductors have virtually touched everyone's life. The amazing pace of the advance in silicon (Si) microelectronics, responsible for the increasingly more powerful computers, shows little signs of slowing down. Optoelectronics relies on compound semiconductors, most commonly consisting of column III and V elements in the periodic table, such as gallium arsenide (GaAs), indium phosphide (InP), and related alloys. The research and development of optoelectronic devices have mainly been driven by the applications needs, e.g., in telecommunications, aerospace, and defense. The story of the quantum well infrared photodetector (QWIP), the subject of this book, has been one of the successful examples. The standard GaAs-based QWIPs cover the infrared (IR) wavelength region longer than about 3 μm. The immediate use of these QWIPs is therefore in the aerospace and defense areas.

A natural question arises concerning the motivation for studying QWIPs. After all, technologies based on HgCdTe and InSb are well developed for IR detection and imaging in the wavelength region of about 3–14 μm. What advantages do QWIPs provide? The first and foremost important advantage relates to the availability of a mature material and processing technology for QWIPs based on GaAs. QWIPs based on Si and InP have also been investigated, but they are far less mature. With the mature technology, it is anticipated that the cost of an imaging device based on a QWIP would be substantially less than that based on HgCdTe or InSb, and that a large volume production capability can be easily established. Other advantages relate to the uniqueness and the flexibility of the QWIP approach, e.g., in high speed and multicolor applications. It is necessary and important to compare QWIPs with other established detectors such as those based on HgCdTe and InSb. It is at least equally or even more important to remind ourselves of the lemma of any new technology, best described by the statement made by Kroemer [1]: "The principal applications of any sufficiently new and innovative technology always have been – and will continue to be – applications created by that technology." It is therefore our belief that the wide spread applications of QWIPs will come in new applications such as in the medical and health areas.

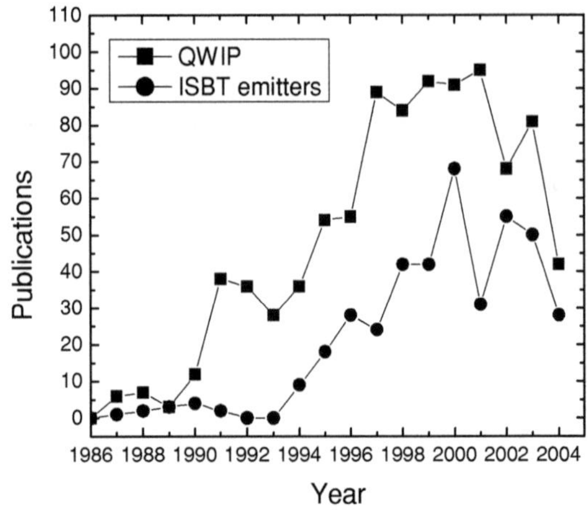

Fig. 1.1. Number of papers related to intersubband transition devices vs. year

The earliest studies of optical intersubband transitions (ISBT) in semiconductors were on two-dimensional (2D) electron systems in metal–oxide–semiconductor inversion layers (see [2] and references therein). Suggestions of using quantum wells for IR devices were first documented in [3,4]. The first experiment on making use of quantum wells for IR detection was reported [5,6]. Proposals for specific embodiments of photodetectors and related theoretical considerations were made [7–11]. The first experiment on ISBT in quantum wells was reported [12]. Subsequently, strong intersubband absorption and Stark shift were observed [13]. The first clear demonstration of QWIPs was made in 1987 [14]; since then tremendous progress was made by the Bell Laboratories' group (see [15] and references therein). Today, large focal plane arrays with excellent uniformity are being fabricated [16–18]. Several review articles [15,16,19,20] relevant to QWIPs have been published. Collections of conference papers related to ISBT in quantum wells can be found [21–23] and to QWIPs specifically [24–26]. A book on QWIPs has been written by Choi [27]. A literature search of the number of papers related to quantum well ISBT devices is graphed in Fig. 1.1. The number of QWIP-related papers has been growing since the initial demonstration. It will be interesting to see how many more years the QWIP research will continue to advance. It is also interesting to contrast this with papers on emitters based on ISBT. The realization of the ISBT laser, the quantum cascade laser (QCL), was made in 1994 [28]. This area will no doubt continue to grow [29,30]. The development of QCLs will likely expand the scope of QWIP applications, e.g., in the areas of free space optical communication and chemical/environmental sensing.

Part I

Physics

2

Basics of Infrared Detection

Before the in-depth analyses of quantum well, ISBT, and QWIP – a specific infrared detector – we first discuss the general concept of how the temperature of an object influences the emission of infrared radiation, and how the detection of this radiation allows us to sense the temperature of this object. We also discuss the general properties of detector signal, noise, and figures of merit such as noise-equivalent power (NEP), detectivity, and noise-equivalent temperature difference (NETD).

2.1 Blackbody Radiation

To fully understand the process of infrared detection, we have to know some basic properties of the signals. The concept of temperature is equivalent to certain energy distributions. The Fermi–Dirac distribution describes the temperature of an ensemble of indistinguishable particles obeying the exclusion principle (fermions), e.g., of carriers in a semiconductor while the thermal energy distribution of particles with unlimited state occupancy (bosons) is given by the Bose–Einstein distribution function

$$f_{\mathrm{B}}(E) = \frac{1}{\exp(E/k_{\mathrm{B}}T) - 1}\,. \tag{2.1}$$

Here E is the energy, T is the temperature, and k_{B} is the Boltzmann constant.

This distribution function describes in particular the energy distribution of a photon field with temperature T. Applying (2.1) to the electromagnetic modes (photon states) in a cavity yields *Planck's radiation law*. In its commonly used form, it states that the *irradiance* $\mathcal{I}_\nu$ (total power per unit surface area) at photon frequency ν is given by [31]

$$\mathrm{d}\mathcal{I}_\nu = \frac{2\pi h}{c^2}\frac{\nu^3\mathrm{d}\nu}{\exp(h\nu/k_{\mathrm{B}}T) - 1}\,. \tag{2.2}$$

Here h is the Planck's constant which connects the frequency with the energy $E = h\nu$ of a photon, and c is the speed of light. Equation (2.2) characterizes the radiation field inside a cavity with the radiation temperature T.

An important property of a thermal radiation field is that the radiation temperature is constant at each position inside the cavity. Considering now the walls of the cavity, thermal equilibrium can only exist if, at each part of the surface, the absorbed radiation power equals the emitted radiation power. The same is of course true for the surface of a small absorbing object inside the cavity. Consequently, at any frequency and incident angle, the emissivity ε of a surface equals the absorptivity (or (1 – reflectivity) if the object is opaque).

Another important consequence of this concept is that the surface of an object with temperature T still emits radiation in the absence of incident radiation or, more generally, in the absence of thermal equilibrium between the object and the radiation field. For a given emissivity $\varepsilon(\nu, \Omega)$, the *radiance* $\mathcal{H}_{\nu,\Omega}$ (radiation power per unit area and steradian) can be expressed as

$$\mathrm{d}\mathcal{H}_{\nu,\Omega} = \varepsilon(\nu, \Omega)\frac{2h}{c^2}\frac{\nu^3\mathrm{d}\nu\cos\vartheta\mathrm{d}\Omega}{\exp(h\nu/k_\mathrm{B}T) - 1}, \tag{2.3}$$

where $\mathrm{d}\Omega = \sin\vartheta\mathrm{d}\vartheta\mathrm{d}\varphi$.

In the simplest case, which is referred to as "blackbody," the emissivity of the surface is $\varepsilon = 1$, such that the emitted radiation is identical to a thermal radiation field. Real objects often show a "greybody" behavior, where ε is constant with a value slightly less than one.

In general, ε can exhibit a complicated angular and frequency dependence. In particular, in the case of structured surfaces, ε may not just depend on the polar angle ϑ but also on the azimuthal angle φ. Intentional modification of these dependencies can be achieved by applying appropriate coatings (reflection/antireflection) and by structuring the surface. Moreover, the reflectivity and thus also the emissivity are strongly affected if diffraction gratings are fabricated at the surface. In addition to diffraction, the emissivity is also influenced by surface plasmons, which can lead to a high degree of spatial coherence and narrow angular dependence of the emitted radiation [32].

For isotropic ε, integration of Eq. (2.3) over ϑ and φ yields the total emitted power density per frequency interval, which has the same value as $\mathcal{I}_\nu$ in (2.2) times an additional factor ε. Substituting for the photon energy E yields the power P_E per energy interval radiated from a surface with area A:

$$\mathrm{d}P_\mathrm{E} = A\varepsilon\frac{2\pi}{h^3c^2}\frac{E^3\mathrm{d}E}{\exp(E/k_\mathrm{B}T) - 1}. \tag{2.4}$$

Integration over the variable E (or ν) results in the Stefan–Boltzmann equation

$$P_\mathrm{tot} = A\varepsilon\sigma T^4 \tag{2.5}$$

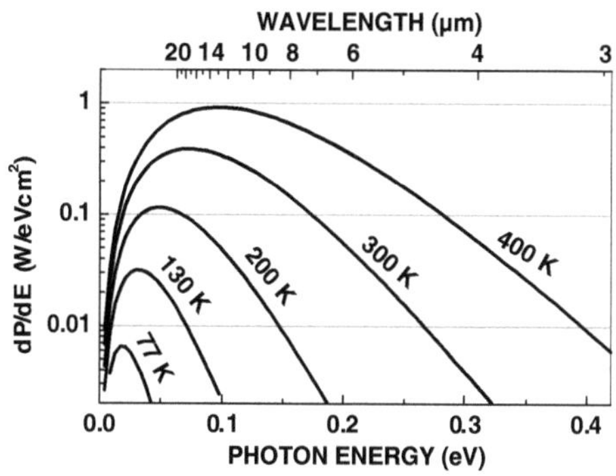

Fig. 2.1. Energy distribution of blackbody radiation vs. photon energy

with the Stefan–Boltzmann constant $\sigma = 2\pi^5 k_B^4/15c^2h^3 = 5.67\times10^{-8}\,\mathrm{W/m^2K^4}$. According to Eq. (2.5), the total emitted radiation power density P_{tot} of a blackbody with $\varepsilon = 1$ at 300 K equals $46\,\mathrm{mW\ cm^{-2}}$.

The radiative energy distribution of a blackbody with $\varepsilon = 1$ is shown in Fig. 2.1. Note that a different functional shape is obtained if the distribution is expressed as a function of the wavelength $\lambda = c/\nu$, since $\mathrm{d}\nu = \mathrm{d}\lambda/\lambda^2$.

The blackbody radiation incident onto a small detector with area A through an optical objective can be expressed in terms of the *f-number* $F_\#$ of the objective, which is the ratio between its focal length f_L and the lens diameter D_L. We thus obtain $\tan(\vartheta/2) = D_\mathrm{L}/2f_\mathrm{L} = 1/2F_\#$, or $\sin^2(\vartheta/2) = 1/(4F_\#^2+1)$. The objective redirects the incident light emitted from a blackbody onto the detector, and it can be shown that the incident photon flux is the same as if the clear aperture of the lens itself were a blackbody with the same temperature. Integrating over the angular variables, (2.3) thus yields

$$\mathrm{d}P_{\mathrm{v,lens}} = A\frac{1}{4F_\#^2+1}\frac{2\pi h\nu^3\mathrm{d}\nu}{c^2(\exp(h\nu/k_\mathrm{B}T)-1)}, \qquad (2.6)$$

which determines the radiation power incident onto the detector. This equation provides the basis for calculating the temperature resolution later in this chapter. We note that ϑ or $F_\#$ also determine the optical field of view.

2.2 Signal, Noise, and Noise-Equivalent Power

We now assume that the power P_S of a *signal* with photon energy $h\nu$, which is equivalent to a photon number $\Phi = P_\mathrm{S}/h\nu$ per unit time, is incident on a photon detector with an area A. There is a probability η, also called the (internal) quantum efficiency, that an incident photon is absorbed in the detector and contributes to the signal current I_S that is flowing in the external circuit. The *photoconductive gain* g_{photo} is defined as the ratio between the statistically averaged number $\overline{n_\text{x}}$ of electrons that are collected in the external circuit and the average number $\overline{n_\text{det}}$ of absorbed or detected photons,

$$g_\text{photo} = \overline{n_\text{x}}/\overline{n_\text{det}}. \tag{2.7}$$

Here the statistical average $\bar{x}$ of a stochastic variable x denotes its average over a large set of samples, each with duration τ_int.

I_S can thus be expressed as $I_\text{S} = e\eta g_\text{photo}\Phi$. The responsivity $\mathcal{R}$ is then defined as the ratio I_S/P_S, which leads to

$$\mathcal{R} = \frac{e}{h\nu}\eta g_\text{photo}. \tag{2.8}$$

In the case of a photoconductor, g_photo can be expressed as the ratio between the mean free path of the photoexcited carriers before recombination and the total thickness of the active region between the contacts; and, equivalently, as the ratio between the excited carrier lifetime and the total transit time. In the case of a photodiode, each detected photon contributes exactly one electron to the signal. The responsivity is thus given by (2.8) with $g = 1$, though the concept of gain is usually not applied to a photodiode. For both cases, we have implicitly assumed that τ_int is much larger than the duration τ_p of the signal pulses associated with individual detected photons. If τ_int approaches τ_p, then the signal (and also the noise associated with the signal) will depend on the sampling time and on the detection frequency. In the context of QWIPs, this case will be discussed in Chap. 10.

The *noise* associated with a stochastic variable is determined by its statistical properties. In mathematical terms, the noise associated with a stochastic variable x is given by its *variance*, defined by $\text{var}(x) = \overline{(x - \overline{x})^2}$.

In the case of a photodiode, each detected photon gives rise to a photocharge of exactly one electron. Thus, the number n_x of electrons collected during the sampling time τ_int obeys a Poisson distribution, meaning that the probability $p(n)$ of collecting n electrons is $p(n) = (\overline{n_\text{x}}^n/n!)\exp(-\overline{n_\text{x}})$. The variance for the specific case of a Poisson distribution is $\text{var}(n_\text{x}) = \overline{n_\text{x}}$.

The time-averaged current $\bar{I} = e\overline{n_\text{x}}/\tau_\text{int}$ is associated with the (squared) noise current

$$i_\text{n}^2 = \text{var}(I) \tag{2.9}$$

which, in the case of a photodiode, gives rise to $i_n^2 = e\bar{I}/\tau_\text{int}$. In practice, noise is measured as a mean square current transmitted through a filter with an effective bandwidth Δf. It can be shown that the sampling or measurement time τ_int is related to the bandwidth by

$$\Delta f = 1/2\tau_\text{int}. \tag{2.10}$$

The noise $i_{n,s}$ associated with a Poisson distribution is called *shot noise.* According to the previous discussion, it can be written as

$$i_\text{n,s}^2 = 2e\bar{I}\Delta f. \tag{2.11}$$

For a given signal, shot noise yields the lowest noise level obtainable for any detector since there is a one-to-one correspondence with the noise already contained in the statistics of the incident photons themselves.

An "ideal" photoconductor exhibits a somewhat more complicated statistics, since the lifetime of photoexcited electrons obeys in turn a Poisson distribution. Assuming again $\tau_{\mathrm{int}} \gg \tau_{\mathrm{p}}$, the signal is thus composed of a sequence of short pulses that are Poisson-distributed in time, and their amplitudes also show a Poisson distribution. It can be shown that these statistics result in the following noise expression:

$$i_{\mathrm{n,gr}}^2 = 4eg_{\mathrm{photo}}\bar{I}\Delta f. \tag{2.12}$$

This expression can be understood by the argument (see Sect. 4.3) that both carrier generation and recombination are associated with Poisson distributions, each of which generates a noise contribution as in (2.11). $i_{\mathrm{n,gr}}$ is thus called *generation-recombination (g-r) noise.* We note that a "real" photoconductor may obtain additional noise contributions, attributable to, e.g., impurity levels, traps, or various scattering events including impact ionization [33,34].

We point out that, for an ideal photoconductor, currents induced by optically or thermally generated carriers and the associated g-r noise are by definition asociated with the same gain. The index of g_{photo} will therefore be omitted in the present context. Due to the discrete microscopic structure of QWIPs, it will sometimes be necessary, however, to distinguish between the gains arising from the responsivity and the noise.

Noise is not only induced by the signal P_{S} itself, but also by a background power P_{B} originating, e.g., from objects adjacent to the signal source, stray light, or emission from the objective, and by the dark current I_{dark} of the detector. Assuming that I_{dark} is associated with the same noise behavior as the optically induced currents, which is usually a good approximation for most detectors, the resulting noise current is readily obtained by substituting $I = \mathcal{R}P_{\mathrm{S}} + \mathcal{R}P_{\mathrm{B}} + I_{\mathrm{dark}}$ into Eqs. (2.11) and (2.12). Depending on the relative magnitudes of $\mathcal{R}P_{\mathrm{S}}$, $\mathcal{R}P_{\mathrm{B}}$, and I_{dark}, we then distinguish between signal-noise-limited, background-noise-limited, and dark-current-limited detection.

Assuming that the dark current is caused by thermal excitation, we define the *thermal generation rate* G_{th} (the number of thermally generated carriers per time and volume $\mathcal{V}$). I_{dark} is thus expressed as

$$I_{\mathrm{dark}} = egG_{\mathrm{th}}\mathcal{V}. \tag{2.13}$$

In the case of QWIPs, G_{th} is obtained by spatial averaging over the detector volume. Background-limited (BL) detection thus refers to the situation that the optical generation rate induced by the radiation exceeds G_{th}.

The NEP is defined as the signal power needed to obtain a unity signal-to-noise ratio. Since the ratio between the signal power and the noise power equals the ratio between the squared currents, the NEP is determined by the condition $i_\mathrm{n}^2 = \mathcal{R}^2 P_\mathrm{S}^2$, such that NEP $= i_\mathrm{n}/\mathcal{R}$. For signal-noise limited (SL) detection, we thus obtain

$$(\mathrm{NEP})_{\mathrm{SL,s}} = \frac{h\nu}{\eta\tau_{\mathrm{int}}} = \frac{2h\nu\Delta f}{\eta} \tag{2.14}$$

in the case of a photodiode, and

$$(\mathrm{NEP})_{\mathrm{SL,gr}} = \frac{2h\nu}{\eta\tau_{\mathrm{int}}} = \frac{4h\nu\Delta f}{\eta} \tag{2.15}$$

for a photoconductor. Similarly, BL detection yields

$$(\mathrm{NEP})_{\mathrm{BL,s}} = \sqrt{\frac{h\nu P_\mathrm{B}}{\eta\tau_{\mathrm{int}}}} = \sqrt{\frac{2h\nu\Delta f P_\mathrm{B}}{\eta}} \tag{2.16}$$

in the case of a photodiode, and

$$(\mathrm{NEP})_{\mathrm{BL,gr}} = \sqrt{\frac{2h\nu P_\mathrm{B}}{\eta\tau_{\mathrm{int}}}} = \sqrt{\frac{4h\nu\Delta f P_\mathrm{B}}{\eta}} \tag{2.17}$$

for a photoconductor. In the dark-limited (DL) case, (2.13) gives rise to

$$(\mathrm{NEP})_{\mathrm{DL,gr}} = \frac{h\nu}{\eta}\sqrt{4G_{\mathrm{th}}\mathcal{V}\Delta f} = \frac{h\nu}{\eta}\sqrt{\frac{4G_{\mathrm{th}}\mathcal{V}\Delta f}{\tau_{\mathrm{int}}}}. \tag{2.18}$$

We point out that the NEP in (2.15), (2.17), and (2.18) is not influenced by the gain of the photoconductor.

2.3 Detectivity and Noise-Equivalent Temperature Difference

The *detectivity* D of a detector is defined as the inverse of NEP. In order to specify the performance of a detector, the *specific detectivity* $D^\star = D\sqrt{A\Delta f}$ is often used. $D^\star$ is the detectivity normalized with respect to the detector area and the bandwidth of the measurement. This definition leads to the general expression

$$D^* = \frac{R\sqrt{A\Delta f}}{i_\mathrm{n}}. \tag{2.19}$$

For BL detection in the presence of the background photon flux density $\Phi_{\mathrm{B,ph}}$ (irradiance $\mathcal{I}$), given by

$$\Phi_{\mathrm{B,ph}} = \frac{P_\mathrm{B}}{h\nu A} = \frac{\mathcal{I}}{h\nu}, \tag{2.20}$$

we thus obtain from (2.16) and (2.17) the specific detectivities

$$D^{\star}_{\mathrm{BL,s}} = \sqrt{\frac{\eta}{2h\nu\mathcal{I}}} \tag{2.21}$$

and

$$D^{\star}_{\mathrm{BL,gr}} = \sqrt{\frac{\eta}{4h\nu\mathcal{I}}}. \tag{2.22}$$

Equations (2.21) and (2.22) are sometimes referred to as the $D^{\star}$ of a photovoltaic and photoconductive detector, respectively.

In the DL case, (2.18) yields

$$D^{\star}_{\mathrm{DL,gr}} = \frac{\eta}{h\nu\sqrt{4G_{\mathrm{th}}L_{\mathrm{Det}}}}, \tag{2.23}$$

where we have introduced the total thickness $L_{\mathrm{Det}} = \mathcal{V}/A$ of the photoconductor. Assuming that the detected radiation is absorbed with the penetration depth α, η is given by $(1 - \exp(\alpha L_{\mathrm{Det}}))$, and η is proportional to L_{Det} for small L_{Det}. Therefore, the ratio α/G_{th} can be used as a figure of merit of the detector material. For larger L_{Det}, this figure of merit can still be used since D^* according to Eq. (2.23) has its maximum for $L_{\mathrm{Det}} = 1.26/\alpha$, where it has a value of $(0.31/h\nu)\sqrt{\alpha/G_{\mathrm{th}}}$ [20].

As expected, the detectivities in Eqs. (2.21)–(2.23) are independent of the measurement bandwidth and detector area. More generally, $D^{\star}$ is a figure of merit that specifies any detector for which i_{n}^2 is proportional to the detector area.

The NETD is defined as the temperature difference ΔT at which the induced change ΔP_{B} of the background power equals NEP, i.e.,

$$\mathrm{NETD} = \frac{\mathrm{NEP}}{\mathrm{d}P_{\mathrm{B}}/\mathrm{d}T}. \tag{2.24}$$

We note that (2.24), if expressed in terms of signal electrons $N_{\mathrm{S}} = P_{\mathrm{B}}\mathcal{R}\tau_{\mathrm{int}}/e$ and noise electrons $N_{\mathrm{N}} = i_{\mathrm{n}}\tau_{\mathrm{int}}/e$, is equivalent to the intuitive relation $N_{\mathrm{N}} = (\mathrm{d}N_{\mathrm{S}}/\mathrm{d}T) \times \mathrm{NETD}$. The NETD is probably the most important figure of merit characterizing detectors and arrays used for passive infrared detection and imaging.

In order to keep the following discussion simple, we consider infrared detection within a narrow spectral band $\Delta\nu$ around the detection energy $h\nu$. In addition, $h\nu/k_{\mathrm{B}}T$ is assumed to be large enough such that the Bose–Einstein distribution in Planck's radiation formula can be approximated by a simple exponential. In this case, we obtain

$$\frac{\mathrm{d}P_{\mathrm{B}}}{\mathrm{d}T} \approx \frac{h\nu}{k_{\mathrm{B}}T^2}P_{\mathrm{B}}, \tag{2.25}$$

or equivalently,

$$\frac{\mathrm{d}N_{\mathrm{S}}}{\mathrm{d}T} \approx \frac{h\nu}{k_{\mathrm{B}}T^2}N_{\mathrm{S}}. \tag{2.26}$$

Equation (2.26) now allows us to relate thermal resolution to the signal-to-noise ratio

$$\mathrm{NETD} = \frac{k_{\mathrm{B}}T^2}{h\nu}\frac{N_{\mathrm{N}}}{N_{\mathrm{S}}}. \tag{2.27}$$

This means in particular that in order to achieve a certain NETD, the required signal-to-noise ratio is proportional to the detection wavelength.

Restricting the discussion to an ideal photoconductor as in Eq. (2.17), the NETD can now be expressed as

$$\mathrm{NETD} = k_{\mathrm{B}}T^2\sqrt{\frac{2}{h\nu\eta P_{\mathrm{B}}\tau_{\mathrm{int}}}}. \tag{2.28}$$

Substituting for P_{B} the accumulated signal electrons N_{S} yields

$$\mathrm{NETD} = \frac{k_{\mathrm{B}}T^2}{h\nu}\sqrt{\frac{2g}{N_{\mathrm{S}}}}. \tag{2.29}$$

To give a typical example, let us assume thermal detection within the spectral band from 8 to 9 µm at 300 K radiation temperature using a detector with $\eta = 10\%$ and $A = 30 \times 30\,\mu\mathrm{m}^2$ through an $F_{\#} = 2$ objective at an integration time $\tau_{\mathrm{int}} = 20\,\mathrm{ms}$. According to Eq. (2.6), we expect an incident power of $P_{\mathrm{B}} = 1.5\,\mathrm{nW}$, and Eq. (2.28) predicts NETD = 7 mK. If $g_{\mathrm{photo}} = 1$, the photo charge amounts to about $N_{\mathrm{S}} = 1.3 \times 10^8$ signal electrons. Since this number is already somewhat larger than the typical storage capacity of a readout integrated circuit, such an NETD is only achievable by increasing the storage capacity or by working at reduced noise levels, which is achievable for lower g_{photo}. We will come back to this point in Chap. 9.

Finally, it is worthwhile to take a closer look at the number of noise electrons $N_{\mathrm{N}} = i_{\mathrm{n,gr}}\tau_{\mathrm{int}}/e$. To this end, we express Eq. (2.12) in terms of N_{S} and N_{N}, which yields

$$N_{\mathrm{N}} = \sqrt{2gN_{\mathrm{S}}}\,. \tag{2.30}$$

The above example then yields $N_{\mathrm{S}}/N_{\mathrm{N}} = 8{,}000$ (see also (2.27)), which imposes additional harsh requirements on the readout noise and on other noise sources associated with the detection electronics. The difficulty arises from the necessity to resolve very small changes in the thermal background, which, according to (2.25), are as low as $\Delta P_{\mathrm{B}}/P_{\mathrm{B}} = 1.3 \times 10^{-4}$ for $\Delta T = 7\,\mathrm{mK}$.

3

Semiconductor Quantum Wells and Intersubband Transitions

3.1 Quantum Wells

The success of QWIPs has been made possible by semiconductor crystal growth technology such as molecular beam epitaxy (MBE) [35,36] and by device concepts using bandgap engineering [37]. Semiconductor structures with atomic layer control are now routinely fabricated by MBEs. Several introductory textbooks on quantum well physics have been written by Bastard [38], Weisbuch and Vinter [39], Shik [40], and Harrison [41].

3.2 Intersubband Transitions

The term intersubband transition (ISBT) refers to the electronic transition between the confined states in quantum wells. The physics related to the optical ISBT in quantum wells is treated in detail by Helm [42]. Other materials of specific relevance to QWIPs can be found in [27,43]. Here, we discuss only ISBTs in the conduction band (CB). Examples presented here are mostly on samples made by GaAs-based MBE. For this system, a polarization selection rule was realized in the early days [7]: only the light polarized in the growth direction can cause ISBTs. (This will be discussed in more mathematical details in Sect. 3.3.) The selection rule is valid for quantum wells where the single band isotropic effective mass approximation holds. Since the argument is based on the effective mass approximation, the selection rule is naturally not rigorous. As band mixing is the cause of the breakdown of the selection rule, physical intuition tells us that the accuracy of the selection rule should be related to the ratio of the energy scales involved. The closest band to the conduction band (at least for the GaAs case) is the valence band (VB). The relevant energy ratio is then E_n/E_g, where E_n $(n = 1, 2, \ldots)$ are the eigenenergies of the confined states in the conduction band quantum well and E_g is the bandgap. (Note that the reference point – zero energy – for E_n is chosen at the conduction bandedge of the well.) For common quantum wells used

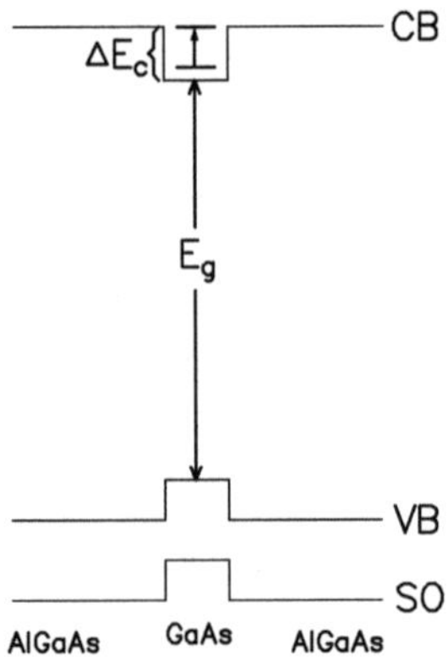

Fig. 3.1. Schematic bandedge profile of a GaAs/AlGaAs quantum well. Conduction band (CB), valence band (VB), and spin–orbit split-off (SO) band are shown. The bandgap energy (E_g) is about ten times the conduction band offset (ΔE_c)

in QWIPs (see, [15]) the ratio is small and therefore the selection rule is expected to be quite accurate, with a deviation of at most 10%, level, as shown experimentally [44].

The bandedge profile of a GaAs/AlGaAs quantum well is depicted in Fig. 3.1. An optical ISBT in the conduction band (CB) is shown by an arrow. The vertical energy scale is drawn roughly in the correct proportion. It is seen that the closest bands, valence band (VB), and spin–orbit split-off (SO) band, are far away in comparison with the ISBT energy involved. It is therefore natural to expect that the effect of mixing to the VB or SO bands should be small, and hence the selection rule holds quite accurately.

Because of the selection rule, a normal incidence geometry (i.e., light incident normal to the wafer and along the growth direction) is not suitable. A commonly employed 45° edge facet geometry is shown Fig. 3.2, as first used [14]. This geometry "throws away" one half of the unpolarized light, but is simple and convenient to realize, and is usually used to obtain a detector performance benchmark. The majority of applications of QWIPs requires large 2D arrays where the facet geometry is not suited. Various diffraction gratings are used for these practical devices, discussed in Chap. 6.

Within the single band effective mass approximation, the Schrödinger equation reads

$$-\nabla \frac{\hbar^2}{2m} \nabla \psi + V\psi = E\psi, \tag{3.1}$$

where m is the effective mass and V is the potential such as the one labeled CB in Fig. 3.1. If we choose the quantum well direction as the z-axis, the potential V and the mass m depend only on z and the wavefunction is separable into a lateral ($x - y$ plane) part ψ_{xy} and a z part ψ_z. The lateral part is simply a planewave with kinetic energy $E_{\|}$. ψ_z is a solution of

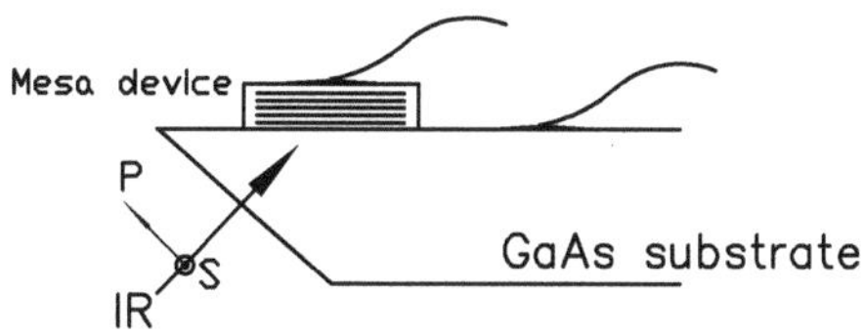

Fig. 3.2. 45° edge facet light-coupling geometry. The figure is not drawn to scale. The semi-insulating substrate thickness is usually in the range of 400–700 μm. For testing individual detector performance, mesa devices of areas from about 100^2–1000^2 μm^2 are used. The IR light is shone normal to the facet surface. The P and S polarizations are defined with respect to the light incidence on the quantum wells

$$-\frac{\mathrm{d}}{\mathrm{d}z}\frac{\hbar^2}{2m}\frac{\mathrm{d}}{\mathrm{d}z}\psi_z + V\psi_z = E_\perp\psi_z, \tag{3.2}$$

such that $E = E_\parallel + E_\perp$. Taking into account that $m = m(z)$ for different barrier and well materials, the operator $\mathrm{d}/\mathrm{d}z$ does not commute with $1/m$. Then (3.2) requires ψ_z and $(1/m)\mathrm{d}\psi_z/\mathrm{d}z$ to be continuous at the interfaces between different materials.

To obtain a simple picture of an ISBT, much of the physics can be illustrated using the simplest model of an infinitely high barrier square quantum well [12], i.e., $V = 0$ for $0 \le z \le L_\mathrm{w}$, $V = \infty$ for $z < 0$ and $z > L_\mathrm{w}$, where L_w is the well width. In this case the eigenstate wavefunction and energy are trivial:

$$\psi_\mathrm{n}(\boldsymbol{k}_{xy}) = \sqrt{\frac{2}{L_\mathrm{w}A}}\sin\left(\frac{\pi n z}{L_\mathrm{w}}\right)\exp(\mathrm{i}\boldsymbol{k}_{xy}\cdot\boldsymbol{x}), \tag{3.3}$$

$$E_n(\boldsymbol{k}_{xy}) = \frac{\hbar^2}{2m}(\pi^2 n^2/L_\mathrm{w}^2 + k_{xy}^2), \tag{3.4}$$

where A is the normalization area in the x–y plane, n is a positive integer, $\boldsymbol{k}_{xy}$ is the in-plane wavevector, and m is the effective mass in the well. Equation (3.4) explains the reason for the term "subband" instead of a single state. For a given quantized state one can put many electrons occupying different in-plane momenta. For the ground state in equilibrium, the occupation of electrons leads to a Fermi energy determined by $n_\mathrm{2D} = (m/\pi\hbar^2)E_\mathrm{f}$, where n_2D is the 2D quantum well carrier density and $m/\pi\hbar^2$ is the 2D density of states.

The dipole matrix moment between any two states (e.g., n and n') with opposite parity is:

$$e\langle z\rangle_{n,n'} = eL_\mathrm{w}\frac{8}{\pi^2}\frac{nn'}{(n^2 - n'^2)^2}. \tag{3.5}$$

Note that the in-plane momentum remains the same between initial and final states.

The oscillator strength is

$$f \equiv \frac{2m\omega}{\hbar}\langle z\rangle_{n,n'}^2 = \frac{64}{\pi^2}\frac{n^2 n'^2}{(n^2 - n'^2)^3}. \tag{3.6}$$

The absorption probability for an IR beam polarized in the plane of incidence and propagating at an angle of θ with respect to the growth axis is

$$\eta = \frac{e^2 h}{4\epsilon_0 n_\mathrm{r} mc}\frac{\sin^2\theta}{\cos\theta} n_\mathrm{2D} f \frac{1}{\pi}\frac{\Delta E}{(E_{n,n'} - \hbar\omega)^2 + (\Delta E)^2}, \tag{3.7}$$

where ϵ_0 is the vacuum permitivity, n_r is the index of refraction, c is the speed of light, $E_{n,n'} \equiv E_n - E_{n'}$, and ΔE is the broadening half width. The lineshape associated with broadening is modeled by a Lorenzian. At the peak ($E_{n,n'} = \hbar\omega$) the absorption is inversely proportional to ΔE. For a given n_2D, the absorption is inversely proportional to m, i.e., the smaller the effective mass the larger the absorption. Note that for a given E_f, however, the absorption is independent of m since $n_\mathrm{2D} = (m/\pi\hbar^2)E_\mathrm{f}$. Note also that the integrated absorption is independent of ΔE. The derivation of (3.7) is straightforward by using a dipole interaction Hamiltonian and Fermi's golden rule (see Sect. 3.3.1 and [42]).

The factor $\sin^2\theta$ in (3.7) comes from the polarization selection rule discussed above. The factor $\cos\theta$ seems to give an unphysical result when $\theta \to 90°$; however, since the meaning of η is the absorption probability of light passing through the well, in this extreme the passing length becomes infinitely long resulting in an infinitely large absorption. If one had considered a quantity of absorption constant α defined by $\eta = \alpha\times$(length), where (length)$= L_\mathrm{w}/\cos\theta$ is the propagation length, this quantity would have always been finite. For a real quantum well with finite barrier height, it would be more physical to choose the length using the quantum well structure thickness (including barriers) instead of only the well width taken here.

Let us put some typical numbers into (3.7) to get a feeling for how strong the ISBT absorption is. For a typical 8–12 μm peaked QWIP, the half width is about $\Delta E = 0.01$ eV. For ground state to first excited state transition, the oscillator strength is $f = 0.961$ (see (3.6) with $n' = 1$ and $n = 2$). For 77 K operation, the carrier density is set to about $n_\mathrm{2D} = 5 \times 10^{11}\,\mathrm{cm}^{-2}$ (see later for the reason). For a GaAs well, the reduced effective mass is $m^* = 0.067$ ($m = m^* \times m_\mathrm{e}$, where m_e is the free electron mass) and the refractive index is about $n_\mathrm{r} = 3.3$. With these values and for a 45° angle ($\theta = 45°$), the absorption for a single quantum well is $\eta = 0.54\%$ (for polarized light). Let us also evaluate the absorption constant for the case of $\theta = 90°$, with other parameters the same as above. Taking a quantum well structure thickness of 50 nm (including barriers), the peak absorption constant is $\alpha = 1{,}520\,\mathrm{cm}^{-1}$.

For a standard QWIP, the optimum well design is the one having the first excited state in resonance with the top of the barrier. This configuration gives *at the same time* both a large peak absorption (similar to the bound state to bound state transition discussed above) and a rapid escape for the excited electrons. The optimum design configuration has been experimentally proven [45,46]. To design a quantum well for a given QWIP wavelength, one needs to know how the barrier height (conduction band offset) relates to heterosystem parameters, i.e., Al fraction (x) in the GaAs/$Al_xGa_{1-x}As$ case. Surveying many samples by comparing the calculated transition energy with the experimental peak absorption, we find a range of values for the conduction band offset $\Delta E_c = (0.87 \pm 0.04) \times x\,\text{eV}$, where x is the Al fraction. The calculation is a simple eigenenergy calculation of a square quantum well. Higher order effects, such as nonparabolicity and many-body effects, influence the precise values of the calculation (as discussed in Sect. 3.4).

For thermal imaging the spectral region of 3–12 μm is most interesting. QWIPs based on GaAs can easily cover this region. Figure 3.3 shows spectral response curves of six individual QWIPs with InGaAs or GaAs wells and AlGaAs barriers. By employing special designs, QWIPs covering wavelengths much shorter than 3 μm may be possible (see [47–49]). The short wavelength is limited by the lack of available high barriers. Using a set of p-type GaAs/AlGaAs QWIPs, a wide spectral coverage was demonstrated [50]. The shortest wavelength device with a high Al fraction (95%) peaks at 1.9 μm and covers a range of 1.4–3 μm. For beyond 12 μm, QWIPs with measured cutoff wavelengths as long as 31 μm have been demonstrated [51–53]. How-

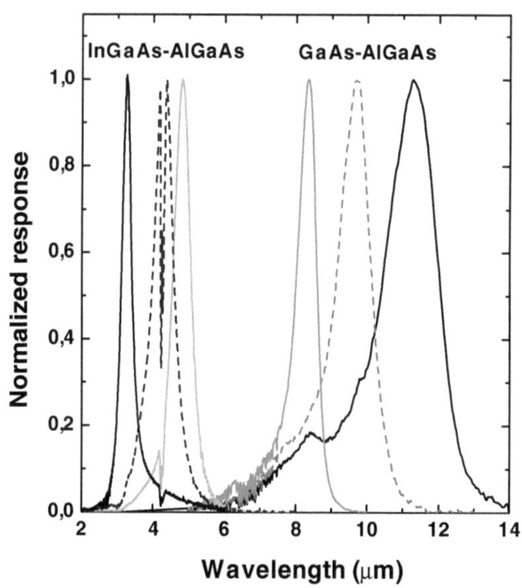

Fig. 3.3. Spectral response curves of six QWIPs covering the two atmospheric transmission windows of 3–5 and 8–12 μm wavelength regions. The dip in the second curve from the left at about 4.2 μm is due to the CO_2 absorption, and the noise from 5.5 to 7.5 μm is due to the water absorption

ever, these longer wavelength QWIPs have small potential energy barriers which require temperatures much lower than 77 K cooling for operation. The region of 30–40 meV (31–41 μm) is masked by the restrahlen band (optical phonon absorption), making it difficult to design QWIPs here. Very recently, far-infrared (FIR) or terahertz QWIPs have been reported and background-limited operation has been achieved for devices up to 100 μm in detection wavelength [54,55]. For long wavelengths (longer than about 20 μm), since free carrier absorption becomes strong and the plasma frequency becomes comparable to the detection frequency, additional design considerations must be made.

3.3 Intersubband Transition: More Details

In this section we present more mathematical details of the physics of ISBTs in a single particle picture. For those readers who do not want to be bored with the math, this section may be skipped. We use the effective mass approximation and deal with a single spherical band which is a good approximation for the conduction band of the GaAs–AlGaAs system. Higher order effects such as scattering-assisted ISBTs [56] and many particle effects [57,58] are not directly included and will be discussed in the following sections.

3.3.1 Basic Formulae

The interaction potential appropriate for calculating the radiative transition absorption rate using Fermi's golden rule is given by

$$H_{\mathrm{rad}} = \frac{e}{m}\left(\frac{\phi\hbar}{2\epsilon_0 n_{\mathrm{r}}\omega c}\right)^{1/2} \hat{\varepsilon}\cdot\boldsymbol{p}\, e^{i\boldsymbol{q}\cdot\boldsymbol{x}}, \tag{3.8}$$

where ϕ is the incident photon flux (number of photons per unit area per unit time), ω is the photon angular frequency, $\hat{\varepsilon}$ is the polarization unit vector, $\boldsymbol{p}$ is the electron momentum operator, n_{r} is the material refractive index, and $\boldsymbol{q}$ is the photon momentum which is perpendicular to $\hat{\varepsilon}$. We assume that the photon flux is sufficiently small so that the interaction in (3.8) can be treated as a perturbation. For high IR intensities, the perturbation approach fails and the higher order term proportional to ϕ (in addition to the $\phi^{1/2}$ term in (3.8)) must be included in the analysis. The derivation of (3.8) is straightforward starting from the dipole interaction Hamiltonian $(e/2m)(\boldsymbol{A}\cdot\boldsymbol{p}+\boldsymbol{p}\cdot\boldsymbol{A})$ and keeping only the contribution for photon absorption, where $\boldsymbol{A}$ is the vector potential.

The envelope wavefunction is written as

$$\psi(\boldsymbol{x}) = \psi_{xy}\psi_z, \tag{3.9}$$

where ψ_{xy} and ψ_z are two separable components of the envelope function ψ in the x–y plane and in the z-direction, respectively. As before, the in-plane envelope function is simply a planewave

$$\psi_{xy} = A^{-1/2}\mathrm{e}^{\mathrm{i}\boldsymbol{k}_{xy}\cdot\boldsymbol{x}}, \tag{3.10}$$

where A is the in-plane normalization area, and the 2D wavevector $\boldsymbol{k}_{xy}$ is $\boldsymbol{k}_{xy} = (k_x, k_y)$. The energy associated with the x–y motion is $E_{\|} = \hbar^2 k_{xy}^2/(2m)$.

The matrix element of interest between the two states ψ and ψ' is

$$M = \langle\psi'|H_{\mathrm{rad}}|\psi\rangle. \tag{3.11}$$

Some general properties can be obtained using (3.11). We first express (3.11) in terms of the z-component and the in-plane envelope function in the quantum well:

$$\begin{aligned}
M &= \langle\psi'|H_{\mathrm{rad}}|\psi\rangle \\
&= \frac{e}{m}\left(\frac{\phi\hbar}{2\epsilon_0 n_{\mathrm{r}}\omega c}\right)^{1/2}\left[\left(\frac{1}{A}\int \mathrm{d}x\mathrm{d}y\mathrm{e}^{-\mathrm{i}\boldsymbol{k}'_{xy}\cdot\boldsymbol{x}}\hat{\varepsilon}\cdot\boldsymbol{p}_{xy}\mathrm{e}^{\mathrm{i}\boldsymbol{k}_{xy}\cdot\boldsymbol{x}}\right)\right. \\
&\quad\times\left(\int \mathrm{d}z\psi^*_{z,n'}\mathrm{e}^{\mathrm{i}q_z z}\psi_{z,n}\right) + \left(\frac{1}{A}\int \mathrm{d}x\mathrm{d}y\mathrm{e}^{\mathrm{i}(\boldsymbol{k}_{xy}-\boldsymbol{k}'_{xy})\cdot\boldsymbol{x}}\mathrm{e}^{\mathrm{i}\boldsymbol{q}_{xy}\cdot\boldsymbol{x}}\right) \\
&\quad\left.\times\left(\int \mathrm{d}z\psi^*_{z,n'}\hat{\varepsilon}\cdot\hat{z}p_z\psi_{z,n}\right)\right], \\
&\approx \frac{e}{m}\left(\frac{\phi\hbar}{2\epsilon_0 n_{\mathrm{r}}\omega c}\right)^{1/2}\left[\hbar(\hat{\varepsilon}\cdot\boldsymbol{k}_{xy})\delta_{\boldsymbol{k}'_{xy},\,\boldsymbol{k}_{xy}}\langle\psi_{z,n'}|\mathrm{e}^{\mathrm{i}q_z z}|\psi_{z,n}\rangle\right. \\
&\quad\left.+\,\delta_{\boldsymbol{k}_{xy}-\boldsymbol{k}'_{xy},\,\boldsymbol{q}_{xy}}(\hat{\varepsilon}\cdot\hat{z})\langle\psi_{z,n'}|p_z|\psi_{z,n}\rangle\right],
\end{aligned} \tag{3.12}$$

where $\hat{z}$ is a unit vector in the z-direction. The factor $e^{iq_z z}$ in the first term of the (3.12) can normally be neglected by setting $e^{iq_z z} = 1$. The photon wavelength (λ) of interest is greater than about $3/n_{\mathrm{r}} \approx 1\,\mu\mathrm{m}$ and the domain of the z-direction integration is about $w \sim 100$ Å (the bound state wavefunction extent), giving $q_z z \sim 2\pi w/\lambda \ll 1$, where $n_{\mathrm{r}} \approx 3$ for GaAs. The first term in (3.12) is proportional to $\delta_{\boldsymbol{k}'_{xy},\,\boldsymbol{k}_{xy}}\delta_{nn'}$ which vanishes because $\boldsymbol{k}'_{xy} = \boldsymbol{k}_{xy}$ and $n = n'$ correspond to a transition between the same state (forbidden by energy conservation). The first term can be shown to vanish for a superlattice, where $\mathrm{e}^{\mathrm{i}q_z z} \approx 1$ is no longer true in general. The physical reason for a vanishing contribution from the first term in (3.12) is that photons cannot cause direct transitions between free carrier states. Note that the in-plane motion of the carriers is free. The commonly referred "free carrier absorption" is in fact a higher order process assisted by scattering events (see [59]). In general, if the incident light is polarized in the x–y plane, the matrix element in (3.12) vanishes. By a similar argument that the photon momentum can be neglected, we can neglect the $\boldsymbol{q}_{xy}$ in the δ-function of the second term in (3.12). The matrix element then becomes

$$M = \frac{e}{m}\left(\frac{\phi\hbar}{2\epsilon_0 n_r \omega c}\right)^{1/2} (\delta_{\boldsymbol{k}_{xy},\, \boldsymbol{k'}_{xy}}) \sin\theta \langle \psi_{z,n'}|p_z|\psi_{z,n}\rangle, \tag{3.13}$$

where $(\hat{\varepsilon}\cdot\hat{z}) = \sin\theta$ and θ is the internal angle of incidence. The polarization selection rule [7] can now be stated as: *optical intersubband transitions associated with a single spherical band are induced by light polarized in the quantum well direction.*

Using Fermi's golden rule

$$W = \frac{2\pi}{\hbar}\sum_{f,\, i} |M|^2 f_i(1-f_f)\delta(E_f - E_i - \hbar\omega), \tag{3.14}$$

the total transition rate (W) is easily calculated by summing over initial and final states, where f_i and f_f are Fermi factors for the initial and final states. Here, for a pure bound-to-bound transition, we sum over only the in-plane 2D states for the initial and final state summations, which is appropriate for transitions between two 2D subbands. If stimulated emisssion (i.e., deexcitation towards lower energies) is taken into account as well, the total transition rate is given by the slightly simpler expression

$$\tilde{W} = \frac{2\pi}{\hbar}\sum_{f,\, i} |M|^2 (f_i - f_f)\delta(E_f - E_i - \hbar\omega), \tag{3.15}$$

since terms proportional to $f_i f_f$ vanish.

At zero temperature and weak excitation, Fermi factors become $f_i = 1$ and $f_f = 0$, assuming that only the ground state subband is occupied. The momentum conserving $\delta_{\boldsymbol{k}_{xy},\, \boldsymbol{k'}_{xy}}$ function in (3.13) takes care of the final state summation in (3.14), and the initial state summation gives simply the 2D density of electrons in the ground state subband (n_{2D}) multiplied by the area (A). Equation (3.14) is then trivially evaluated (defining the absorption quantum efficiency η):

$$\eta \equiv W/(\phi A\cos\theta) = \frac{e^2 h}{4\epsilon_0 n_r m c}\frac{\sin^2\theta}{\cos\theta} n_{2D} f\ \delta(E_2 - E_1 - \hbar\omega), \tag{3.16}$$

where the oscillator strength [12,60] is defined by

$$f \equiv \frac{2}{m\hbar\omega}|\langle\psi_2|p_z|\psi_1\rangle|^2, \tag{3.17}$$

where ψ_2 and ψ_1 are the first excited and the ground state z-direction envelope functions, and E_2 and E_1 are the corresponding eigenenergies. An equivalent expression for the oscillator strength is given by

$$f = \frac{2m\omega}{\hbar}|\langle\psi_2|z|\psi_1\rangle|^2. \tag{3.18}$$

The equivalence between (3.17) and (3.18) is shown by considering the matrix element of commutator $\langle\psi_2|[H_z, z]|\psi_1\rangle$ and noting $E_2 - E_1 = \hbar\omega$, where H_z is the quantum well effective mass Hamiltonian in the z-direction. For finite temperatures, as long as $E_{21} - E_{\rm F} \gg k_{\rm B}T$, (3.16) gives an excellent approximation, where $E_{21} \equiv E_2 - E_1$ and $E_{\rm F}$ is the Fermi energy related to the 2D electron density $n_{\rm 2D}$ in the ground state subband by

$$E_{\rm F} = (\pi\hbar^2/m)n_{\rm 2D}. \tag{3.19}$$

The expression $m/(\pi\hbar^2)$ is the 2D density of states. Relevant energy scales of interest are $E_{21} - E_{\rm F} \sim 120$ meV and $k_{\rm B}T < 10$ meV. (Note that mid-infrared (MIR) detectors are normally cooled to $T < 100$ K.) Taking into account the finite lifetime of the excited state (usually limited mainly by optical phonon emission), the δ-function in (3.16) becomes a Lorentzian:

$$\eta = \frac{e^2h}{4\epsilon_0 n_{\rm r} mc}\frac{\sin^2\theta}{\cos\theta}n_{\rm 2D}f\frac{1}{\pi}\frac{\Delta E}{(\hbar\omega - E_{21})^2 + \Delta E^2}, \tag{3.20}$$

where ΔE is the half width. Ideally, in the absence of other elastic broadening mechanisms (e.g., by interface roughness and well width fluctuations), ΔE is related to the lifetime by $\tau_{\rm life} = \hbar/(2\Delta E)$. For convenience, we will use (3.20) as a model, even when elastic broadening is not negligible as in the following discussions. For GaAs at 10 µm wavelength, the constant $e^2h/4\epsilon_0 n_{\rm r} mc$ in (3.20) equals 5.2×10^{-16} eV cm^2. Taking $n_{\rm 2D} = 10^{12}$ cm^{-2} and $\Delta E \sim 0.01$ eV, peak absorption (at $\hbar\omega = E_{21}$) of about a percent is expected, which is consistent with measured results for absorption per quantum well [61].

3.3.2 Calculations for a Symmetric Quantum Well

The original proposal by Coon and Karunasiri [7] to use ISBTs for IR detection predicted large absorption quantum efficiencies (up to 50% for only one well) when the upper state (the ISBT absorption final state) is nearly in resonance with the top of the barrier. This large absorption has never been observed experimentally: an absorption of a fraction of a percent per well per IR path at internal angles less than 45° is commonly observed [12,14,61,62]. Asai and Kawamura studied the well width dependence of intersubband absorption using a series of samples which included the situation where the upper state is nearly in resonance with the top of the barrier, but no absorption enhancement was observed [62]. Some of our own detector samples had the upper state very close to the top of the barrier, but we only observed absorptions of less than 1% per well per IR path [61]. The work reported [46] has been motivated by the question of whether one can obtain an extremely large absorption by positioning the upper state in resonance with the top of the barrier. This regime of having the upper state close to the top of the barrier is of practical importance, since it corresponds to the optimum detector design [45].

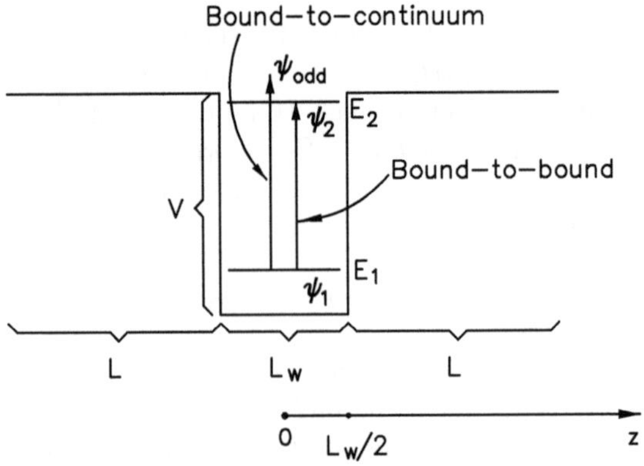

Fig. 3.4. The quantum well potential profile. The ground and the first excited states are labeled by E_1 and E_2, respectively. For small well widths (L_w), the first excited state is "pushed out" of the well and only the ground state exits. Both bound-to-bound and bound-to-continuum intersubband transitions are considered

We reproduce the analytical results of [46] for ISBT absorption including both bound-to-bound and bound-to-continuum contributions and including Lorentzian broadening for the finite final state lifetime. Using these analytical results, the physical reason for the large predicted absorption in the limit of zero broadening and for the case when the upper (final) state is in resonance with the top of the barrier is discussed. Calculated absorption spectra are compared with the experiments of Asai and Kawamura [62].

The bound-to-bound ISBT is discussed first. The symmetric quantum well potential profile is shown schematically in Fig. 3.4. The ground (ψ_1) and the first excited (ψ_2) state wavefunctions are

$$\psi_1 = C_1 \begin{cases} \mathrm{e}^{\kappa_1(z+L_{\mathrm{w}/2})}\cos k_1 L_{\mathrm{w}/2} & \text{if } z < -L_{\mathrm{w}/2} \\ \cos k_1 z & \text{if } -L_{\mathrm{w}/2} \le z \le L_{\mathrm{w}/2} \\ \mathrm{e}^{-\kappa_1(z-L_{\mathrm{w}/2})}\cos k_1 L_{\mathrm{w}/2} & \text{if } z > L_{\mathrm{w}/2} \end{cases} \tag{3.21}$$

$$C_1 = \frac{1}{\sqrt{L_{\mathrm{w}/2} + (V/\kappa_1 E_1)\cos^2 k_1 L_{\mathrm{w}/2}}} \tag{3.22}$$

with

$$\cos k_1 L_{\mathrm{w}/2} - \frac{m_\mathrm{b} k_1}{m\kappa_1}\sin k_1 L_{\mathrm{w}/2} = 0, \tag{3.23}$$

and

$$\psi_2 = C_2 \begin{cases} -\mathrm{e}^{\kappa_2(z+L_{\mathrm{w}/2})}\sin k_2 L_{\mathrm{w}/2} & \text{if } z < -L_{\mathrm{w}/2}, \\ \sin k_2 z & \text{if } -L_{\mathrm{w}/2} \le z \le L_{\mathrm{w}/2}, \\ \mathrm{e}^{-\kappa_2(z-L_{\mathrm{w}/2})}\sin k_2 L_{\mathrm{w}/2} & \text{if } z > L_{\mathrm{w}/2}, \end{cases} \tag{3.24}$$

$$C_2 = \frac{1}{\sqrt{L_{\mathrm{w}/2} + (V/\kappa_2 E_2)\sin^2 k_2 L_{\mathrm{w}/2}}} \tag{3.25}$$

with

$$\cos k_2 L_{\mathrm{w}/2} + \frac{m\kappa_2}{m_{\mathrm{b}} k_2} \sin k_2 L_{\mathrm{w}/2} = 0, \tag{3.26}$$

where $k_{1,2} = \sqrt{2mE_{1,2}}/\hbar$, $\kappa_{1,2} = \sqrt{2m_{\mathrm{b}}(V - E_{1,2})}/\hbar$, E_1 (E_2) is the ground (first excited) state eigenenergy, m (m_{b}) is the well (barrier) effective mass, V is the barrier height, and $L_{\mathrm{w}/2}$ is the half well width (i.e., the well width is $L_{\mathrm{w}} = 2L_{\mathrm{w}/2}$). Equations (3.23) and (3.26) determine the eigenenergies E_1 and E_2. In fact, all even parity bound states satisfy (3.21), (3.22), and (3.23), and all odd parity bound states satisfy (3.24), (3.25), and (3.26). We have set the origin ($z = 0$) at the center of the well.

The above-barrier continuum eigenstates can be chosen to have even or odd parity. For ISBTs from the (even parity) ground state, only the odd parity continuum states are allowed as final states:

$$\psi_{\mathrm{odd}} = \frac{1}{\sqrt{L}} \begin{cases} \sin[k'(z + L_{\mathrm{w}/2}) - \beta] & \text{if } z < -L_{\mathrm{w}/2}, \\ \left(\sin^2 kL_{\mathrm{w}/2} + \frac{m_{\mathrm{b}}k}{mk'} \cos^2 kL_{\mathrm{w}/2}\right)^{-1/2} \sin kz & \text{if } -L_{\mathrm{w}/2} \le z \le L_{\mathrm{w}/2}, \\ \sin[k'(z - L_{\mathrm{w}/2}) + \beta] & \text{if } z > L_{\mathrm{w}/2}, \end{cases} \tag{3.27}$$

where $k = \sqrt{2mE_z}/\hbar$, $k' = \sqrt{2m_{\mathrm{b}}(E_z - V)}/\hbar$, E_z is the energy associated with only the z-direction motion, β is given by $\tan\beta = (mk'/m_{\mathrm{b}}k)\tan kL_{\mathrm{w}/2}$, and L is a normalization length on either side of the well (see Fig. 3.4). We have used a box-normalization scheme ($L \gg L_{\mathrm{w}}$), and the above-barrier continuum states are normalized in a length $2L$.

The ISBT oscillator strength (see (3.17) and (3.17)) is given by

$$f \equiv \frac{2m\omega}{\hbar} |\langle z \rangle|^2 = \frac{2\hbar}{m\omega} |\langle \partial/\partial z \rangle|^2, \tag{3.28}$$

where $\langle \cdots \rangle$ represents a matrix element between wavefunctions.

After some algebraic manipulation, the oscillator strength for the bound-to-bound transition (using ψ_1 and ψ_2 in (3.28)) is found to be

$$f_{B-B} = \frac{8\hbar C_1^2 V^2}{m\omega(E_2 - E_1)^2} \cos^2 k_1 L_{\mathrm{w}/2} \frac{E_2 \kappa_2 \sin^2 k_2 L_{\mathrm{w}/2}}{E_2 \kappa_2 L_{\mathrm{w}/2} + V \sin^2 k_2 L_{\mathrm{w}/2}} \tag{3.29}$$

and that for the bound-to-continuum (using ψ_1 and ψ_{odd}) is

$$f_{B-C} = \frac{8\hbar C_1^2 V^2}{m\omega L(E_z - E_1)^2} \cos^2 k_1 L_{\mathrm{w}/2} \frac{(E_z - V)\tan^2 kL_{\mathrm{w}/2}}{E_z + (E_z - V)\tan^2 kL_{\mathrm{w}/2}}. \tag{3.30}$$

For analytical clarity, we have set $m = m_{\mathrm{b}}$ in deriving the results given in (3.29) and (3.30). This is a good approximation since ψ_1 is localized mainly in the well, with a very small exponential tail into the barrier. In determining the eigenenergies, however, we do include the difference between m and m_{b}, which is important especially for the excited state E_2.

We consider the two example measurement geometries shown in Fig. 3.5a and b. The first geometry is relevant to the 45° facet detectors and to the

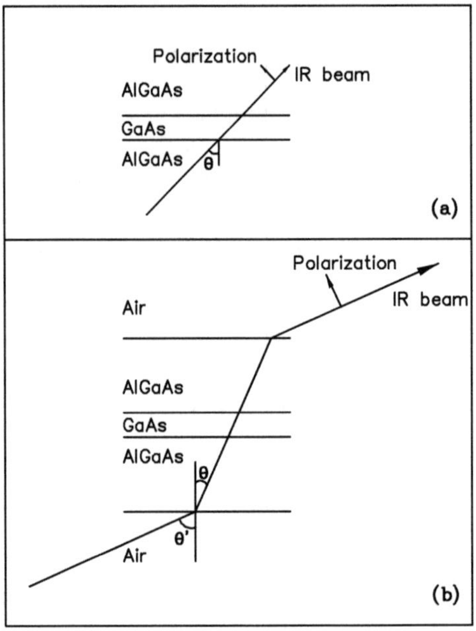

Fig. 3.5. Two measurement geometries considered in the model calculation

45° zigzag waveguides [42]. The second is commonly used for the Brewster angle transmission measurements [62]. From (3.20), the absorption quantum efficiency for Fig. 3.5a is explicitly given by

$$\eta = \frac{e^2 h}{4\epsilon_0 n_r mc} \frac{\sin^2\theta}{\cos\theta} n_{2D} \sum_{\text{final}} \frac{1}{\pi \Delta E} \frac{1}{1 + [(E_{\text{final}} - E_1 - \hbar\omega)/\Delta E]^2} f, \quad (3.31)$$

where E_{final} is the final state energy associated with the z-direction motion, the polarization of the light is in the plane of incidence (p-polarized), and the summation is over the final states. The contribution from the bound-to-bound transition is trivially evaluated, and the contribution from bound-to-continuum transitions involves converting the sum into a integral by $\sum \rightarrow \int_0^\infty L\,\mathrm{d}k'/\pi$. We then have

$$\eta = \frac{e^2 h}{4\epsilon_0 n_r mc} \frac{\sin^2\theta}{\cos\theta} n_{2D} \frac{1}{\pi \Delta E} \left[\frac{1}{1 + [(E_2 - E_1 - \hbar\omega)/\Delta E]^2} f_{B-B} + \right.$$
$$\left. + \frac{L\sqrt{2m}}{2\pi\hbar} \int_V^\infty \frac{\mathrm{d}E_z}{\sqrt{E_z - V}} \frac{1}{1 + [(E_z - E_1 - \hbar\omega)/\Delta E]^2} f_{B-C} \right]. \quad (3.32)$$

Note that the above is independent of L, as it should be, because $f_{B-C} \propto L^{-1}$ (see (3.30)).

Because of the divergence in the one-dimensional (1D) density of states at $E_z = V$ (the $1/\sqrt{E_z - V}$ factor in the second term in (3.32)), the bound-to-continuum contribution can be large when the broadening is small. In fact, in the limit of $\Delta E \to 0$, one obtains a divergent result. We will discuss this divergence explicitly to show that it is an effect of the density of states together with a special arrangement of the quantum well. In the $\Delta E \to 0$ limit, the Lorentzian lineshape becomes a δ-function, and the second term in (3.32), i.e., $\sum f_{B-C}$, becomes

$$\sum_{\text{final}} f_{B-C} = \frac{8C_1^2 V^2 \sqrt{2m}}{2\pi^2 \Delta E m\omega (E_z - E_1)^2} \cos^2(k_1 L_{\text{w}/2}) \times \frac{1}{\sqrt{E_z - V}} \frac{(E_z - V)\tan^2(kL_{\text{w}/2})}{E_z + (E_z - V)\tan^2(kL_{\text{w}/2})}. \tag{3.33}$$

The conservation of energy gives $E_z = E_1 + \hbar\omega$. Let us concentrate on the following factor from (3.33):

$$\begin{aligned} &\frac{1}{\sqrt{E_z - V}} \frac{(E_z - V)\tan^2 kL_{\text{w}/2}}{E_z + (E_z - V)\tan^2 kL_{\text{w}/2}} \\ &= \frac{1}{\sqrt{\hbar\omega - (V - E_1)}} \frac{[\hbar\omega - (V - E_1)]\tan^2 kL_{\text{w}/2}}{(E_1 + \hbar\omega) + [\hbar\omega - (V - E_1)]\tan^2 kL_{\text{w}/2}}. \end{aligned} \tag{3.34}$$

One can see immediately that the divergence in the density of states at $\hbar\omega = V - E_1$ is normally canceled by the factor $\hbar\omega - (V - E_1)$ in the numerator from the oscillator strength f_{B-C} *except* when $\tan^2 kL_{\text{w}/2} = \infty$, i.e., $kL_{\text{w}/2} = \pi/2 + P\pi$, where $P = 0, 1, 2, \ldots$.

In order to clarify the physical correspondence of this "selection rule" for $kL_{\text{w}/2}$, we rewrite the upper state eigenenergy condition (3.26 in the following form:

$$\tan k_2 L_{\text{w}/2} = -\frac{m_{\text{b}} k_2}{m \kappa_2}. \tag{3.35}$$

When $\kappa_2 \to 0$, we have $\tan^2 k_2 L_{\text{w}/2} \to \infty$ for which $k_2 L_{\text{w}/2} = \pi/2 + P\pi$. This implies that the condition $k_2 L_{\text{w}/2} = \pi/2 + P\pi$ corresponds to an odd parity "bound" state which is exactly in resonance with the top of the barrier. The physical reason for the divergent absorption is now clear: it arises from the combination of the (1) 1D density of states and (2) having the final state in resonance with the top of the barrier. One therefore might expect that a very large absorption could be obtained in a quantum well specially designed to this situation [7]. However, this does not occur in reality, because of the broadening factor. This is shown below, where calculated examples with realistic values for the broadening are given and are compared with experiments.

Figure 3.6 shows calculated absorption spectra for a single path through one quantum well. The measurement geometry corresponds to that of Fig. 3.5a. The internal angle of incidence is 45°, and the IR light is p-polarized. Parameters appropriate for an $Al_{0.33}Ga_{0.67}As$–GaAs quantum well were used. The GaAs well width was varied from 35 to 65 Å in 3 Å steps. The well reduced effective mass is 0.067 and that for the barrier is 0.094. The barrier height used

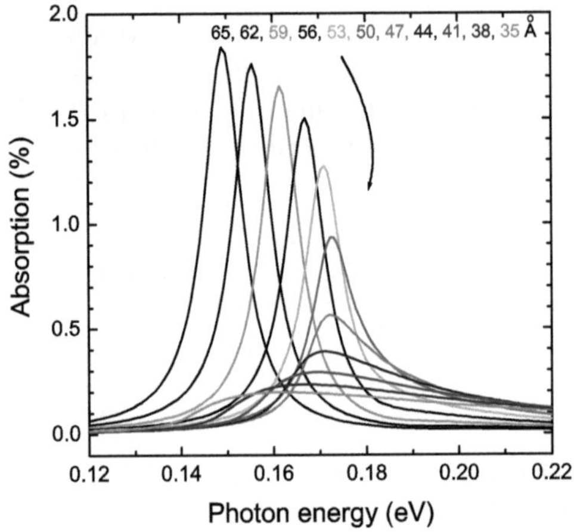

Fig. 3.6. Calculated absorption quantum efficiency vs. photon energy for one well for different well well widths from 35 to 65 Å. Parameters appropriate to $Al_{33}Ga_{67}As$–GaAs were used. The electron density in the well is 9×10^{11} cm^{-2}, and the internal angle of incidence is 45° (p-polarized)

was 0.25 eV. The broadening full width $2\,\Delta E$ was taken to be 10 meV. The 2D electron density in the well n_{2D} is 9×10^{11} cm^{-2}. The crossover from one bound state to two bound states occurs in the range of 48–49 Å. From Fig. 3.6, the absorption spectra become narrower in lineshape and higher in peak strength when the well width is increased. There is no abrupt change in the spectra when crossing from the pure bound-to-continuum case ($L_w \leq 47$ Å) to the case where both bound-to-bound and bound-to-continuum ISBTs contribute to the absorption ($L_w \geq 50$ Å).

To compare with experiments, including the interesting situation of crossing from one bound state to two bound states, one needs to measure a series of samples having either different well widths or different barrier heights. The only systematic study of the well width dependence of ISBTs in this crossover regime found in the literature is by Asai and Kawamura [62]. Some of their results are reproduced in Fig. 3.7 (dashed lines). The calculated absorption spectra, using parameters appropriate for their 70-period InGaAs–InAlAs quantum well structure grown on InP, are shown in Fig. 3.7. The InGaAs well width is varied from 35 to 100 Å. The reduced effective masses are 0.042 for the well and 0.075 for the barrier with 0.47 eV height. The broadening $2\,\Delta E$ is taken to be 20 meV and the 2D electron density in the well $n_{2D} = L_w \times (1.5 \times 10^{18}\ \text{cm}^{-3})$. The Brewster angle measurement geometry as shown in Fig. 3.5b uses an external (internal) angle of incidence of 73° (17°) and p-polarized IR light. Note that for the Brewster geometry, the factor $\sin^2\theta/(n_r \cos\theta)$ reduces [12] to $1/(n_r^2\sqrt{n_r^2+1})$. Since the bandgap of the well material (InGaAs) is much narrower than that of GaAs, it is necessary to include the effect of band nonparabolicity (see Sect. 3.4.4) to obtain an agreement of the calculated absorption peak positions with experiment. The

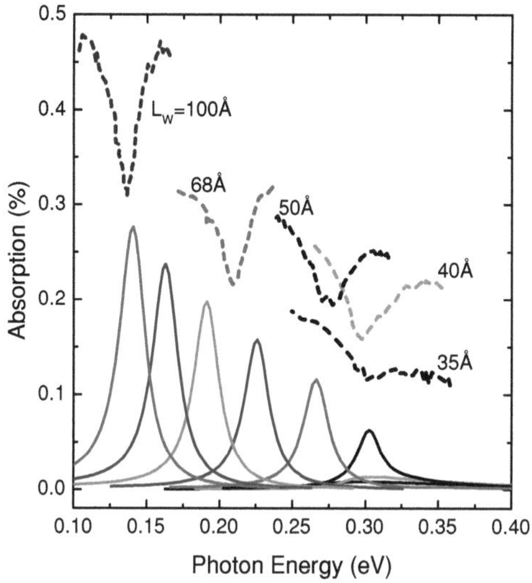

Fig. 3.7. Experimental transmittance data (dashed lines) of Asai and Kawamura [62], and calculated absorption quantum efficiency vs. photon energy (solid lines) for 70 wells for different well well widths of 35, 40, 50, 60, 70, 80, 90, and 100 Å. Parameters appropriate to InGaAs–InAlAs quantum wells grown in InP were used. The electron density in the well is $L_w \times (1.5 \cdot 10^{18}\,\mathrm{cm}^{-3})$, and the internal angle of incidence is 17° (p-polarized), where L_w is the well width.

calculated curves in Fig. 3.7 were obtained using the same formalism for the calculation of Fig. 3.6, but included an energy dependent effective mass [63] with parameter $\alpha = 1.24\,\mathrm{eV}^{-1}$. The general agreement between our calculations and the measurements of Asai and Kawamura is good.

To end this section, we point out that ISBTs have been investigated in many other structures and materials, though the conduction band n-type QWs are the most successful one for QWIPs. When tunnel coupling between adjacent QWs is nonnegligible, the multi-quantum well (MQW) is called a superlattice. Transport and ISBT in superlattices have been reviewed by Helm [64]. ISBTs can be also observed and potentially made use of in the valence band QWs [50,65–67]. Various material systems have been investigated recently, notably GaN/AlN QWs [68–71].

3.3.3 Transfer-Matrix Method

In order to calculate subband energies and associated wave functions not just for simple quantum wells but for potential structures involving a larger number of material layers, the transfer-matrix method is often applied. Here the boundary conditions between subsequent layers result in a matrix multiplication, thus giving rise to a "transfer matrix" describing the propagation across the whole multilayer.

We restrict ourselves here to piecewise constant potentials. A similar calculation involving Airy functions can in principle be executed for the case of piecewise linear potentials. Let us define a piecewise constant potential $V(z) = V_\nu$ and effective mass $m(z) = m_\nu$ within the intervals $(z_{\nu-1}, z_\nu)$ for $\nu = 1, ..., N$, respectively, $V(z) = V_0$ for $z < z_0$, and $V(z) = V_N$ for $z > z_N$. Let us further assume $V_0 = V_N$, $m_0 = m_N$. Our goal is to solve the Schrödinger equation (3.2) in order to determine the wave function ψ. We use a basis representation, where the wavefunction $\psi(z)$ within the interval $(z_{\nu-1}, z_\nu)$ is given by

$$\psi(z) = A_\nu \mathrm{e}^{\mathrm{i}k_\nu(z-z_\nu)} + B_\nu \mathrm{e}^{-\mathrm{i}k_\nu(z-z_\nu)} \quad \text{for } z_{\nu-1} \le z < z_\nu. \tag{3.36}$$

Equation (3.36) is a solution of (3.2) inside the interval $(z_{\nu-1}, z_\nu)$ if the momentum parameter (or wavenumber) k_ν is given by

$$k_\nu = \begin{cases} (1/\hbar)\sqrt{2m_\nu(E - V_\nu)} & \text{if } E > V_\nu, \\ -i(1/\hbar)\sqrt{2m_\nu(V_\nu - E)} & \text{if } E < V_\nu. \end{cases} \tag{3.37}$$

The parameters A_ν and B_ν are now determined by the requirement that the functions ψ and $\mathrm{d}/\mathrm{d}z(\psi/m)$ are continuous at the positions $z = z_\nu$. These boundary conditions can be expressed as

$$\begin{pmatrix} A_{\nu+1} \\ B_{\nu+1} \end{pmatrix} = \mathbf{T}_{\nu+1,\nu} \begin{pmatrix} A_\nu \\ B_\nu \end{pmatrix}. \tag{3.38}$$

The matrix $\mathbf{T}_{\nu+1,\nu}$ is given by

$$\mathbf{T}_{\nu+1,\nu} = \frac{1}{2} \begin{pmatrix} (1+\alpha_\nu)\mathrm{e}^{\mathrm{i}k_{\nu+1}d_{\nu+1}} & (1-\alpha_\nu)\mathrm{e}^{\mathrm{i}k_{\nu+1}d_{\nu+1}} \\ (1-\alpha_\nu)\mathrm{e}^{-\mathrm{i}k_{\nu+1}d_{\nu+1}} & (1+\alpha_\nu)\mathrm{e}^{-\mathrm{i}k_{\nu+1}d_{\nu+1}} \end{pmatrix}, \tag{3.39}$$

where we have introduced $d_{\nu+1} = z_{\nu+1} - z_\nu$ and

$$\alpha_\nu = \frac{k_\nu m_{\nu+1}}{k_{\nu+1} m_\nu}. \tag{3.40}$$

Equation (3.38) represents a recursion relation between the coefficients A_ν, B_ν associated with adjacent intervals. The iteration of (3.38) thus gives rise to an equation of the form

$$\begin{pmatrix} A_N \\ B_N \end{pmatrix} = \mathbf{T} \begin{pmatrix} A_0 \\ B_0 \end{pmatrix}, \tag{3.41}$$

where we have defined the transfer matrix[1]

$$\mathbf{T} = \mathbf{T}_{N,N-1}\mathbf{T}_{N-1,N-2} \cdot \ldots \cdot \mathbf{T}_{1,0}. \tag{3.42}$$

[1] Some authors define the transfer matrix $\widetilde{\mathbf{T}}$ via $\begin{pmatrix} A_0 \\ B_0 \end{pmatrix} = \widetilde{\mathbf{T}} \begin{pmatrix} A_N \\ B_N \end{pmatrix}$, which is related to the present definition via $\widetilde{\mathbf{T}} = \mathbf{T}^{-1} = \frac{1}{\det \mathbf{T}} \begin{pmatrix} T_{11} & -T_{12} \\ -T_{21} & T_{22} \end{pmatrix}$.

By construction, the determinant of T is given by $\det \mathbf{T} = m_N k_0 / k_N m_0$. In particular, we have $\det \mathbf{T} = 1$ if the potential and mass are identical for $\nu = 0$ and $\nu = N$.

This transfer matrix $\mathbf{T} \equiv \begin{pmatrix} T_{11} & T_{12} \\ T_{21} & T_{22} \end{pmatrix}$ now provides access to various types of numerical calculations:

- For bound states, we have $E < V_0$ and $E < V_N$. According to (3.36) and (3.37), we need $A_0 = 0$ and $B_N = 0$ in order to obtain an asymptotically decaying ψ. This condition is equivalent to $T_{22} = 0$, such that the associated subband energies can be calculated by finding the zeroes of $T_{22}(E)$.
- In order to determine tunnel probabilities (or transmission probabilities) of continuum states with $E > V_0$ and $E > V_N$, we can think of a particle wave coming in from the left, which is partially transmitted. For $A_0 = 1$, we thus have $B_N = 0$, which yields $A_N = (\det \mathbf{T})/T_{22}$. The transmission probability is thus given by $\mathcal{T} = |(\det \mathbf{T})/T_{22}|^{-2}$.
- Conducting the analogous calculation with a periodic potential allows us to evaluate quasiperiodic (or superlattice) states. Here quasiperiodicity refers to the condition that $\psi(z + d) = \exp(\mathrm{i}\phi)\psi(z)$ after the period d, which implies that $\mathbf{T}$ should have an eigenvalue of $\exp(\mathrm{i}\phi)$, and the Bloch-phase ϕ determines the momentum-wavevector $k = 2\pi\phi/d$ of the superlattice dispersion relation. Since we have $\det \mathbf{T} = 1$ for a periodic potential, the other eigenvalue is $\exp(-\mathrm{i}\phi)$, and the trace $\mathrm{tr}(\mathbf{T}) \equiv T_{11} + T_{22}$ satisfies $\mathrm{tr}(\mathbf{T}) = 2\cos\phi$. The latter relation can be used to calculate the miniband structure; in particular, miniband states are characterized by the condition $|\mathrm{tr}\ \mathbf{T}| \leq 2$.

3.4 Corrections to the Intersubband Energy and Lineshape

The previous considerations focus on the main contribution to the ISBT energy, magnitude, and lineshape based on the single-particle Schrödinger equation. Here we will treat some additional interactions and give criteria for when these interactions need to be taken into account.

3.4.1 Coulomb Interaction

Since carriers (electrons or holes) and ionized dopants (donors or acceptors) all have charges, the Coulomb interaction is probably the first that occurs to one to include in addition to the simple quantum well potential.

In fact, the Coulomb interaction is the origin of the commonly observed asymmetry between the forward and reverse I–V characteristics of GaAs/AlGaAs QWIPs, where the segregation of the Si dopant atoms occurring during growth leads to an asymmetry in the doping profile and hence in the quantum well potential. As a consequence, the effective barrier heights seen by electrons in the well in the forward and reverse directions will be different. Modeling the potential profile of n-type GaAs/AlGaAs QWIPs and comparing with experiments will be discussed in more detail in Sect. 7.3. Since the current depends exponentially on the barrier height, I–V measurements are very sensitive to the amount of asymmetry in the barrier heights.

As a simple illustrative example, the electrostatic potential V due the ionized dopants and electrons in the ground state satisfies the Poisson equation:

$$\frac{\mathrm{d}^2V}{\mathrm{d}x^2} = \frac{e^2}{\epsilon}[N_\mathrm{D}^+(x) - n_{2\mathrm{D}}\psi^*(x)\psi(x)], \tag{3.43}$$

where ϵ is the vacuum permitivity multiplied by the material dielectric constant, and $\psi(x)$ is the ground state wavefunction. Choosing $V|_{x=-\infty} = 0$ and $\mathrm{d}V/\mathrm{d}x|_{x=-\infty} = 0$, we get

$$V(x) = \frac{e^2}{\epsilon}\int_{-\infty}^{x} \mathrm{d}x_1(x - x_1)[N_\mathrm{D}^+(x_1) - n_{2\mathrm{D}}\psi^*(x_1)\psi(x_1)]. \tag{3.44}$$

The resulting quantum well potential is the sum of the perfect square well potential due to the bandedge profile and the electrostatic potential given by (3.44). We assume that all dopants are ionized, so that $\int_{-\infty}^{\infty} \mathrm{d}x N_\mathrm{D}^+(x) = n_{2\mathrm{D}}$; and we use the ground state wavefunction of the perfect square well for simplicity (see Sect. 3.3.2).

Obviously when the potential profile is modified, the eigenenergy and eigenfunction would change as a result. To include the leading order correction, one could perform a textbook first-order perturbation calculation for the position of the ground state eigenenergy. A change in wavefunction then leads to a change in potential profile. A rigorous solution should then be done self-consistently by numerically solving the Poisson and Schrödinger equations. To show an example of the effect discussed here, we plot in Fig. 3.8 a calculated self-consistent potential for a quantum well designed for 9-µm peak detection. We have purposely used a higher-than-optimum doping density of $10^{12}\,\mathrm{cm}^{-2}$ to show clearly the band bending. Even with this higher doping, the correction to the ISBT energy is only about 2%. The effect is even smaller for the case of a uniform doping in the well because the spread out of the ionized dopant charges mostly balances the carrier charge determined by the wavefunction shape.

Helm [42] calculated an example of a 8-nm well with 5-nm center region uniformly doped to $2 \times 10^{18}\,\mathrm{cm}^{-3}$ giving rise to a 2D electron density of $10^{12}\,\mathrm{cm}^{-2}$. The effect of the Coulomb interaction is still very small; the correction to the intersubband energy is about 1%. In addition, he also calculated the same well and electron density but for a modulation doping case. Modulation doping refers to the scheme of doping the barrier and having electrons transferred into the well. Modulation doped QWs exhibit substantially enhanced lateral mobility which leads to a narrower intersubband linewidth.

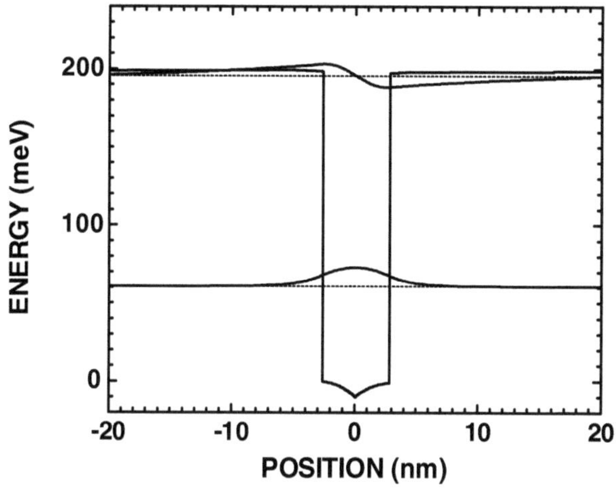

Fig. 3.8. Self-consistently calculated potential profile for a quantum well with 5.4-nm well width, $x = 0.24$ Aluminum fraction, and center delta doping of $10^{12}\,\mathrm{cm}^{-2}$. The ground and the first excited states are at 60.8 and 195.9 meV

The band bending, however, is substantial as a result of having electrons separated from the ionized donors. In his calculated example, the intersubband energy was changed from 111.1 to 104.4 meV (about 6% change).

3.4.2 Many-Particle Effects

Electrons or holes are identical spin-1/2 particles obeying Fermi statistics and Pauli principle, which implies that the wavefunction of the system must be anti-symmetric. There are effects associated with these special characteristics, which do not have classical analogies. Perhaps the most commonly noted one is the exchange correction to the energy. We briefly discuss the cause of this exchange correction using a two-electron example.

Suppose we have two electrons occupying two single-particle states ψ_a and ψ_b, the wavefunction of the system (before including the Coulomb interaction between them) is

$$\psi(\boldsymbol{x}_1, \boldsymbol{x}_2) = \frac{1}{\sqrt{2}}(\psi_a(\boldsymbol{x}_1)\psi_b(\boldsymbol{x}_2) - \psi_a(\boldsymbol{x}_2)\psi_b(\boldsymbol{x}_1)). \tag{3.45}$$

This is the Slater determinant of a two-particle wavefunction. We now "add" the Coulomb interaction $V(\boldsymbol{x}_1 - \boldsymbol{x}_2) = e^2/4\pi\epsilon|\boldsymbol{x}_1 - \boldsymbol{x}_2|$ and evaluate the leading order correction to the energy of the system. The first-order perturbation energy correction is:

$$\begin{aligned}\langle\psi|V|\psi\rangle = \frac{1}{2}\int & \mathrm{d}\boldsymbol{x}_1\mathrm{d}\boldsymbol{x}_2[\psi_a^*(\boldsymbol{x}_1)\psi_b^*(\boldsymbol{x}_2)V\psi_a(\boldsymbol{x}_1)\psi_b(\boldsymbol{x}_2)\\ & + \psi_a^*(\boldsymbol{x}_2)\psi_b^*(\boldsymbol{x}_1)V\psi_a(\boldsymbol{x}_2)\psi_b(\boldsymbol{x}_1) - \psi_a^*(\boldsymbol{x}_1)\psi_b^*(\boldsymbol{x}_2)V\psi_a(\boldsymbol{x}_2)\psi_b(\boldsymbol{x}_1)\\ & - \psi_a^*(\boldsymbol{x}_2)\psi_b^*(\boldsymbol{x}_1)V\psi_a(\boldsymbol{x}_1)\psi_b(\boldsymbol{x}_2)].\end{aligned} \tag{3.46}$$

The first two terms are identical as the Coulomb interaction depends only on $|\boldsymbol{x}_1-\boldsymbol{x}_2|$, i.e., that the subscripts 1 and 2 are interchangeable. This is precisely the direct Coulomb correction discussed in Sect. 3.4.1.

$$E_{\mathrm{Coul}} = \langle\psi|V|\psi\rangle_{\mathrm{Coulomb}} = \int \mathrm{d}\boldsymbol{x}_1\mathrm{d}\boldsymbol{x}_2\frac{\rho_a(\boldsymbol{x}_1)\rho_b(\boldsymbol{x}_2)}{4\pi\epsilon|\boldsymbol{x}_1-\boldsymbol{x}_2|}, \tag{3.47}$$

where the charge densities are $\rho_a(\boldsymbol{x}) = e|\psi_a(\boldsymbol{x})|^2$ and $\rho_b(\boldsymbol{x}) = e|\psi_b(\boldsymbol{x})|^2$.

The last two terms in (3.46) are the "exchange" correction:

$$E_{\mathrm{ex}} = \langle\psi|V|\psi\rangle_{\mathrm{Exchange}} = \int \mathrm{d}\boldsymbol{x}_1\mathrm{d}\boldsymbol{x}_2\mathrm{Re}[\psi_a^*(\boldsymbol{x}_1)\psi_b^*(\boldsymbol{x}_2)V\psi_a(\boldsymbol{x}_2)\psi_b(\boldsymbol{x}_1)]. \tag{3.48}$$

The term "exchange" can now be understood: similar to a Coulomb interaction integral but with the wavefunctions (or the particle position indices) exchanged. Clearly the exchange interaction does not have a classical analogy and is purely a quantum mechanical effect of spin-1/2 identical particles or Fermions.

For a many-particle system, one could proceed with a similar procedure of constructing a Slater determinant wavefunction and calculating the correction (or diagonalizing the full Hamiltonian). One can imagine there will be many more terms than in (3.46). In addition to the direct Coulomb interaction, the leading term correction is still the exchange and what is left (all the rest) is called "correlation." A full treatment of exchange-correlation is beyond the scope of this book, and in the following we list some results and discuss the qualitative features. Using a scheme called Hartree–Fock equation (including the direct Coulomb and exchange contributions), Bandara et al. [72] derived an approximate expression for a single QW with infinitely high potentials:

$$E_{\mathrm{ex}} \approx -\frac{e^2}{4\pi\epsilon}k_{\mathrm{F}}[(2/\pi)\boldsymbol{E}(k/k_{\mathrm{F}}) - 0.32(k_{\mathrm{F}}/k_{\mathrm{L}})], \tag{3.49}$$

where $\boldsymbol{E}(k/k_{\mathrm{F}})$ is the complete elliptical integral of the second kind, and $k_{\mathrm{L}} = \pi/L_{\mathrm{w}}$. For $k = 0$, the above becomes:

$$E_{\mathrm{ex}}(k=0) \approx -\frac{e^2}{4\pi\epsilon}k_{\mathrm{F}}[1 - 0.32(k_{\mathrm{F}}/k_{\mathrm{L}})] \tag{3.50}$$

and for $k = k_{\mathrm{F}}$:

$$E_{\mathrm{ex}}(k=k_{\mathrm{F}}) \approx -\frac{e^2}{4\pi\epsilon}k_{\mathrm{F}}[2/\pi - 0.32(k_{\mathrm{F}}/k_{\mathrm{L}})]. \tag{3.51}$$

Including only the exchange effect tends to overestimate the correction because the correlation is usually of opposite sign. A well-known approach that includes both exchange and correlation contributions is the local density approximation (LDA) within the well-celebrated density functional theory. A popular form of the LDA potential is given in [73] and refinements have been made to improve the accuracy [74]. The LDA formalism is very convenient in numerical calculations because the full exchange-correlation "potential" depends only on the local electron density.

So far we have only discussed corrections to the eigenenergy when the static Coulomb interaction is included. There are also dynamic effects, i.e., the observed ISBT resonance position is not simply the difference between the subband energies. There has been a body of work in connection with the Si inversion layers where ISBT was also studied; for a review see Ando et al. [2]. It has been shown that the ISBT resonance occurs at

$$\tilde{E}_{21}^2 = E_{21}^2(1+\alpha-\beta), \tag{3.52}$$

where $E_{21} = E_2 - E_1$ is the difference between subband 1 and 2, and α and β describe the depolarization and exciton-like shifts, respectively. The heuristic physical picture of the depolarization effect is as follows. Under the AC field associated with the incident photons, the electron that is trying to make the transition feels the time-dependent change of all other electrons. The depolarization shift essentially comes from direct Coulomb (Hartree) interaction with the oscillating electrons. Similarly, the exciton-like correction is due to the time-dependent Coulomb interaction with the "hole" left behind in the lower subband. The expressions for the two constants α and β are given by

$$\alpha = \frac{2e^2 n_{2\mathrm{D}}}{\epsilon E_{21}} S \tag{3.53}$$

with

$$S = \int_{-\infty}^{\infty} \mathrm{d}z \left[\int_{-\infty}^{z} \mathrm{d}z' \psi_2(z')\psi_1(z')\right]^2, \tag{3.54}$$

and

$$\beta = -\frac{2n_{2\mathrm{D}}}{E_{21}} \int_{-\infty}^{\infty} \mathrm{d}z \psi_2(z)^2 \psi_1(z)^2 \frac{\partial V_{xc}[n(z)]}{\partial n(z)}, \tag{3.55}$$

where $V_{xc}[n(z)]$ is the exchange-correlation energy in the LDA. Equations (3.53)and (3.54) have been derived in several ways, while (3.55) applies only under the LDA. Since both α and β are linearly related to the electron density, a higher density clearly leads to a larger correction.

There is a simpler way in estimating the depolarization correction [75,76]. The result is still written in the form $\tilde{E}_{21}^2 = E_{21}^2(1+\alpha)$. For $\alpha \ll 1$, the depolarization correction is given by

$$E_{\mathrm{depol}} = \tilde{E}_{21} - E_{21} \approx \frac{e^2\hbar^2 n_{3\mathrm{D}} f_{12}}{2\epsilon E_{21} m^*}, \tag{3.56}$$

where f_{12} is the oscillator strength.

Helm [42] gave a more detailed discussion of all these many-particle effects. We give again his results on the calculated example of a well-doped ($2 \times 10^{18}\,\mathrm{cm}^{-3}$ over 5 nm) example of an 8-nm GaAs well with $\mathrm{Al_{0.3}Ga_{0.7}As}$ barriers. The ground to first excited state separation under the ideal square well is 111.2 meV. The direct Coulomb (Hartree) changes it to 112.3 meV, and including exchange-correlation results in 114.6 meV. The depolarization effect moves the intersubband resonance to 123.4 meV, and the exciton-like effect leads to 120.3 meV. The net effect of all interactions is to move the resonance to a slightly larger energy, consistent with experiments.

3.4.3 Further Interactions

The linewidth of ISBT relates to various scattering effects by interactions of electrons with interface roughness, phonons, alloy disorder, and impurities. In principle, since ISBT is a fully collective phenomenon, the linewidth and resonance position should be calculated using many-particle theory. As an approximation, however, we discuss the linewidth in a single particle picture. The theoretical justification relies on the work of Nikonov et al. [77] who showed that in the limit of small band nonparabolicity the many-particle effects only cause blue shifts in the absorption spectra (depolarization shift) and that the linewidth is solely determined by the single particle relaxation rate. Nevertheless, experimental [78] and recent theoretical [79] results indicate that collective effects have a certain influence on the intersubband linewidth. In QWIPs, this contribution is not essential because of their specific device parameters (low carrier densities), and will therefore be neglected.

Unuma et al. [80] carried out a systematic investigation on the linewidth due to scattering by interface roughness, phonons, alloy disorder, and impurities. They found that interface roughness is a major factor because it broadens the width of the excited state by a large amount. The physical origin of this broadening is easy to understand by simply considering the eigenenergy of a QW with infinitely high barriers:

$$E_n = \frac{\hbar^2}{2m} \frac{\pi^2 n^2}{L_\mathrm{w}^2}. \tag{3.57}$$

A fluctuation in the well width δL_w leads to a broadening in eigenenergy of

$$\delta E_n = \frac{\hbar^2 \pi^2 n^2}{2m} \frac{2}{L_\mathrm{w}^3} \delta L_\mathrm{w} \tag{3.58}$$

or

$$\delta E_n = 2E_n \frac{\delta L_\mathrm{w}}{L_\mathrm{w}}. \tag{3.59}$$

Clearly a larger subband index n leads to a larger δE_n. For a 10-nm GaAs well, if there is a one-monolayer (0.3 nm) well width fluctuation, the broadening of the ground state ($n = 1$) is 3.3 meV, and that for the first excited state ($n = 2$)

is 13.4 meV. This should give rise to an observed linewidth of at least 10 meV, comparable to experimental results.

Depending on the correlation length Λ of these thickness fluctuations, two different kinds of broadening exist. If Λ exceeds the coherence length l_c of the electrons, then electrons at lateral positions in the QW plane can be considered as independent. In this case, the observed broadening can be significantly larger than the "intrinsic" broadening associated with the scattering lifetime of the transition. This situation is refered to as "inhomogeneous broadening." On the other hand, if $\Lambda < l_c$, then the interface fluctuations increase the homogeneous broadening (or lifeime broadening) of the electrons. This case leads to a reduced scattering time of the in-plane momentum (intrasubband scattering), which takes into account of the fact that the lateral electron momentum does not have a strict physical meaning any more. A detailed theoretical treatment of this situation including numerical calculations can be found in [80]. Since the contribution from intrasubband scattering in the first excited subband is much larger than that in the ground subband, interface roughness scattering has a much stronger influence (by about one order of magnitude) on the homogenous lifetime of the transition than on the momentum relaxation time [80].

Interface roughness scattering plays a particularly important role in GaAs and InGaAs quantum wells, such that ISBT are mostly homogeneously broadened. While homogeneous and inhomogeneous broadening cannot be distinguished in standard absorption measurements, prominent differences exist in ultrafast spectroscopy and nonlinear optics (see also Chap. 10).

An additional influence on the ISBT energy and spectrum is caused by the Coulomb potential of individual dopant atoms and the associated binding energy. At the carrier densities usually used in QWIPs, the associated donor states, (confined in the QW plane by the impurity potential) couple into impurity bands. In addition to the usual ISBT of "free" carriers without in-plane confinement, there are in addition transitions from the ground impurity band (located slightly below E_1) to excited impurity bands which are analogous to higher subbands but share the in-plane confinement. In most cases, the impurity-related absorption band (which also depends on the precise impurity positions in the growth direction) cannot be observed experimentally since it overlaps with the "free" intersubband absorption, in particular for bound-to-continuum QWIPs. Nevertheless, recent investigations resulted in a clear evidence of two spectral bands split by about 6 meV for a $E_1 \rightarrow E_2$ transition energy around 150 meV. Increasing the carrier density from 2×10^{11} to $4 \times 10^{11}\,\mathrm{cm}^{-2}$ gave rise to a Mott transition, where, due to the increase of the coupling, the impurity band merges with the conduction band [81]. The impurity potential is supposed to be of stronger influence in THz-QWIPs (discussed in Sect. 4.4), which operate not only at smaller transition energies, but also at lower operating temperatures and carrier densities. At low carrier densities and temperatures, changes in the absorption spectrum can be caused by carrier freeze-out into the impurity band. Pronounced effects have been demonstrated recently in superlattices where the parity selection rule was affected by the impurity potential [82].

3.4.4 Band Nonparabolicity

Much of the discussion in this book uses the simplest effective mass model of one parabolic band, i.e., the energy vs. wavevector dispersion relation is described by $E = \hbar^2 k^2/2m$, with the effective mass m taken to be constant. This ideal situation is a good approximation only if the energy of the electron is close to the bandedge. Far away from the bandedge the dispersion becomes different from a parabolic band. Within the $k \cdot p$ formalism [83], this nonparabolicity is explained by mixing of the conduction band with additional (spin split-off and valence) bands. The 8-band $k \cdot p$ model, originally proposed by E.O. Kane [83], uses an 8×8-matrix Hamilton operator acting on wavefunctions of electron, heavy-hole, light-hole, and split-off bands, each with two spin values. Taking into account also the second conduction band, located about 3 eV above E_C in GaAs, results in a 14-band model [84].

Two-Band Model

As a simpler approach, we discuss here an effective two-band model, which can be regarded as a good approximation to an n-type QWIP. In fact, the three valence bands used in the original Kane model are energetically far enough away from the conduction band and can be merged into one effective valence band. The free Hamiltonian H_0 is given here by the 2×2-matrix [85–87]

$$H_0 = \begin{pmatrix} V_\mathrm{C}(z) & -\mathrm{i}\widetilde{\gamma}\frac{\partial}{\partial z} \\ -\mathrm{i}\widetilde{\gamma}\frac{\partial}{\partial z} & V_\mathrm{V}(z) \end{pmatrix}, \tag{3.60}$$

where $V_\mathrm{C}(z)$, $V_\mathrm{V}(z)$ are the spatial distributions of the conduction and valence bandedge, respectively. This Hamiltonian acts on the wave function $\psi = (\psi_\mathrm{C}, \psi_\mathrm{V})^t$ which comprises a conduction part ψ_C and a valence part ψ_V. The wave equation $H_0\psi = E\psi$ now implies that ψ_C is accompanied by the valence part

$$\psi_\mathrm{V} = \frac{-i\widetilde{\gamma}}{E - V_\mathrm{V}} \frac{\partial}{\partial z} \psi_\mathrm{C}. \tag{3.61}$$

In particular, a bound state ψ with real-valued ψ_C contains an imaginary ψ_V.

The meaning of the parameter $\widetilde{\gamma}$, which is related to the $k{\cdot}p$ matrix element π_{cv} via $\widetilde{\gamma} = \sqrt{2/3}\hbar\pi_{cv}$ [85–87], becomes clear when solving the wave equation for ψ_C,

$$\left[V_\mathrm{C} - \frac{\partial}{\partial z} \frac{\widetilde{\gamma}^2}{E - V_\mathrm{V}} \frac{\partial}{\partial z} \right] \psi_\mathrm{C} = E\psi_\mathrm{C}. \tag{3.62}$$

Comparing this equation with an effective mass approximation suggests a correspondence between $\hbar^2/2m$ and $\widetilde{\gamma}^2/(E - V_\mathrm{V})$, which can be interpreted as an energy dependence of the effective mass,

$$m = \frac{\hbar^2}{2\tilde{\gamma}^2}(E - V_\mathrm{V}). \tag{3.63}$$

For constant potentials V_C and V_V, (3.62) leads to a hyperbolic dispersion of the electron momentum k, given by the relation $\tilde{\gamma}^2 k^2 = (E - V_\mathrm{C})(E - V_\mathrm{V})$, or equivalently,

$$E - V_\mathrm{C} = -\frac{E_\mathrm{g}}{2} \pm \sqrt{\frac{E_\mathrm{g}^2}{4} + \tilde{\gamma}^2 k^2} \tag{3.64}$$

with the bandgap energy $E_\mathrm{g} = E_\mathrm{C} - E_\mathrm{V}$. From (3.64), the group velocity $v_\mathrm{g} = \hbar^{-1}|\partial E/\partial k|$ at the energy $E > V_\mathrm{C}$ is found to be

$$v_\mathrm{g} = \frac{2\tilde{\gamma}}{\hbar} \frac{\sqrt{E - V_\mathrm{C}}\sqrt{E - V_\mathrm{V}}}{2E - V_\mathrm{C} - V_\mathrm{V}}. \tag{3.65}$$

We note that $v_\mathrm{g} \to \tilde{\gamma}/\hbar$ for $E \to \infty$, which indicates a mathematical analogy between the two-band model and relativistic quantum mechanics.

The probability density associated with a state $\psi = (\psi_\mathrm{C}, \psi_\mathrm{V})^t$ is given by $\rho = \overline{\psi_\mathrm{C}}\psi_\mathrm{C} + \overline{\psi_\mathrm{V}}\psi_\mathrm{V} = \psi^*\psi$, and the current density by $j = \frac{\tilde{\gamma}}{\hbar}(\overline{\psi_\mathrm{C}}\psi_\mathrm{V} + \overline{\psi_\mathrm{V}}\psi_\mathrm{C}) = \psi^* v_\mathrm{op}\psi$. Here we have introduced the velocity operator $v_\mathrm{op} = (i/\hbar)(H_0 z - zH_0) = \frac{\tilde{\gamma}}{\hbar}\begin{pmatrix} 0 & 1 \\ 1 & 0 \end{pmatrix}$. Sufficiently close to the conduction bandedge, we typically have $|\psi_\mathrm{C}| \gg |\psi_\mathrm{V}|$, so ψ_C can be used for plotting the wavefunction.

The parameter $\tilde{\gamma}$ is now determined by the condition that the correct effective mass is obtained at the conduction bandedge. For GaAs, we use $E_\mathrm{C} - E_\mathrm{V} = 1.519\,\mathrm{eV}$ and (3.63) yields $\tilde{\gamma} = 0.9294\,\mathrm{eV\ nm}$. For the conduction bandedge as a function of the Al-content x, we use the expression [88] $E_\mathrm{C} = 0.57(1.594\,x + x(1 - x)(0.127 - 1.310\,x))\,\mathrm{eV}$, where the prefactor 0.57 is the band offset parameter. In order to ensure the correct reduced effective mass value also at higher Al-contents, i.e., $m^* = 0.067 + 0.083\,x$, we define the *effective* valence bandedge E_V (rather than the real one) via the relation [87] $E_\mathrm{V} = E_\mathrm{C} - 1.519(1 + \frac{0.083}{0.067}\,x)\,\mathrm{eV}$.

We note that a two-band formalism yields a significantly better description of the observed subband structure than the usual one-band Hamiltonian. This also becomes clear when comparing the momentum dispersions predicted by different models, as shown in Fig. 3.9. Significant deviations exist between the predictions of the effective-mass approximation and the two-band model, and the latter already behaves quite similar to the 14-band model of [84]. These deviations directly influence the accuracy of calculated subband energies, in particular for excited subbands. Figure 3.9 also indicates the necessity to distinguish between the dispersions of the barrier ($\mathrm{Al_{0.3}Ga_{0.7}As}$) and quantum well (GaAs) materials.

For multilayers, a transfer matrix calculation in the two-band approximation essentially works as already described in Sect. 3.3.3. Using the potentials $V_\mathrm{C}(z) = V_{\mathrm{C}\nu}$ and $V_\mathrm{V}(z) = V_{\mathrm{V}\nu}$ inside the interval $[z_{\nu-1}, z_\nu]$, (3.38), (3.39), (3.41), and (3.42) still hold after defining the momentum parameters (instead of (3.37))

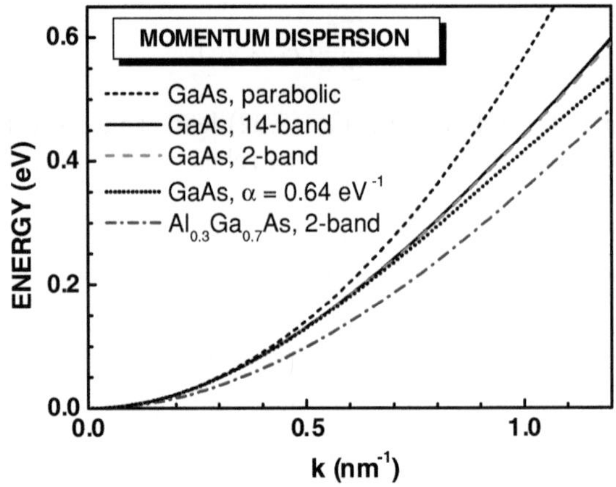

Fig. 3.9. Momentum dispersions as obtained from the "parabolic" effective mass model (3.1), the two-band model (3.64), the model of (3.69) with $\alpha = 0.64\,\mathrm{eV}^{-1}$ for GaAs, and the 14-band calculation of [84]

$$k_\nu = \begin{cases} (1/\widetilde{\gamma})\sqrt{(E - V_{\mathrm{C}\nu})(E - V_{\mathrm{V}\nu})} & \text{if } E > V_{\mathrm{C}\nu} \\ (-\mathrm{i}/\widetilde{\gamma})\sqrt{(V_{\mathrm{C}\nu} - E)(E - V_{\mathrm{V}\nu})} & \text{if } E < V_{\mathrm{C}\nu} \end{cases} \tag{3.66}$$

and the parameter (instead of (3.40))

$$\alpha_\nu = \frac{k_\nu(E - V_{\mathrm{V},\nu+1})}{k_{\nu+1}(E - V_{\mathrm{V}\nu})}. \tag{3.67}$$

Results of subband calculations using the transfer-matrix method in the two-band approximation are shown in Figs. 3.8 and 5.3.

Simple Model for Nonparabolicity

Finally, as a simpler way to include the effect of nonparabolicity, one could use the following expression:

$$E = \frac{\hbar^2 k^2}{2m}(1 - \gamma k^2), \tag{3.68}$$

where γ is a nonparabolicity parameter or, more conveniently,

$$E = \frac{\hbar^2 k^2}{2m}(1 - \alpha E), \tag{3.69}$$

where α is a nonparabolicity parameter with the dimension of an inverse energy. To determine the eigenenergies, one needs to numerically solve the appropriate transcendental equations (see Sect. 3.3.2 for the case of a symmetrical single QW). In computer programs, one scans the energy and looks for roots. The expressions in the transcendental equations require relations between E and k. Using (3.69), the band nonparabolicity is taken into account in this simple way (used in Sect. 3.3.2).

Applying this scheme to the case of a quantum well with infinitely high barriers, one easily gets (instead of (3.4)):

$$E_n = \frac{\hbar^2}{2m}\left(\frac{\pi n}{L_\mathrm{w}}\right)^2 \Big/ \left(1 + \frac{\hbar^2}{2m}\left(\frac{\pi n}{L_\mathrm{w}}\right)^2 \alpha\right). \tag{3.70}$$

Taking the reduced effective mass $m^* = 0.067$ for GaAs and a well width of 10 nm, the value of the quantity $\frac{\hbar^2}{2m}(\frac{\pi}{L_\mathrm{w}})^2$ is 0.0558 eV. The parameter α for GaAs amounts to less than $1\,\mathrm{eV}^{-1}$. The nonparabolicity correction for the ground state $n = 1$ is very small and for the excited states can be nonneglibible. For InGaAs lattice matched to InP, the value taken in the calculation of Sect. 3.3.2 is $\alpha = 1.24\,\mathrm{eV}^{-1}$, and the results compare well with experiments.

In the in-plane directions, nonparabolicity causes upper and lower subbands to be nonparallel in their E vs. $\boldsymbol{k}_{xy}$ relations (see also Fig. 3.9). This was expected to cause a broadening in the observed ISBT linewidth [89]. However the situation is complicated by the collective nature of the ISBT. Warburton et al. [78] showed that the ISBT linewidth in InAs/AlSb QWs (with highly nonparabolic band structure) is dominated by the collective effect resulting in a narrow and almost temperature-independent linewidth.

3.5 Intersubband Relaxation and Carrier Capture

The photoconductive gain in QWIPs is strongly influenced by the time constant for relaxation of the photoexcited carriers back into the ground subband. We therefore provide in this section some basic concepts on how to treat theoretically the relevant interactions leading to carrier capture, namely electron–phonon, electron–impurity, and electron–electron scattering. We restrict ourselves here to n-type QWIPs (i.e., to electrons), since the bandstructure parameters of the conduction band are known reasonably well, and since the coupling between heavy and light hole states in the valence band produces additional complications.

The standard theoretical approach uses Fermi's golden rule

$$\frac{1}{\tau_i} = \frac{2\pi}{\hbar}\sum_\mathrm{f} |\langle i|H_\mathrm{scatt}|f\rangle|^2 \delta(E_\mathrm{f} - E_i), \tag{3.71}$$

to calculate the lifetime τ_i of the initial state i at the energy E_i [90–92]. The scattering processes into the final states f at energy E_f are induced by the scattering Hamiltonian H_scatt. We note that the energies E_i and E_f of the initial and final states include not only the electrons but also other excitations (e.g., phonons) participating in the interaction.

Using the interaction Hamiltonians treated below, typical applications of (3.71) include the determination of time constants for intersubband relaxation, carrier capture, and tunneling. In many cases, only one initial state $\langle i|$ is relevant for the relaxation process (particularly if $\langle i|$ is a bound state), or it is a good approximation to take into account one representative initial state.

The most relevant interaction here is the Fröhlich interaction between electrons and longitudinal optical (LO) phonons. At large carrier densities ($\sim 10^{12}\,\mathrm{cm}^{-2}$ and greater), electron–impurity and electron–electron scattering also have a substantial influence. Radiative relaxation governed by the Hamiltonian of (3.8) is orders of magnitude less efficient than electron–photon interaction and can be neglected in this context.

3.5.1 Electron-Phonon Interaction

In general, the elecron–phonon interaction can be written as [92]

$$H_{\mathrm{ep}} = \sum_{\mathbf{q},\mu} C_\mu(\mathbf{q}) \mathrm{e}^{\mathrm{i}\mathbf{q}\mathbf{r}} \hat{b}^{\mu}_{\mathbf{q}} + C^*_\mu(\mathbf{q}) \mathrm{e}^{-\mathrm{i}\mathbf{q}\mathbf{r}} \hat{b}^{\mu\dagger}_{\mathbf{q}}, \tag{3.72}$$

where μ is the band index, $\mathbf{q}$ the phonon wavevector, $\hat{b}^{\mu\dagger}_{\mathbf{q}}$, $\hat{b}^{\mu}_{\mathbf{q}}$ are the phonon creation and annihilation operators, and $C_\mu(\mathbf{q})$ the coupling constant. The asterisk denotes the complex conjugate. The scattering process is then described using electron states $|\Psi_i\rangle$ and a boson representation of the phonon states (e.g., $\hat{b}^{\mu\dagger}_{\mathbf{q_f}}|0\rangle$ for a one-phonon state with final phonon wavevector $\mathbf{q}_f$).

In the following, we consider bulk-like LO phonons and neglect their momentum dispersion. This approximation has proven to yield realistic time constants both in the contexts of intersubband relaxation [91] and carrier capture in QW structures [90,93]. In this case, the *Fröhlich interaction* between electrons and LO phonons can be written as

$$C_\mu(\mathbf{q}) = \frac{\mathrm{i}e}{|\mathbf{q}|} \sqrt{\frac{\pi E_{\mathrm{LO}}}{\mathcal{V}} \left(\frac{1}{\epsilon_\infty} - \frac{1}{\epsilon_{\mathrm{S}}}\right)}, \tag{3.73}$$

where E_{LO} is the LO phonon energy, ϵ_∞ (ϵ_{S}) the high-frequency (static) dielectric constant, and $\mathcal{V}$ the normalization volume.

The resulting scattering rate for Fröhlich scattering by LO phonons from subband Ψ_i to subband Ψ_f with the effective mass m is then given by [91,92]

$$\frac{1}{\tau_i} = \frac{me^2 E_{\mathrm{LO}}}{4\hbar^3} \left(\frac{1}{\epsilon_\infty} - \frac{1}{\epsilon_{\mathrm{S}}}\right) \int\!\!\int \mathrm{d}z\mathrm{d}z' \Psi^*_f(z)\Psi_f(z')\Psi_i(z)\Psi^*_i(z') \times \int_0^{2\pi} \mathrm{d}\varphi \frac{\exp(-a_\varphi |z-z'|)}{a_\varphi}, \tag{3.74}$$

where we have defined the quantitiy $a_\varphi = |\mathbf{k}_f - \mathbf{k}_i|$. Here φ denotes the angle between the initial and final electron wavevectors $\mathbf{k}_i$ and $\mathbf{k}_f$. For the present discussion, we assume $\mathbf{k}_i = 0$, so that $a_\varphi = k_f$. At high temperatures, (3.74) should be multiplied by $N_{\mathrm{LO}} + 1/2 \pm 1/2$, with the Bose-Einstein occupation factor $N_{\mathrm{LO}} = (\exp(\hbar\omega_{\mathrm{LO}}/k_{\mathrm{B}}T) - 1)^{-1}$, with the plus and minus signs referring to phonon emission and phonon absorption, respectively. This correction can be neglected for temperatures at which IR detectors are usually operated.

Equations (3.71)–(3.74) show the most prominent property of LO phonon-assisted relaxation – the interaction is inversely proportional to the phonon wavevector $\boldsymbol{q}$. For intersubband energies $E_2 - E_1$ higher than the LO phonon energy E_{LO}, τ_i thus increases roughly proportional to $E_2 - E_1 - E_{\mathrm{LO}}$. For $E_2 - E_1 \gg E_{\mathrm{LO}}$, (3.74) can be approximated (in the case of bound-to-continuum QWIPs) by the relation [20,94]

$$\frac{1}{\tau_i} = \frac{e^2 \lambda_{\mathrm{c}} E_{\mathrm{LO}} I_1}{4h^2 c L_{\mathrm{p}}} \left(\frac{1}{\epsilon_\infty} - \frac{1}{\epsilon_{\mathrm{S}}} \right), \tag{3.75}$$

where λ_{c} is the cutoff wavelength, L_{p} a characteristic length (taken as the QWIP period), and $I_1 \approx 2$ a dimensionless integral. Equation (3.75) is quite useful for a quick estimate of the capture time, and yields about 5 ps for typical QWIP parameters.

For $E_2 - E_1 \to E_{\mathrm{LO}}$, $\boldsymbol{q}$ in (3.74) goes to zero, but τ_i remains finite since the number of final states decreases. For energy spacings below E_{LO}, the emission of LO-phonon becomes irrelevant; in this case, scattering by acoustic phonons (not discussed here) becomes important which leads to lower associated decay rates. Intersubbband relaxation times in excess of 500 ps have thus been observed at low temperatures [95].

In addition to these general trends, carrier capture in QWIPs also depends on the actual manifold of continuum states over which the electrons are distributed prior to their capture. Here the situation rapidly becomes involved by the influence of resonance effects, due to electric field-induced localization and resonant couplings of minibands in the continuum and due to the enhancement associated with the quasibound nature of the excited subband. We therefore refer to [41,91] for a more detailed theoretical treatment.

Figure 3.10a shows calculated LO phonon-mediated intersubband scattering times $\tau_{2\to1}$ from the second into the first subband for GaAs/AlGaAs quantum wells between 5.5 nm (where the second subband is quasibound) and 17.8 nm (where $E_2 - E_1 = E_{\mathrm{LO}}$) at $T = 0$ K. The increase of $\tau_{2\to1}$ with decreasing well width (increasing intersubband energy) is clearly seen. The steep increase of $\tau_{2\to1}$ at well widths below ~6 nm arises from the reduced overlap when the second subband turns into a quasibound state.

3.5.2 Electron-Impurity and Electron-Electron Scattering

Both electron–impurity and electron–electron scattering are based on the Coulomb interaction $H_{\mathrm{C}} = e^2/4\pi\epsilon r$. Since the scattering potential for both

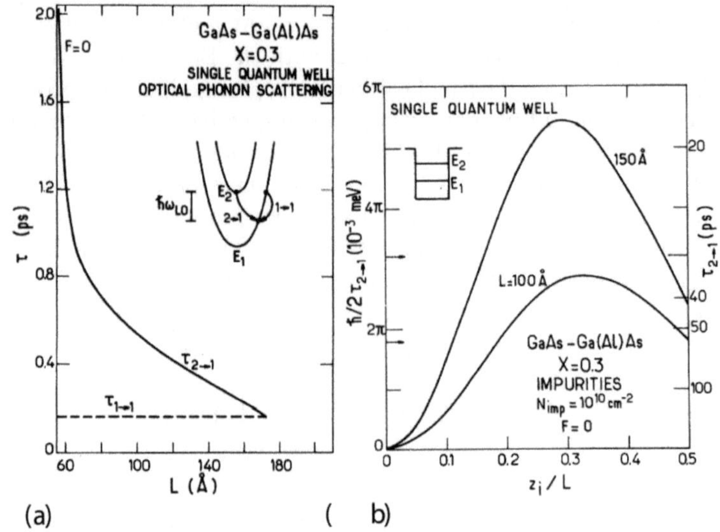

Fig. 3.10. (**a**) Inter subband ($\tau_{2\to1}$) and intrasubband ($\tau_{1\to1}$) relaxation times associated with LO phonon emission in GaAs/$Al_{0.3}Ga_{0.7}As$ single-quantum-wells at $T = 0\,K$ vs. well thickness. The inset depicts the schematics of the scattering process. (**b**) $\tau_{2\to1}$ (right scale) and broadening $1/2\tau_{2\to1}$ due to electron–impurity scattering vs. dimensionless impurity position z_i/L (on-center: $z_i/L = 0$, on-edge: $z_i/L = 1/2$). Horizontal arrows correspond to homogeneous impurity distributions in the GaAs quantum well (From [91])

processes is basically identical, the associated relaxation times are of the same order of magnitude.

Scattering by Ionized Impurities

Scattering by ionized (i.e., unscreened) impurities has been investigated theoretically by Ferreira and Bastard [91]. An ionized impurity located at $\boldsymbol{R}_\nu = (\boldsymbol{R}_{\parallel\nu}, z_\nu)$ gives rise to the scattering potential

$$H_\nu = \frac{2\pi e^2}{\epsilon_s \mathcal{V}^{2/3}} \sum_{\boldsymbol{Q}_\parallel} \frac{1}{Q_\parallel} \exp\left(-Q_\parallel |z - z_\nu| + \mathrm{i}\boldsymbol{Q}_\parallel \cdot (\boldsymbol{r}_\parallel - \boldsymbol{R}_{\parallel\nu})\right), \qquad (3.76)$$

where the summation is over the wavevector $\boldsymbol{Q}_\parallel = (Q_x, Q_y)$. $\mathcal{V}$ is the volume of the sample and ϵ_s the static dielectric constant.

The associated scattering rate τ_i has been evaluated [91] for the case of a sheet of impurities with the areal density N_{imp} at the position z_0. The result is given by

$$\frac{1}{\tau_i} = \frac{2\pi^2 e^4}{\hbar \epsilon_S^2} N_{\mathrm{imp}} \sum_{\substack{f \\ E_f < E_i}} \frac{1}{E_i - E_f}$$

$$\times \left| \left\langle \Psi_f \left| \exp\left(-|z - z_0| \sqrt{\frac{2m}{\hbar^2}(E_i - E_f)} \right) \right| \Psi_i \right\rangle \right|^2 , \tag{3.77}$$

which holds for an initial electron momentum $\mathbf{k}_i = 0$. For impurities distributed on several layer planes, (3.77) has to be calculated and summed up for each impurity sheet.

Associated intersubband relaxation times $\tau_{2\to 1}$ are plotted in Fig. 3.10b for 10 and 15 nm wide GaAs quantum wells as a function of the impurity position. For an impurity density of $\sim 10^{10}\,\mathrm{cm}^{-2}$, typical time constants are a few 10 ps. Note however that impurity scattering is parity forbidden according to (3.77), if the impurities happen to be located exactly at the well center.

Electron–Electron Scattering

In the case of electron–electron scattering, at least two carriers with different initial and final states participate in the interaction, thus giving rise to matrix elements of the form

$$\langle f|H_\mathrm{C}|i\rangle = \langle \psi_a(\boldsymbol{x}_1))\psi_b(\boldsymbol{x}_2)| \frac{e^2}{4\pi\epsilon|\boldsymbol{x}_1 - \boldsymbol{x}_2|} |\psi_c(\boldsymbol{x}_1))\psi_d(\boldsymbol{x}_2)\rangle. \tag{3.78}$$

The summation thus involves four independent states $\psi_a, \ldots, \psi_d$; a more complicated situation than the one covered in Sect. 3.4.2. According to (3.78), one or two electrons per scattering event are transferred into a different subband. For further evaluation of this scattering process, see [41].

At higher carrier densities, the electron–electron interaction in a quantum well is screened due to the presence of other carriers, i.e., the probability for electron–electron scattering is reduced since many other carriers respond to the changes induced by such a scattering event. For example, at $n_\mathrm{2D} \sim 10^{12}\,\mathrm{cm}^{-2}$, scattering rates are a few times $10^{12}\,\mathrm{s}^{-1}$ without screening and about 50% less if screening is taken into account [41]. The influence of screening is negligible at low carrier densities $n_\mathrm{2D} \sim 10^{10}\,\mathrm{cm}^{-2}$, where the calculated scattering rate is a few $10^{10}\,\mathrm{s}^{-1}$ [41].

Ando [96] has taken a different approach, where electron–electron scattering is treated in terms of "collective excitations." This approach allows for some computational simplifications, since intersubband relaxation in the presence of many-particle effects leads to "effective" lifetimes and broadenings. In this way, additional interactions (including optical-phonon scattering) can still be taken into account in the presence of electron–electron interaction [80]. Intersubband scattering rates using different screening models have been compared by Lee and Galbraith [97]. They found that intersubband scattering can be considerably enhanced by collective effects, and that screening – in spite of a sometimes negligible quantitative influence – gives rise to a qualitative difference at small momentum transfer, where scattering rates remain finite (in contrast to the single-particle picture discussed in Sect. 3.5.1).

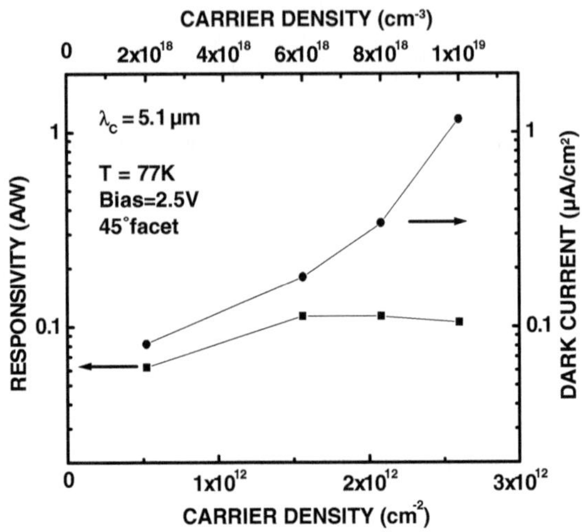

Fig. 3.11. Peak responsivity and dark current of $In_{0.3}Ga_{0.7}As/Al_{0.32}Ga_{0.68}As$ MWIR QWIPs vs. carrier density (from [18])

Figure 3.11 shows the measured responsivity of InGaAs/AlGaAs QWIPs with 2.6 nm well width, as a function of the carrier density [18], which can give a feeling for the practical influence of these scattering processes. While the responsivity is approximately linear with the carrier density at the usual $n_{2D} = 1 \ldots 5 \times 10^{11}\,\mathrm{cm}^{-2}$, the figure clearly shows that $\mathcal{R}$ saturates above $10^{12}\,\mathrm{cm}^{-2}$, and even indicates a slight trend towards decreasing $\mathcal{R}$ (which we attribute to a broadening of the spectra). At low n_{2D}, carrier capture mainly occurs via LO phonon scattering, which leads to the linear behavior since τ_c (and therefore g) remains approximately unchanged. The saturation at high n_{2D} is understood by a scattering rate proportional to n_{2D}, such that the increase in quantum efficiency is compensated by the resulting decrease in the gain, thus leading to a constant responsivity.

4

Photoconductive QWIP

The simplest picture of a QWIP made of n-type GaAs/AlGaAs is given in Fig. 4.1. The detector operation is based on photoemission of electrons from the quantum wells. The contacts on both sides are of the same n-type. The device is essentially a unipolar photoconductor. Usually many wells (10–100) are required for sufficient absorption.

4.1 Dark Current

A good understanding of the dark current is crucial for design and optimization of QWIPs because dark current contributes to the detector noise and dictates the operating temperature. We first present two simple physical models. As such, the results provide only an order-of-magnitude estimate; however, the physical principles are clear. A brief discussion of numerical models suited for guiding the fine-tuning of the device parameters and for interpreting higher order effects is then given.

4.1.1 Simple Models

There are several common assumptions or approximations made to define the physical regime for all the discussions in this section. These are: (a) the interwell tunneling contributes negligibly to the dark current, (b) the electron density in each well remains constant, (c) the heavily doped emitter serves as a perfectly injecting contact, and (d) mainly one bound state is confined in the quantum well, including the case where the upper state (final state of the ISBT) is in resonance or very close to the top of the barrier. Assumption (a) is satisfied by requiring the barriers to be sufficiently thick. Assumption (b) is a good approximation, but is not strictly valid especially at large bias voltages as shown experimentally [58]. Assumption (c) is expected to be valid for QWIPs with a large number of quantum wells, consistent with experimental results [98]. The effect of contacts becomes important for QWIPs with a small number

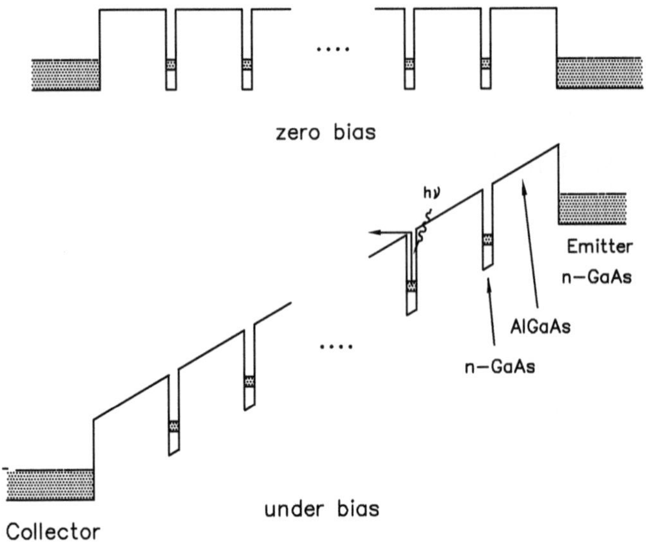

Fig. 4.1. Schematic conduction bandedge profile of a GaAs/AlGaAs QWIP under zero (*above*) and finite (*below*) bias. The electron population in the n-type wells is provided by doping using silicon. The emitter and collector contact layers are doped with silicon. Photons ($h\nu$) excite electrons from quantum wells, causing a photocurrent

of quantum wells as shown in simulations [99]. To produce good detectors, condition (d) is required, as stated in Sect. 3.1.

Having defined the physical regime, the dark current in a typical and standard photoconductive QWIP is controlled by the flow of electrons above the barriers, and by the emission and capture of electrons in the wells. Figure 4.2 presents pictorially the electron distribution (top) and the processes controlling the dark current (bottom).

The top part of the figure indicates that at finite temperatures electrons are not only bound in the well, but are also distributed outside of the well and on top of the barriers. The energy region for electrons contributing to the dark current is indicated by the large brace bracket.

The lower part of Fig. 4.2 shows the dark current paths. In the barrier regions (on top of the barriers), the current flows in a three-dimensional (3D) fashion, and the current density is labeled as j_{3D} which *equals* the dark current J_{dark}. In the vicinity of each well, the emission of electrons from the well (j_e) contributes to the dark current. This current, which tends to lower the electron density in the well, *must* be balanced by the trapping or capture of electrons into the well under steady state ($j_c = j_e$). Since the dark current is the same throughout the structure, j_{3D} and j_e (or j_c) are related. If we define a trapping or capture probability p_c for an electron traversing a well with energy larger than the barrier height, we must have $j_c = p_c\, j_{3D}$, and the sum

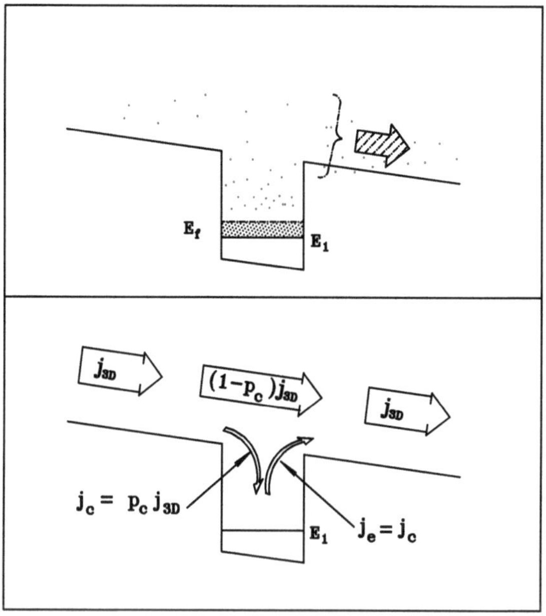

Fig. 4.2. Schematic representation of the electron distribution (*top*) and the processes controlling the dark current (*bottom*). Symbols are defined in the text

of the captured and uncaptured fractions must equal the current in the barrier region: $j_{3D} = j_c + (1 - p_c)j_{3D} = j_e + (1 - p_c)j_{3D}$. With this physical picture, one can model the dark current J_{dark} by calculating either j_{3D} directly or j_e, and in the latter case $J_{dark} = j_e/p_c$.

3D Carrier Drift Model

The first physical model calculates J_{dark} by directly estimating j_{3D}. A 3D electron density on top of the barriers N_{3D} is estimated with only the drift contribution taken into account (diffusion is neglected). The model was first presented in a very clear and concise paper by Kane et al. [100]. The dark current density is given by

$$J_{dark} = eN_{3D}v(F), \tag{4.1}$$

where $v(F)$ is the drift velocity as a function of electric field F. The 3D density is calculated by treating the barriers as a bulk semiconductor. Superlattice band structure effects are neglected, justified because the barriers are thick (much thicker than the wells) and the resulting superlattice miniband gaps are less than the thermal energy k_BT at device operating temperatures. The only 2D quantum well effect comes into the picture for the evaluation of the Fermi energy E_f. Usually QWIPs are degenerately doped in the wells, i.e., the top of the Fermi sea is higher than the energy of the lowest subband. Assuming a complete ionization, the 2D doping density N_D equals the electron density within a given well, as a good approximation. The Fermi energy is then calculated by $N_D = (m/\pi\hbar^2)E_f$. With all above assumptions, a simple calculation yields

$$N_{3D} \approx 2(m_b k_B T/2\pi\hbar^2)^{3/2} \exp(-E_{act}/k_B T), \tag{4.2}$$

where m_b is the barrier effective mass, and E_{act} is the thermal activation energy which equals the energy difference between the top of the barrier and the top of the Fermi sea in the well. We have assumed that $E_{act}/k_B T \gg 1$, appropriate for most practical cases.

The drift velocity takes the usual form

$$v(F) = \frac{\mu F}{[1 + (\mu F/v_{sat})^2]^{1/2}}, \tag{4.3}$$

where μ is the low field mobility and v_{sat} is the saturated drift velocity. While (4.3) is sufficient for the present considerations, to take into account the negative differential mobility phenomenon a more elaborate form of $v(F)$ will be used in Sect. 7.2.2.

A note on the validity of the model is needed. The key step is the evaluation of N_{3D}, by taking the equilibrium value at zero bias with the Fermi level determined by the well doping. It is therefore expected that the result is only valid for low electric field. As an extension and perhaps an improvement, Man and Pan [101] and Chu et al. [102] proposed a model involving different carrier temperatures or Fermi levels for the 3D barrier electrons and the 2D well electrons. Man and Pan adopted an empirical expression relating the barrier hot electron temperature to the 2D well electron temperature and applied electric field, whereas Chu et al. calculated the 3D barrier electron Fermi level by balancing the tunneling escape rate and the capture rate due to electron–phonon scattering. For both cases better fits to experiments were obtained.

The simple model compares well with experiments in the expected regime of low applied electric field. We show a comparison with three samples designed to have high absorption for heterodyne detection, having high well doping and sacrificing dark current. The main difference between the samples is the doping density in the well. The three samples were grown by MBE on semi-insulating GaAs substrates. The period of the 100-repeat multiple quantum well structure consists of a GaAs well and $Al_xGa_{1-x}As$ barriers, the GaAs well center region is doped with Si, and the top and bottom GaAs contact layers are 400 and 800 nm thick, doped with Si to $2 \times 10^{18}\,cm^{-3}$. The quantum well parameters are listed in Table 4.1.

Values of $\mu = 1{,}000\,cm^2\ Vs^{-1}$, for the mobility and $v_{sat} = 10^7\,cm\,s^{-1}$ for the saturated velocity were used for fitting. The final parameter needed is the activation energy, which is $E_{act} = V_b - (E_1 - E_{ex}) + E_f$, where V_b is the barrier height and E_1 is the ground state eigenenergy. The energy shift in ground state due to the exchange effect is taken into account using the

Table 4.1. Quantum well parameters of the 100 well QWIPs

N_D (cm^{-2})	x	L_w (nm)	L_b (nm)	V_b (eV)	E_1 (eV)	E_{ex} (eV)
1×10^{12}	0.200	6.6	25.0	0.18	0.049	0.015
1.5×10^{12}	0.192	6.6	25.0	0.17	0.048	0.017
2×10^{12}	0.197	5.9	24.0	0.18	0.056	0.020

Symbols: x – Al fraction, L_w – well width, L_b – barrier width, N_D – doping density in the well, V_b – barrier height, E_1 – ground state energy, and E_{ex} – exchange energy

expression given in Sect. 3.4. With these parameters, we obtain an adequate fit to the experimental data in Fig. 4.3 (dashed curves) in the low-field region (below a few kilovolt per centimeter), as expected for the model. The fit can be improved if the effect of barrier lowering due to the applied bias field is included. The dash-dotted curves in Fig. 4.3 are calculated using $E_{act} = V_b - (E_1 - E_{ex}) + E_f - eFL_w/2$; the last term takes into account the effective barrier lowering because E_1 (referenced to the center of the well) is approximately independent of F.

In fact, this simple 3D carrier drift model can be applied to more complicated structures, as the assumption of having only one bound state can be relaxed. As an example, we apply this model to a set of three p-type QWIPs [103] where several bound states (including both heavy and light hole states) are confined in the well. The main difference between the samples is the barrier height (or the x value). The period of the 100-repeat multiple quantum well structure consists of a GaAs well and $Al_xGa_{1-x}As$ barriers. The

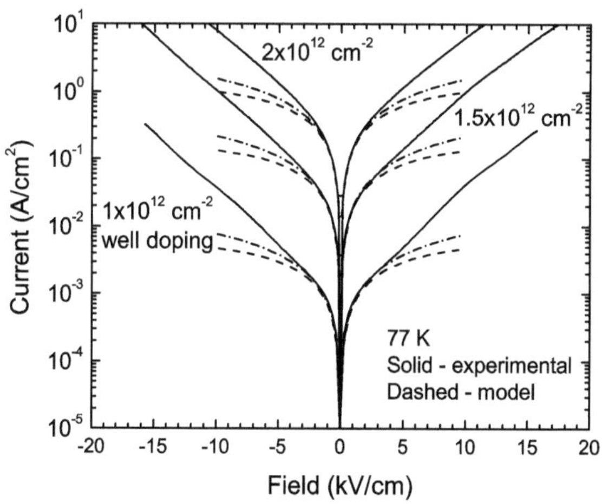

Fig. 4.3. Dark current characteristics at 77 K for a set of n-type QWIPs. Solid curves are experimental results, while dashed and dash-dotted are calculated using the simple 3D drift model for low field. The three samples differ mainly by the doping in the well

Table 4.2. Quantum well parameters of the 100 well p-type QWIPs

x	L_w (nm)	V_b (eV)	HH1 (eV)	HH2 (eV)	LH1 (eV)
0.215	4.0	0.114	0.029	0.101	0.057
0.245	4.1	0.130	0.029	0.106	0.059
0.290	4.1	0.154	0.031	0.114	0.064

The symbols: x – Al fraction, L_w – well width, V_b – barrier height, HH – heavy hole, LH – light hole, and the number after HH or LH is the eigenlevel index. All barriers are 20 nm thick, and all well center regions are doped with Be to 10^{12} cm^{-2}

top and bottom GaAs contact layers are 200 and 600 nm thick, doped with Be to 8×10^{18} cm^{-3}. The quantum well parameters are listed in Table 4.2; for completeness, the calculated eigenenergies are given as well.

The 3D density of free holes in the barrier region is estimated using the Fermi energy determined by the well doping. Both heavy and light hole densities are included. The hole effective mass in the barrier was obtained by linearly interpolating the values for GaAs and AlAs. The hole mobility values are not known for this type of structures, and we have used a value of $\mu_h = 100\,\text{cm}^2\,\text{Vs}^{-1}$ as an approximation. The activation energy is $E_{act} = V_b - (E_{HH1} - E_{ex}) + E_f$, where E_{HH1} is the heavy hole ground state eigenenergy. Again using the expression in Sect. 3.4, the calculated exchange energy E_{ex} is 26 meV. The eigenenergies and the Fermi energy are calculated using the 8×8 envelope function approximation [104], which gives $E_f = 9$ meV, where E_f is referenced to the HH1 eigenlevel. Using these parameters, we obtain an adequate fit to the experimental data in Fig. 4.4 in the low-field region (below a few kilovolt per centimeter) as expected for the model.

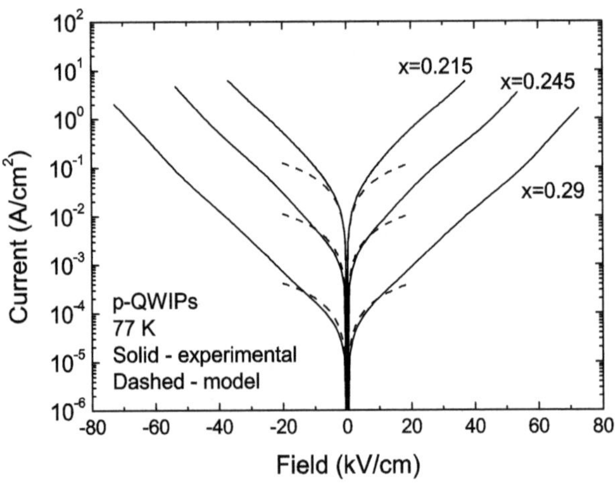

Fig. 4.4. Dark current characteristics at 77 K for a set of p-type QWIPs. Solid curves are experimental results, while dashed are calculated using the simple 3D drift model for low field. The three samples differ mainly by the barrier Al fraction (x) which determines the barrier height

Emission-Capture Model

The second approach [105] calculates j_e and then $J_{dark} = j_e/p_c$ (see the lower part of Fig. 4.2).

We first review and comment on several other published models in the literature. The model given by Levine et al. [106] has been widely used [105,107,108]. However, they [106] did not discuss the process of trapping or capture to balance the emission or escape. Andrew and Miller [107] included an image charge effect using the usual formula for the image potential, but this is incorrect for a quantum well with mainly an occupied ground state; thus the image charge effect is expected to be small for QWIPs [109]. In a well-cited paper, Kinch and Yariv [110] formulated a model by estimating the 2D electrons distributed above the barriers, but they did not discuss the process of trapping or capture. The model by Petrov and Shik [111] estimates j_e and takes this as the total dark current. The estimate was done by integrating the product of a velocity v_z and the transmission coefficient, where the z-coordinate is in the direction of the current flow. This approach models the electron escape by direct tunneling only, and neglects the scattering-assisted escape process.

Scattering-assisted escape is the dominant process for a typical QWIP, especially at low fields. Electrons associated with the confined ground state in the well and distributed on the 2D in-plane dispersion curve undergo a scattering event to get out of the well and then become a 3D mobile carrier in the barrier. This physical picture is illustrated in Fig. 4.5.

We now construct the dark current expression. Under certain approximations, the result from considering j_e and then $J_{dark} = j_e/p_c$ should be consistent with the previous expression (4.1). Referring to Fig. 4.5, the escape

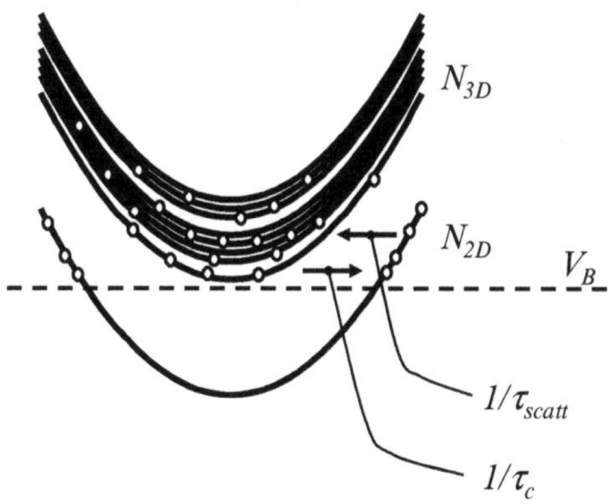

Fig. 4.5. Processes of scattering of electrons from the ground state subband into the continuum states and capture of electrons from the continuum back to the subband. The main contribution to the scattering escape is from electrons with their energy higher than the barrier height

current density can be written as

$$j_e = eN_{2D}/\tau_{scatt}, \tag{4.4}$$

where N_{2D} is a 2D electron density which only includes electrons on the upper part (with energy greater than the barrier height) of the ground state subband and τ_{scatt} is the scattering time to transfer these electrons from the 2D subband to the nonconfined continuum on top of the barrier. The standard QWIPs having their barriers much wider than wells, we neglect any superlattice miniband effects.

The capture probability is related to the relevant time constants by

$$p_c = \frac{\tau_{trans}}{\tau_c + \tau_{trans}}, \tag{4.5}$$

where τ_c is the capture time for an excited electron back into the well, and τ_{trans} is the transit time for an electron across *one* quantum well region including the surrounding barriers. (Note that the capture time τ_c is equivalent to the excited electron lifetime τ_{life}, more widely used in the discussion of photoconductors. Also note that τ_{trans} defined here is associated with only *one* period of the multiple quantum well device in contrast to, e.g., [112], where the same symbol was used for the transit time across the entire detector.)

In the limit of $p_c \ll 1$, i.e., $\tau_c \gg \tau_{trans}$, as is true for actual devices at operating electric fields, the dark current becomes

$$J_{dark} = j_e/p_c = e\frac{N_{2D}}{\tau_{scatt}}\frac{1}{p_c} = e\frac{N_{2D}\tau_c}{\tau_{scatt}\tau_{trans}} = e\frac{N_{2D}}{L_p}\frac{\tau_c}{\tau_{scatt}}v, \tag{4.6}$$

where L_p is the period length of the multiple quantum well structure, which is the sum of the well and barrier widths $L_p = L_w + L_b$. This gives some physical insights into the physical processes involved in the dark current. The quantity N_{2D}/τ_{scatt} represents the thermal escape or generation of electrons from the quantum well, and $1/p_c$, as shown later, is directly proportional to the photoconductive gain. This implicit dependence of J_{dark} on the photoconductive gain will be conceptually important when discussing the detectivity of a QWIP.

We can establish a relationship between N_{2D} and N_{3D} due to the balance in scattering escape and capture (see Fig. 4.5):

$$\frac{N_{2D}}{\tau_{scatt}} = \frac{N_{3D}L_p}{\tau_c}, \tag{4.7}$$

This is consistent with the expectation that τ_{scatt} is independent of L_b for large barrier widths ($L_b \to \infty$), whereas τ_c should be proportional to L_b.

Substituting (4.7) into (4.6), we get

$$J_{dark} = eN_{3D}\frac{L_p}{\tau_{trans}} = eN_{3D}v, \tag{4.8}$$

with the drift velocity $v = L_p/\tau_{trans}$, exactly the same as in (4.1).

It is also easy to establish an approximate relationship between τ_{scatt} and τ_{c}. The 3D density has been given in (4.2). Similarly by counting all electrons distributed on the ground state subband above the activation energy, we get

$$N_{\text{2D}} \approx \frac{m}{\pi\hbar^2} k_{\text{B}} T \exp(-E_{\text{act}}/k_{\text{B}}T). \tag{4.9}$$

Using (4.2), (4.7), and (4.9), we get

$$\frac{\tau_{\text{c}}}{\tau_{\text{scatt}}} \approx \frac{m_{\text{b}}}{m} L_{\text{p}} \left(\frac{m_{\text{b}} k_{\text{B}} T}{h\hbar} \right)^{1/2}. \tag{4.10}$$

Equation (4.10) yields $\tau_{\text{c}}/\tau_{\text{scatt}} \approx 2$ for typical GaAs/AlGaAs QWIPs with $L_{\text{p}} = 50\,\text{nm}$ at $T = 77\,\text{K}$, while experimentally, τ_{scatt} is in the range of 1–2 ps and τ_{c} is 5–10 ps. The emission-capture and 3D carrier drift models thus yield quite similar results.

Another popular model already mentioned earlier in this section, given by Levine et al. [106], gives rise to the formula

$$J_{\text{dark}} = e \frac{N_{\text{2D}}}{L_{\text{p}}} v. \tag{4.11}$$

Comparing with (4.6), the above is equivalent to setting $\tau_{\text{c}} = \tau_{\text{scatt}}$ in (4.6). Although (4.11) has been widely used by many authors, it ignores the implicit dependence of J_{dark} on the photoconductive gain, and it implies an unrealistic proportionality between J_{dark} and $1/L_{\text{p}}$. Nevertheless, both approaches yield the correct order of magnitude for practical dark current calculations.

Calculation of dark current now relies on obtaining a good estimate of N_{2D}. So far we calculated N_{2D} by including only those electrons with their energies above the barrier height. Electrons near the barrier height (but below) can also contribute by scattering-assisted tunneling. This contribution can be accounted for in the following way. We use the important theoretical result of Meshkov [113] which can be stated as follows: including scattering processes, an electron tunneling rate in a 1D potential is controlled by the *total* energy rather than the energy associated with the tunneling direction. For thick barriers, scattering- assisted processes determine the tunneling transmission probability. We use the following expression, which in view of Meshkov effectively includes the scattering escape process [106]:

$$N_{\text{2D}} = \int_{E_1}^{\infty} \frac{m}{\pi\hbar^2} \mathcal{T}(E,F) \left[1 + \exp\left(\frac{E - E_{\text{f}}}{k_{\text{B}} T} \right) \right]^{-1} \mathrm{d}E, \tag{4.12}$$

where $\mathcal{T}(E,F)$ is the transmission coefficient which is taken to be unity for E higher than the barrier. According to (4.6), the final expression for dark current is then given by

$$\begin{aligned} J_{\text{dark}} &= e\, N_{\text{2D}} \tau_{\text{c}} / L_{\text{p}}\, v\, \tau_{\text{scatt}} \\ &= \frac{ev\tau_{\text{c}}}{\tau_{\text{scatt}}} \int_{E_1}^{\infty} \frac{m}{\pi\hbar^2 L_{\text{p}}} \mathcal{T}(E,F) \left[1 + \exp\left(\frac{E - E_{\text{f}}}{k_{\text{B}} T} \right) \right]^{-1} \mathrm{d}E. \end{aligned} \tag{4.13}$$

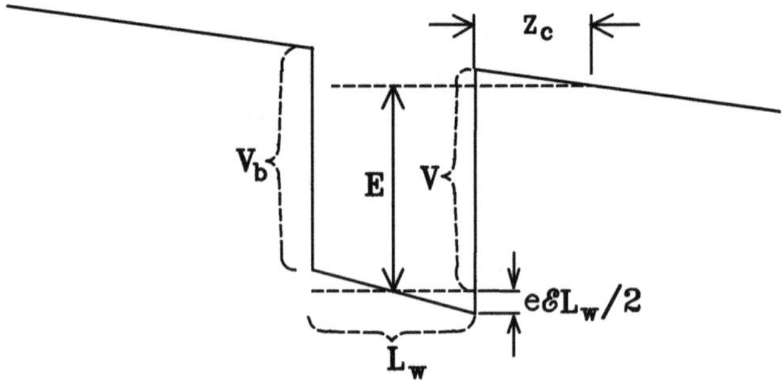

Fig. 4.6. Schematic illustration of relevant energies and the effective barrier lowering. The electron energy E is referenced to the center of the well

As before, the v vs. F relationship is given by (4.3). The electric field dependence in (4.13) is explicitly through (4.3) and implicitly through $\mathcal{T}(E, F)$. Using the WKB approximation, $\mathcal{T}(E, F)$ is given by

$$\mathcal{T}(E,F) = \exp\left[-2\int_0^{z_c} dz\sqrt{2m_b(V-E-eFz)}/\hbar\right], \tag{4.14}$$

for energy less than the barrier height $E < V$, where m_b is the barrier mass and $V = V_b - eFL_w/2$ includes the barrier lowering by applied bias field, and $z_c = (V - E)/eF$ defines the classical turning point. Figure 4.6 shows schematically the relevant quantities. The WKB approximation compares well with more exact calculations using the transfer matrix approach.

Equation (4.13) can be simplified in the pure thermionic emission regime, i.e., the tunneling contribution can be neglected. This is equivalent to setting $\mathcal{T}(E, F) = 0$ for E below the barrier.

Furthermore, the condition $(E - E_f)/k_BT \gg 1$ for $E > V$ generally holds. Equation (4.13) then becomes

$$J_{dark} = \frac{ev\tau_c}{\tau_{scatt}} \frac{m}{\pi\hbar^2 L_p} k_BT \exp(-E_{act}/k_BT). \tag{4.15}$$

This expression closely resembles (4.1) together with (4.2), and corresponds exactly to (4.6).

The model discussed here (with τ_c/τ_{scatt} set to 1) has been extensively compared with experiments (see [43,106]). Given the simplicity of the model, the good agreement obtained exceeds expectations. Figure 4.7 shows an example of model-experiment comparison. The samples have nominally the same parameters other than the number of wells. The well width is 6 nm, the barrier width is 25 nm, the barrier x value is 0.25, and the well doping is 9×10^{11} cm^{-2}. The measured curves in Fig. 4.7 display an asymmetry between positive and negative bias polarities, which was found to be due to the dopant segregation during growth [114,115] (to be discussed in Sect. 7.3).

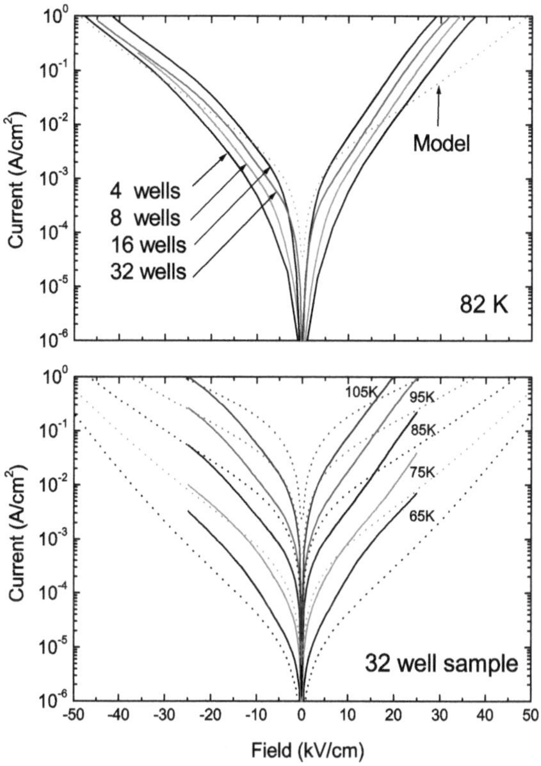

Fig. 4.7. Dark current of a set of nearly identical samples varying only the number of wells. The *dashed curve* is calculated using Levine's formula (4.13) (τ_c/τ_{scatt} set to 1)

4.1.2 Self-Consistent and Numerical Models

Self-Consistent Drift-Diffusion Model

The model developed by Ershov et al. [99,116] calculates the QWIP characteristics by self-consistently solving three equations: (1) Poisson equation, (2) continuity equation for electrons in the barriers, and (3) rate equation for electrons in the quantum wells. The inclusion of the Poisson equation is especially important for QWIPs with a small number of wells (<10) because the field can be substantially different (often higher) for the first few periods starting from the emitter in comparison with the rest of the wells (see Fig. 4.8). The continuity equation involves the current (expressed in the standard drift-diffusion form) and rates of thermal and optical generation and of recombination.

Using this numerical model, we were able to account for the observed unusual capacitance behavior [117] and explain the nonlinear photoconductivity at high excitation power using a CO_2 laser [118]. In addition, the model has

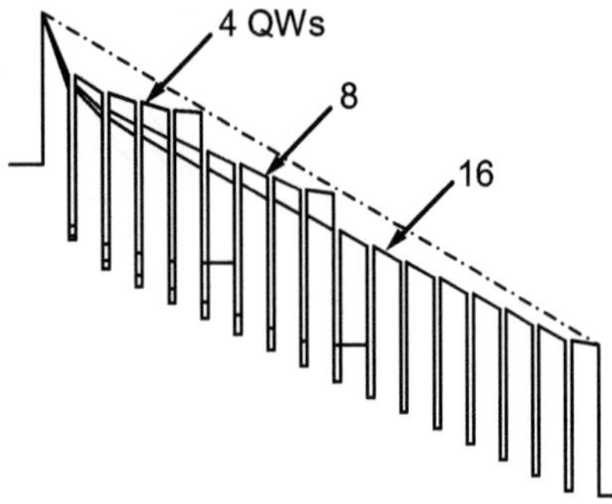

Fig. 4.8. Self-consistently calculated profiles of QWIPs with 4, 8, and 16 wells. The average applied field ($10\,\mathrm{kV\,cm^{-1}}$) is shown by a dash-dotted line (Courtesy of M. Ershov)

the capability to predict transient and hence frequency characteristics [119], as well as photoresponse under localized IR excitation [120].

Self-Consistent Emission-Capture Model

Thibaudeau et al. [121] presented a numerical model which extends the simple emission-capture model presented before. The model allows the electric field to be nonuniform, self-consistently determined by Gauss' law. The authors obtained better agreement with experiments than the simple model.

Ryzhii [122] constructed an analytical model by solving Poisson's equation and an equation governing the electron balance in the quantum well. Interesting functional dependencies of the responsivity on the number of wells and the photon excitation power were found.

Jovanović et al. [123] constructed a quantum mechanical model considering all scattering processes including emission and capture. The model results were compared with experimental data on a GaAs/AlGaAs device and good agreement was found.

Numerical Monte Carlo Model

Ryzhii et al. [124,125] carried out Monte Carlo simulations on QWIPs, in particular their ultrafast electron transport properties. Cellek et al. [126,127] also performed such simulations, analyzing the effects of material properties on the device characteristics. They found the evidence that the L-valley in GaAs/AlGaAs QWIPs plays an important role in determining the responsivity vs. voltage behavior. Monte Carlo simulations shed light on the hot electron distribution on top of the barriers, and should provide guidance to the optimization of QWIPs.

To end this section, although several models have been established, with varying degree of complexity, and good agreement between models and experiments has been obtained, to formulate a true first-principle QWIP model is a highly nontrivial task. This is because the QWIP is a complicated and "dirty" system. Given the wide barriers and narrow wells, the transport mechanism falls between ballistic and drift-diffusion; and due to the high doping and high field, realistic calculations of scattering or trapping rates are extremely complicated and have not been performed so far. The situation becomes even more complicated to model for p-type structures [65,128].

4.2 Photocurrent

Photoconductivity phenomena in solids are well known, and many texts have been written on the subject [31,33,129,130]. The device operation of the photoconductive QWIPs is similar to that of extrinsic semiconductor detectors [131]. The distinct feature of QWIPs in contrast with the conventional intrinsic and extrinsic photoconductors is the discreteness, i.e., incident photons are only absorbed in discrete quantum wells which are normally much narrower than the inactive barrier regions. In this section, we discuss the photocurrent caused by intersubband excitations in a QWIP and introduce the concept of photoconductive gain. Here we consider only the case of positive photoconductivity, i.e., the effect of the incident IR light is to make the device resistance smaller. Negative photoconductivity is possible, e.g., if one has a device with a negative differential resistance region [132–136].

4.2.1 Photoconductive Gain

Photoconductive gain is defined as the number of electrons flowing through the external circuit for each photon absorbed. A model [112] specifically for photoconductive QWIPs has been constructed, which answers exactly what constitutes the mechanism of photoconductive gain and how it depends on device parameters such as the number of wells. The model also explains observations of large ($\gg$1) photoconductive gains [61,137]. In this section, we first present the physical picture, and then derive an expression for the gain. A comparison with experiments and an estimate of relevant time scales follow.

To visualize the physical process and the gain mechanism, a simple one-well structure is given in Fig. 4.9. The top part shows the dark current paths (same as in the bottom part of Fig. 4.2). All these dark current paths remain *unchanged* when IR light is shone on the detector. The additional processes as a result of the IR radiation are shown in the bottom part of Fig. 4.9. There is a *direct* photoemission of electrons from the well, and this, of course, contributes to the observed photocurrent in the collector. The photoconductive gain is a result of the extra current injection from the contact necessary to balance the loss of electrons from the well due to photoemission. The amount of the extra injection must be sufficiently large that its fraction trapped in the well equals the direct photoemission current. On the other hand, the fraction of

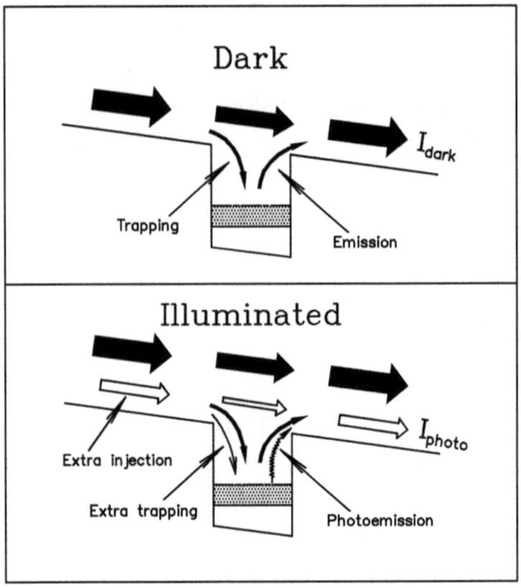

Fig. 4.9. The photoconductive gain mechanism. The *top* part shows the dark current paths, while the *bottom* indicates the direct photoemission and the extra current injection from the contact to balance the loss of electrons from the well. The dark current paths remain the same under illumination. The collected total photocurrent is the sum of the direct photoexcited and the extra injection contributions

the extra injected current that reaches the collector contact is in fact indistinguishable from the direct photoemitted current, and therefore contributes to the observed photocurrent. The total photocurrent consists of contributions from the direct photoemission and the extra current injection. Note that the physical mechanism given here is the same as for a conventional photoconductor [33] though this simple physical picture was presented only in the context of QWIP physics [112]. The common physical picture to explain larger than unity gain states that photoexcited electrons circulate around the circuit several times. This seems plausible, and appears in textbooks (see, p. 97, [130]), but is misleading since a collector "absorbs" all electrons and the excess energy of the electrons gives rise to the Ohmic heating.

An important fact, perhaps counter-intuitive at first sight, is that the magnitude of photocurrent is independent of the number of wells if the absorption for each well is the same. To understand this pictorially, Fig. 4.10 shows the photoccurrent paths for a two-well structure. If the first well is next to the emitter, one can see that for the second well the processes of photoemission and refilling are identical to those in the first well. The only difference is that for the first well the extra injection comes directly from the emitter whereas for the second well the injection is the resulting total photocurrent in the barrier separating the two wells. The same argument can be made for any subsequent wells. This means that the magnitude of the photocurrent is unaffected by adding more wells as long as the magnitude of absorption and hence

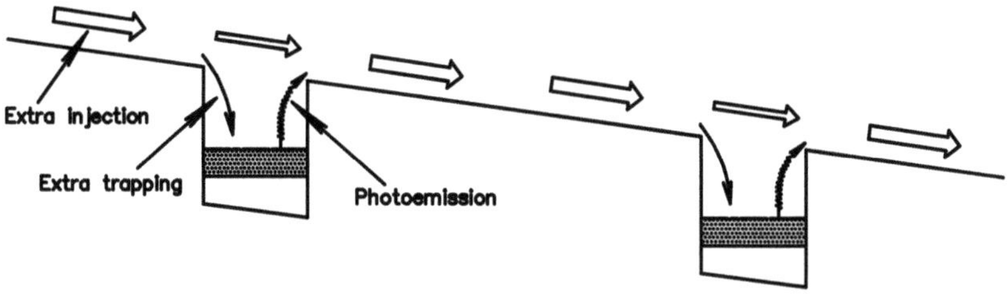

Fig. 4.10. A two-well case which illustrates the independence of photocurrent on number of wells. The left well is next to the emitter. The extra injection is the observed photocurrent, and the same injection balances the photoemission from both wells

photoemission from all the wells remains the same. A further discussion of the photocurrent paths will be given in Sect. 10.1.1.

We present a derivation of the photoconductive gain expression explicitly for QWIPs. The model is constructed under the same assumptions (a)–(d) as in Sect. 4.1.1. A clarification of how the assumption of an injecting contact (assumption (c) in Sect. 4.1.1) can be fulfilled is in order, because this is an important point both for the dark current model and for the concept of photoconductive gain. To have a good injecting contact, the barrier between the emitter contact and the multiquantumwell (MQW) region must not be large. In most cases this barrier is the same as the barrier separating wells in the MQW region. If needed, an extra injection of electrons is achieved by increasing the electric field at the emitter–MQW junction. This is a self-consistent process: e.g., if one adds an extra emission channel of electrons from the wells (e.g., by photoemission), the wells will tend to become slightly charged, which increases the electric field at the emitter–MQW junction and hence increase injection to balance the loss of electrons in the wells [138]. The mechanism is shown schematically in Fig. 4.11.

Having discussed the two pictures of one- and two-well cases (Figs. 4.9 and 4.10), a simple derivation can be constructed. We first calculate the emission

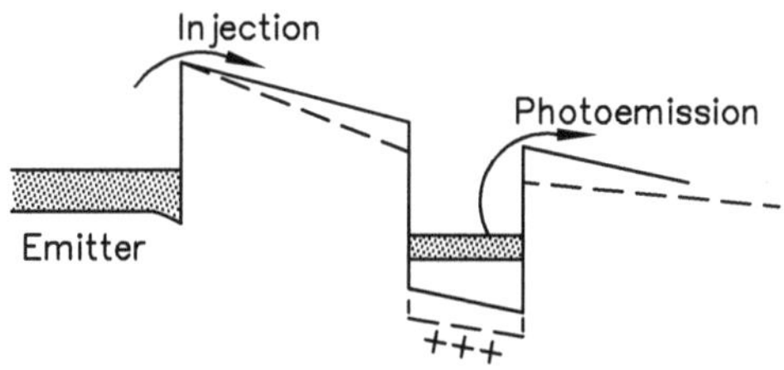

Fig. 4.11. The mechanism of injection. The bandedge profile with and without the photoemission is shown in *dashed* and *solid lines*, respectively. Due to photoemission of electrons, the well becomes slightly charged, leading to an increase in the electric field and injection

current as a consequence of *direct* excitation of electrons into the continuum (the zigzag arrow shown in the bottom part of Fig. 4.9). The photoemission current directly ejected from *one* well is

$$i_{\mathrm{photo}}^{(1)} = e\Phi\eta^{(1)}\frac{\tau_{\mathrm{relax}}}{\tau_{\mathrm{relax}}+\tau_{\mathrm{esc}}} \equiv e\Phi\eta\frac{p_{\mathrm{e}}}{N}, \tag{4.16}$$

where Φ is the incident photon number per unit time, the superscript (1) indicates quantities for one well, τ_{esc} is the escape time, τ_{relax} is the intersubband relaxation time, $\eta \equiv N\eta^{(1)}$ is the total absorption quantum efficiency, N is the number of wells, and the escape probability for an excited electron from the well is given by

$$p_{\mathrm{e}} \equiv \frac{\tau_{\mathrm{relax}}}{\tau_{\mathrm{relax}}+\tau_{\mathrm{esc}}}. \tag{4.17}$$

We have assumed that the amount of absorption is the same for all the wells, i.e., $\eta \equiv N\eta^{(1)}$. The photon flux could depend on the location of the well, but this is very specific to the detector light coupling geometry (e.g., 45° facet coupling [14] or grating coupling [139]). The derivation of (4.16) is straightforward from a rate equation consideration: letting n_{ex} be the number of the excited electrons, we have

$$\frac{\mathrm{d}n_{\mathrm{ex}}}{\mathrm{d}t} = \Phi\eta^{(1)} - \frac{n_{\mathrm{ex}}}{\tau_{\mathrm{esc}}} - \frac{n_{\mathrm{ex}}}{\tau_{\mathrm{relax}}}. \tag{4.18}$$

Under steady state, $\mathrm{d}n_{\mathrm{ex}}/\mathrm{d}t = 0$, we solve for n_{ex} from (4.18). Then the photoemitted current from one well is $en_{\mathrm{ex}}/\tau_{\mathrm{esc}}$ which gives (4.16).

As shown in Figs. 4.9 and 4.10, for each well, the injection current ($i_{\mathrm{photo}}^{(1)}/p_{\mathrm{c}}$) which refills the well to balance the loss due to emission ($i_{\mathrm{photo}}^{(1)}$) *equals* the observed photocurrent. The photocurrent is then given by

$$I_{\mathrm{photo}} = i_{\mathrm{photo}}^{(1)}/p_{\mathrm{c}}. \tag{4.19}$$

Using (4.16), we immediately get

$$I_{\mathrm{photo}} = e\Phi\eta\frac{p_{\mathrm{e}}}{Np_{\mathrm{c}}} \equiv e\Phi\eta g_{\mathrm{photo}} \tag{4.20}$$

and

$$g_{\mathrm{photo}} \equiv \frac{p_{\mathrm{e}}}{Np_{\mathrm{c}}} \tag{4.21}$$

is the photoconductive gain. As before in the discussion of dark current, the capture probability is given by

$$p_{\mathrm{c}} = \frac{\tau_{\mathrm{trans}}}{\tau_{\mathrm{c}}+\tau_{\mathrm{trans}}}, \tag{4.22}$$

where τ_{trans} is the transit time for an electron across *one* quantum well region or the period of the structure. Note that Beck [140], using the same physical model, obtained $g_{\mathrm{photo}} = 1/(Np_{\mathrm{c}})$. The difference from (4.21) is purely due to the difference in the definition of quantum efficiency η. Beck's definition of quantum efficiency is not photon *absorption* quantum efficiency. His quantum efficiency is equivalent to $p_{\mathrm{e}}\eta$ here, which results in the missing p_{e} factor in his photoconductive gain expression in comparison with (4.21). A similar photoconductive gain expression for QWIPs was given [141], valid for $p_{\mathrm{c}} \ll 1$. Heuristically, since the processes of escape and capture are complimentary, the quantities of p_{e} and p_{c} should be correlated in a standard photoconductive QWIP with simple quantum wells. Looking at the times involved in the definitions, (4.17) and (4.22), an approximation $p_{\mathrm{e}} \sim 1 - p_{\mathrm{c}}$ holds, but only as a rough guide.

The photoconductive gain expression for QWIP should correspond to the expression given by the conventional theory of photoconductivity [33], $g_{\mathrm{photo}} = \tau_{\mathrm{c}}/\tau_{\mathrm{trans,tot}}$, where $\tau_{\mathrm{trans,tot}} = (N+1)\tau_{\mathrm{trans}}$ is the total transit time across the detector active region. Under the approximation $p_{\mathrm{e}} \approx 1$, $p_{\mathrm{c}} \approx \tau_{\mathrm{trans}}/\tau_{\mathrm{c}} \ll 1$ and $N \gg 1$, the gain expressions given by (4.21) and the conventional theory become the same:

$$g_{\mathrm{photo}} \approx \frac{1}{Np_{\mathrm{c}}} \approx \frac{\tau_{\mathrm{c}}}{\tau_{\mathrm{trans,tot}}} = \frac{\tau_{\mathrm{c}}v}{NL_{\mathrm{p}}}. \tag{4.23}$$

For QWIPs, the lifetime (capture time) τ_{c} is associated *only* with those processes that scatter an electron into the ground state subband in the well (trapping). For a simple square well, the condition $p_{\mathrm{e}} \approx 1$ is met for a bound-to-continuum case (i.e., only one bound state is confined in the well); while for a bound-to-bound case (two bound states) this is no longer true [142]. For structures where p_{e} and p_{c} can be designed independently [143] (see Chap. 5), (4.23) cannot be used. If the absorption is proportional to N, as is a good approximation for a lot of practical cases, the photocurrent is independent of N since g_{photo} is inversely proportional to N. This was shown experimentally [61] (see Fig. 4.12). Photocurrent independence of N is equivalent to its independence of device length in the conventional theory. This independence does not mean that the detector performance is independent of the number of wells because of noise considerations (see Sect. 4.3).

Using (4.21) for different values of p_{c} and $p_{\mathrm{e}} = 1$, the calculated photoconductive gain is plotted vs. the number of wells in Fig. 4.13, together with some experimental data (at high fields) from [61,100,106,137,144]. The NRC data [61] were obtained on samples with comparable parameters (grown one after the other) except the number of wells. A capture probability of about 0.07 is inferred from these data. Most of the reported detector samples have 50 quantum wells. It is seen from Fig. 4.13 that a range of gain values from about 0.27–0.80 for 50 well samples have been observed, and hence the gain is quite sample dependent. The difference in τ_{trans} between samples is a possible reason for the spread of the observed gain values. The transit time τ_{trans} is mainly determined by the high-field drift velocity of an excited electron in the barrier region and varies somewhat depending on the field strength

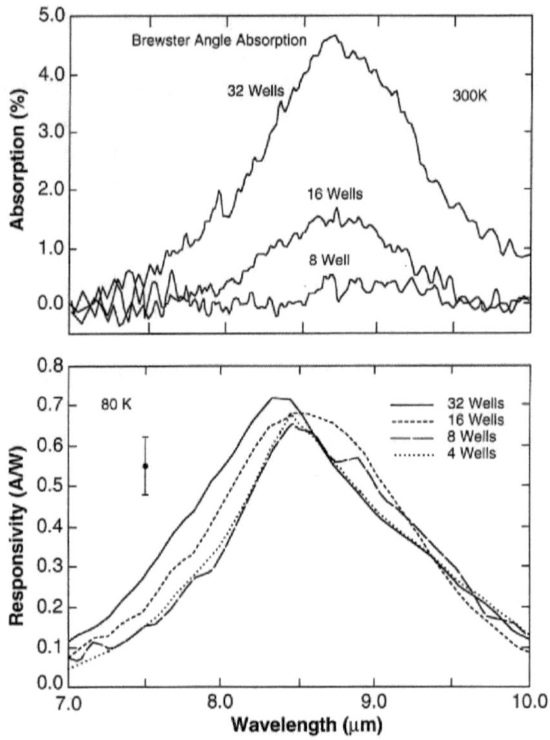

Fig. 4.12. Absorption (*top*) and responsivity (*bottom*) spectra of four samples with nominally identical parameters but the number of wells. This clearly evidences that the responsivity is independent of the number of wells although the absorption is increased proportionally. See [61] for experimental details

and barrier materials. The other possibility is the variation in τ_c. Processes that result in capture are due to scattering by impurities and electrons in the well region, phonons, and interface roughness. Experiments of Gunapala et al. [144] suggest that the impurity and electron–electron scattering may not be the dominant mechanism because the observed gain values did not decrease systematically with increasing well doping density. Phonon scattering would probably result in comparable values of capture probability for similar structures, and may not explain the strong sample dependence. Interfaces between AlGaAs and GaAs could be very different from sample to sample and from one crystal growth facility to another. Finally, the resonant enhancement of the density of states close to the barrier bandedge – usually present in order to obtain efficient intersubband or bound-to-continuum absorption – is expected to cause also a resonantly enhanced capture probability and thus a reduced τ_c.

We can make some estimates of the time scales involved. From the measured intersubband absorption line width (normally no narrower than about

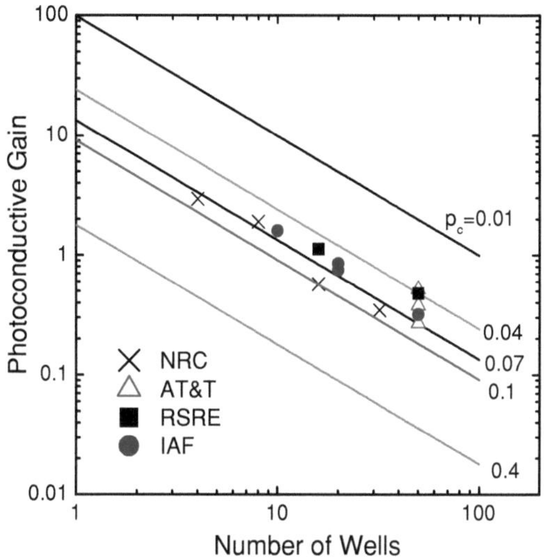

Fig. 4.13. Calculated photoconductive gain vs. the number of wells for capture probability $p_\mathrm{c} = 0.01$, 0.04, 0.07, 0.1, and 0.4, respectively. Experimental data are taken from Liu et al. (*cross*), Levine et al. (*triangle*), Kane et al. (*square*), and Schneider et al. (*circle*)

5 meV), we know that τ_relax is longer than 100 fs for a typical QWIP, i.e., 100 fs is a lower bound for τ_relax. From time-resolved experiments, Tatham et al. [145] projected an upper bound on the intersubband relaxation time of about 500 fs and Baier et al. [146] deduced this time to be approximately 1 ps, both for a bound-to-bound state relaxation process. Using an excited state differential absorption technique, Faist et al. [147] inferred a relaxation time of slightly less than 1 ps, again, for a bound-to-bound situation. Andersson [94] performed calculations of τ_c for QWIPs [see (3.74)] and gave $\tau_\mathrm{c} = 5.5$ ps for a typical structure. There is a difference between the bound-to-bound and continuum-to-bound relaxation processes. Since in the latter case the excited electron is extended in the barrier region, the relaxation should take more time. We therefore expect that τ_c for a typical GaAs/AlGaAs QWIP should be approximately 5 ps. This value is consistent with findings from time-resolved photocurrent measurements [148], which deduced $\tau_\mathrm{c} < 7$ ps. For InGaAs/GaAs QWIPs, capture times are about twice as long (see Sects. 10.1.2 and 10.2.2).

The transit time can be estimated by $\tau_\mathrm{trans} \approx L_\mathrm{p}/v$, where L_p is the quantum well period and v is the drift velocity. For typical parameters of $v = 10^7\,\mathrm{cm\,s^{-1}}$ and $L_\mathrm{p} = 30$–50 nm, τ_trans is estimated to be in the range of 0.3–0.5 ps. One therefore expects a capture probability $[p_\mathrm{c} = \tau_\mathrm{trans}/(\tau_\mathrm{c} + \tau_\mathrm{trans}) \approx \tau_\mathrm{trans}/\tau_\mathrm{c}]$ to be in the range of 0.06–0.1 consistent with existing experiments (see Fig. 4.13).

To quantitatively calculate τ_c is a highly nontrivial problem [149,150]. This problem is also of key importance to the operation of (interband) quantum well lasers [151] as well as QCL [28,29]. Large variations of carrier relaxation time from about 1 to 20 ps have been predicted, and experimental studies using many samples with varying well width have resulted in similar variations [151–153].

4.2.2 Detector Responsivity

In this section, we discuss factors that control the responsivity. The magnitude of responsivity is controlled by both quantum efficiency and photoconductive gain. A high absorption does not necessarily result in a high detector responsivity. There must not only be high absorption but the photoexcited electrons must also escape the wells efficiently to give rise to a large photocurrent. Both experiments [45] and calculations [46] have been performed to address the optimal design of quantum well for maximizing the responsivity. It was shown that the optimum occurs when the excited state is in close resonance with the top of the barrier. In general, the spectral current responsivity is given by

$$\mathcal{R}_i = I_{\mathrm{photo}}/(h\nu\Phi) = \frac{e}{h\nu}\eta g_{\mathrm{photo}}, \tag{4.24}$$

where ν is the photon frequency. For QWIPs, g_{photo} is given by (4.21), which is

$$g_{\mathrm{photo}} = \frac{\tau_{\mathrm{relax}}}{\tau_{\mathrm{relax}} + \tau_{\mathrm{esc}}} \frac{\tau_c + \tau_{\mathrm{trans}}}{\tau_{\mathrm{trans}}} \frac{1}{N}, \tag{4.25}$$

where the first two fractions come from p_e and $1/p_c$, respectively.

Let us discuss ways to maximize the responsivity. Under the assumption of $\eta \propto N$, there is nothing that can be done about the number of wells since $\mathcal{R}_i$ is independent of N. The escape probability must be made close to unity. This is done by ensuring $\tau_{\mathrm{esc}} \ll \tau_{\mathrm{relax}}$. For the bound-to-continuum case, the process of escape takes little time, i.e., once an electron is excited, it is already in the continuum. In this case, $\tau_{\mathrm{esc}} \approx 0$ and $p_e \approx 1$. For the bound-to-bound case, τ_{relax} is shorter than τ_c. From various experiments and calculations (see Sect. 4.2.1), the value of τ_{relax} is about 1 ps (5 ps for τ_c). In order to ensure $p_e \sim 1$, we must have $\tau_{\mathrm{esc}} \ll 1$ ps. This implies that if a bound-to-bound design is employed, one must have the excited state close to the top of the barrier so that the tunneling escape time is much less than 1 ps. For a typical 10 µm GaAs/AlGaAs QWIP under a typical field of 10 kV cm^{-1}, this dictates that the excited state should not be lower than about 10 meV below the top of the barrier. The upper state (E_2) tunneling escape time is estimated easily by considering an "attempt frequency" $v_2/2L_w$ and the transmission probability $\mathcal{T}$, and is given by $\tau_{\mathrm{tunnel}} \approx (2L_w/v_2)\mathcal{T}^{-1}$, where $v_2 = \sqrt{2E_2/m}$. This estimate is semiclassical, but does produce excellent results in comparison with rigorous calculations. The electric field dependent transmission probability is

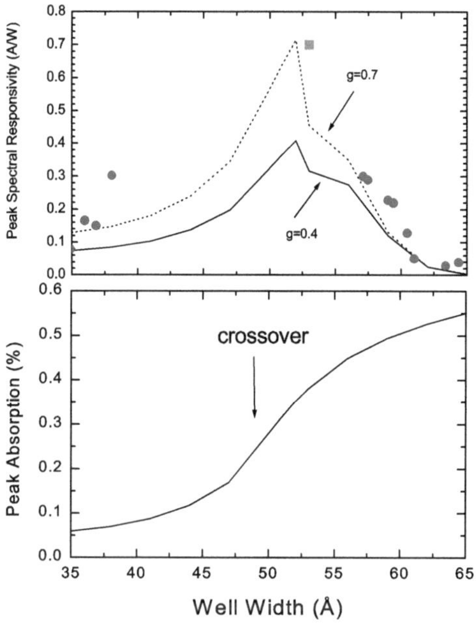

Fig. 4.14. (**a**) Estimated peak spectral responsivity vs. well width for photoconductive gain values of 0.4 (*solid line*) and 0.7 (*dashed line*), and experimental results (*bullets and square*), and (**b**) calculated peak absorption quantum efficiency vs. well width for one well and one infrared path at an internal angle of incidence of 45°. The arrow indicates the crossover from the bound-to-continuum to the bound-to-bound case

easily estimated using a WKB method. A calculated example together with experimental data is shown in Fig. 4.14.

4.3 Detector Performance

4.3.1 Detector Noise

In general, a photoconductor has several sources of noise: $1/f$ noise, Johnson noise, dark current noise, and photon noise (noise associated with the current induced by incident photons). The physical mechanism of $1/f$ noise is very complicated and is an ongoing research topic. For GaAs QWIPs, experiments show that $1/f$ noise seldom limits the detector performance. We therefore neglect the contribution of $1/f$ noise. Johnson noise is inherent to all resistive devices and the noise mean square current is

$$i^2_{\mathrm{n,J}} = \frac{4k_{\mathrm{B}}T}{R}\Delta f, \tag{4.26}$$

where Δf is the measurement bandwidth and R is the device differential resistance. Johnson noise is easily calculated once the device I–V curve is known, and the contribution is usually small in a photoconductive QWIP [142]. Contributions from the dark current noise and the photon noise usually limit the detector ultimate performance in QWIPs. Here we concentrate on these two mechanisms.

In deriving noise expressions, we use the simple physical picture given by Rose (see pp. 97–99 of [33]). One identifies the source α_{n} of the noise and the magnification factor F in observation. Then the noise (square average) is

$$I_{\mathrm{n}}^2 = 2F^2\alpha_{\mathrm{n}}\Delta f. \tag{4.27}$$

Given the transport mechanism (see Figs. 4.2 and 4.9 and related discussions), the dark current noise is generation-recombination (g-r) in nature. The noise current should be given by the standard g-r noise expression [33]

$$i_{\mathrm{n,dark}}^2 = 4eg_{\mathrm{noise}}I_{\mathrm{dark}}\Delta f, \tag{4.28}$$

where g_{noise} is the noise gain, and I_{dark} is the device dark current. If we label the emission current (see the top part in Fig. 4.9) from one well as $i_{\mathrm{e}}^{(1)}$ the dark current is

$$I_{\mathrm{dark}} = i_{\mathrm{e}}^{(1)}/p_{\mathrm{c}} = i_{\mathrm{e}}/(Np_{\mathrm{c}}), \tag{4.29}$$

where $i_{\mathrm{e}} \equiv N \times i_{\mathrm{e}}^{(1)}$ is the total emission current from all N wells. Equivalently, one can express the dark current in an alternative form $I_{\mathrm{dark}} = i_{\mathrm{c}}^{(1)}/p_{\mathrm{c}}$, where $i_{\mathrm{c}}^{(1)}$ is the capture current per well and $i_{\mathrm{c}}^{(1)} = i_{\mathrm{e}}^{(1)}$. The "g-r" (emission and capture here) noise therefore consists of two contributions: fluctuations in i_{e} and i_{c}. The magnification factor is $1/(Np_{\mathrm{c}})$ according to (4.27). Then we have

$$\begin{aligned} i_{\mathrm{n,dark}}^2 &= 2e\left(\frac{1}{Np_{\mathrm{c}}}\right)^2 (i_{\mathrm{e}} + i_{\mathrm{c}})\Delta f \\ &= 4e\left(\frac{1}{Np_{\mathrm{c}}}\right)^2 i_{\mathrm{e}}\Delta f \\ &= 4e\frac{1}{Np_{\mathrm{c}}}I_{\mathrm{dark}}\Delta f \\ &\equiv 4eg_{\mathrm{noise}}I_{\mathrm{dark}}\Delta f, \end{aligned} \tag{4.30}$$

where the noise gain is defined by $g_{\mathrm{noise}} \equiv 1/(Np_{\mathrm{c}})$. In a conventional photoconductor, the noise gain equals the photoconductive gain $g_{\mathrm{noise}} = g_{\mathrm{photo}}$ (at least as a very good approximation for all practical purposes). Here we see that g_{noise} is different [154] from the photoconductive (4.21) $g_{\mathrm{photo}} = p_{\mathrm{e}}/(Np_{\mathrm{c}})$. Experiments [15,142,155] reported a gain derived from the ratio of the measured current responsivity and absorption (i.e., the photoconductive gain), and a gain derived from direct noise measurements (the noise gain). The ratio of the two measured gains gave the escape probability p_{e} which approached unity as the bias voltage was increased.

In the limit of $p_c \to 1$, (4.30) does not give the expected N full shot noise sources in series. Beck [140] extended the model and derived a more general expression $i_n^2 = 4eg_{\text{noise}}I(1 - p_c/2)\Delta f$ which does give the expected result when $p_c \to 1$ (see also Sect. 5.2.3). Several other related discussions of gain in QWIPs are given in [27,156,157].

An expression for photon noise current can be easily obtained by replacing I_{dark} in (4.30) with photocurrent. One of the most important sources of photon noise is caused by the background photons absorbed by a detector, as already discussed in detail in Chap. 2. The background photon noise usually determines a detector operating temperature.

4.3.2 Detectivity and Blip Condition

In this section we provide a simple analytical estimate that displays the key physical parameters. In doing so, we can easily identify the key parameters and point out possible directions for improvement. The two most important IR photon detector characteristics are detectivity ($D^\star$) and background infrared limited performance blip temperature (T_{blip}). $D^\star$ is the signal (per unit incident power)-to-noise ratio appropriately normalized by the detector area and the measurement electrical bandwidth. The relevant noise contributions are from (1) the detector itself (i.e., dark current) and (2) the fluctuation of the photocurrent induced by background photons incident on the detector. The blip regime is defined as the regime where the dominant noise is caused by the background photons. It is always desirable to operate a detector under blip condition for maximal sensitivity.

As already discussed in Chap. 2, the current noise power spectral density $S_i \equiv i_{n,\text{dark}}^2/\Delta f$ is related to dark current by

$$S_i = 4egI_{\text{dark}}, \tag{4.31}$$

where $g \approx 1/Np_c$ is the gain. We assume that we are in the regime where noise and photoconductive gains are the same. This is valid for structures where $p_c \ll 1$ and $p_e \approx 1$, i.e., a nearly ideal photoconductor. The dark current can be estimated (4.1)

$$I_{\text{dark}} = eN_{3D}vA, \tag{4.32}$$

where v is the drift velocity and A is the device area, or by (4.6)

$$I_{\text{dark}} = e\frac{N_{2D}}{L_p}\frac{\tau_c}{\tau_{\text{scatt}}}vA. \tag{4.33}$$

The 3D and 2D above-barrier electron densities N_{3D} and N_{2D} (see (4.2) and (4.9)) can be approximated by

$$N_{3D} = 2\left(\frac{m_b k_B T}{2\pi\hbar^2}\right)^{3/2} \exp(-hc/\lambda_c k_B T + E_f/k_B T), \tag{4.34}$$

$$N_{2D} = \frac{m}{\pi\hbar^2}k_B T \exp(-hc/\lambda_c k_B T + E_f/k_B T), \tag{4.35}$$

with m (m_b) – effective mass in the well (barrier), T – temperature, λ_c – cutoff wavelength, and E_f – Fermi energy. There is a simple relationship between the Fermi energy and the well 2D doping density $N_\mathrm{D} = (m/\pi\hbar^2)E_\mathrm{f}$ (assuming doping is in the well and the dopants are completely ionized).

As before (4.24), the current responsivity of a QWIP is defined as

$$\mathcal{R}_i = e\frac{\lambda}{hc}\eta g. \tag{4.36}$$

The detector dark current limited $D^\star$ is given by

$$D^\star_\mathrm{det} = \frac{\mathcal{R}_i}{\sqrt{S_i/A}}. \tag{4.37}$$

Using the expression $g = \tau_\mathrm{c}/N\tau_\mathrm{trans}$, where τ_trans (as before) is the transit time across one period, and substituting (4.31), (4.32), and (4.36) into (4.37), the 3D drift model yields the detector noise limited $D^\star$

$$D^\star_\mathrm{det} = \frac{\lambda}{2hc}\frac{\eta}{\sqrt{N}}\sqrt{\frac{\tau_\mathrm{c}}{N_\mathrm{3D}L_\mathrm{p}}}. \tag{4.38}$$

Here λ is the wavelength, η is the absorption quantum efficiency, N is the number of quantum wells, and L_p is the quantum well period. Similarly, the 2D emission-capture model (4.6) yields the detectivity

$$D^\star_\mathrm{det} = \frac{\lambda}{2hc}\frac{\eta}{\sqrt{N}}\sqrt{\frac{\tau_\mathrm{scatt}}{N_\mathrm{2D}}}. \tag{4.39}$$

An alternate instructive derivation of (4.39) is based on (2.23). Here the relation $G_\mathrm{th}\mathcal{V} = Nj_\mathrm{e}A$ between the emission current j_e and the 3D thermal generation rate G_th (see (2.13)) gives rise to the simple expression $D^* = \eta/(h\nu\sqrt{4Nj_\mathrm{e}})$. (4.39) then follows by inserting (4.4).

In view of the balance between scattering escape and capture (4.7), (4.38) and (4.39) are actually equivalent (and yield in fact similar numerical values if typical band parameters and time constants are inserted). However, they do show the physical process from different perspectives. While (4.38) relates the detectivity to the 3D effective carrier concentration and the capture process, (4.39) addresses a 2D effective carrier concentration and a scattering (or emission) process. We point out that, by distinguishing between carriers in the 3D continuum and carriers in the 2D subband at high kinetic energy, $D^\star_\mathrm{det}$ (4.39) becomes conceptually different from Levine's model (which leads to the same result except that τ_scatt is replaced by τ_life or τ_c). In fact, if we think of a hypothetical QWIP with extremely thick barriers, $D^\star_\mathrm{det}$ according to (4.38) and (4.39) will remain constant (since $\tau_\mathrm{c} \sim L_\mathrm{p}$), while $D^\star_\mathrm{det} \sim \sqrt{L_\mathrm{p}}$ in Levine's model.

Some limitations of QWIPs are seen from (4.38) and (4.34), and (4.39) and (4.35). Since the lifetime (and scattering time) is short for QWIPs, a lower detector dark current limited $D^{\star}_{\rm det} \propto \sqrt{\tau_{\rm c}/L_{\rm p}}$ ($D^{\star}_{\rm det} \propto \sqrt{\tau_{\rm scatt}}$) is anticipated. In addition, an "extra" thermal excitation factor appears ($E_{\rm f}/k_{\rm B}T$ in the exponent of (4.34) and (4.35)). This causes a larger dark current and hence results in a lower $D^{\star}_{\rm det}$. Note that this discussion is for a QWIP with its noise coming solely from its dark current (i.e., no background); a similar discussion is given below for background photon noise limited detectivity.

From (4.38) and (4.39), the expected general behavior for a photoconductor is seen, such as (1) a higher η, longer $\tau_{\rm c}$ (or $\tau_{\rm scatt}$), shorter $\lambda_{\rm c}$ or lower T lead to a higher $D^{\star}$, and (2) $\lambda_{\rm c}$ and T are the most sensitive parameters, being on the exponent.

Noting that η is proportional to the doping density and hence the Fermi energy ($\eta \propto E_{\rm f}$), there is an optimum value for $E_{\rm f}$. Since $D^{\star} \propto E_{\rm f}\exp(-E_{\rm f}/2k_{\rm B}T)$, the optimum value is found by

$$\frac{\rm d}{{\rm d}E_{\rm f}} E_{\rm f}\exp(-E_{\rm f}/2k_{\rm B}T) = 0, \tag{4.40}$$

which gives the maximum $D^{\star}$ when $E_{\rm f} = 2k_{\rm B}T$ [100]. This condition dictates an optimum value for $N_{\rm D}$ for maximizing $D^{\star}$. Figure 4.15 shows calculated $D^{\star}$ values vs. wavelength and temperature. Typical values are used: $\eta = 25\%$, $N = 50$, $\tau_{\rm c} = 5\,{\rm ps}$, and $m^* = 0.067$ for GaAs reduced effective mass ($m = m_{\rm e}m^*$, where $m_{\rm e}$ is the free electron mass). The optimum doping or Fermi energy value $E_{\rm f} = 2k_{\rm B}T$ is assumed. The calculation uses (4.38) for a quantum well period $L_{\rm p} = 40\,{\rm nm}$. Identical results would of course have been obtained with (4.39), provided that the values chosen for $\tau_{\rm c}$ and $\tau_{\rm scatt}$ satisfy (4.38).

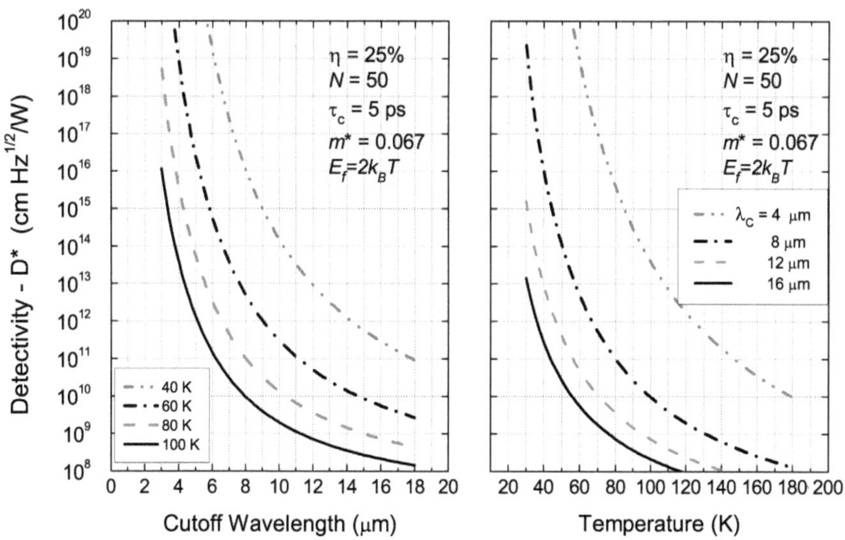

Fig. 4.15. Calculated detector dark current limited detectivity vs. (*left*) wavelength for different temperatures, and vs. (*right*) temperature for different wavelengths

Useful empirical D^* values of GaAs-based QWIPs with 45° facet coupling at around 77 K, as reported by Levine et al. [15,158], are given by

$$D_{\mathrm{e}}^{\star} = 1.1 \times 10^{6} \exp(E_{\mathrm{c}}/2k_{\mathrm{B}}T)\ \mathrm{cm}\sqrt{\mathrm{Hz}}/\mathrm{W} \tag{4.41}$$

for n-type devices, and by

$$D_{\mathrm{h}}^{\star} = 2 \times 10^{5} \exp(E_{\mathrm{c}}/2k_{\mathrm{B}}T)\ \mathrm{cm}\sqrt{\mathrm{Hz}}/\mathrm{W} \tag{4.42}$$

for p-type. As a matter of fact, (4.41) is in nice agreement with the results of Fig. 4.15.

From (4.38) and (4.39), a point worth noting is that *provided* η can be made high, say close to 100%, one should use the *least* number of quantum wells in a QWIP. The limiting and the best case is a single well ($N = 1$) QWIP with $\eta = 100\%$. In this limit, a QWIP would in fact have a comparable performance to that of HgCdTe or InSb detectors. This limit is not completely impractical, e.g. it may be achievable by a waveguide-grating coupler with a high quality factor (see related discussion in Sect. 6.3).

For comparison, the background-limited spectral peak $D^\star$ is given by (see also (2.22))

$$D_{\mathrm{blip}}^{\star} = \frac{\lambda_p}{2hc}\sqrt{\frac{\eta_{\mathrm{p}}}{\phi_{\mathrm{B,ph}}}}, \tag{4.43}$$

where λ_{p} is the peak detection wavelength, η_{p} is the peak absorption, and $\phi_{\mathrm{B,ph}}$ is the integrated background photon number flux (per unit area) incident on the detector. Equation (4.43) is obtained from the same definition (4.37) by replacing I_{dark} in (4.31) to obtain the noise spectral density caused by the background photocurrent.

One can see that for a given wavelength and if a detector is blip, $D^\star$ only depends on the absorption quantum efficiency and the background photon flux. The lifetime becomes irrelevant in this regime. Unlike a broadband photon detector with a cutoff, a QWIP spectral response is peaked and can be approximated by a Lorentzian lineshape. In evaluating $\phi_{B,\mathrm{ph}}$ an integration is therefore performed using the blackbody function and a Lorentzian with unity height peaked at λ_p. Another parameter, the full width of the response curve $\Delta\lambda$, must be specified for the integration. For a standard QWIP, the range is approximately $\Delta\lambda/\lambda_p$ =10–30% [15].

The blip condition is defined when the photocurrent caused by the background equals the dark current. (This condition is equivalent to when background photon noise equals the dark current noise.) For operations at and lower than T_{blip}, the detector is said to be under blip condition. Using (4.32) and (4.36), the blip temperature is found to be determined by the following equations (using 3D drift model):

$$\eta^{(1)}\tau_{\mathrm{c}}\phi_{\mathrm{B,ph}} = 2\left(\frac{m_{\mathrm{b}}k_{\mathrm{B}}T}{2\pi\hbar^2}\right)^{3/2} L_{\mathrm{p}} \exp(-hc/\lambda_{\mathrm{c}}k_{\mathrm{B}}T + E_{\mathrm{f}}/k_{\mathrm{B}}T) \tag{4.44}$$

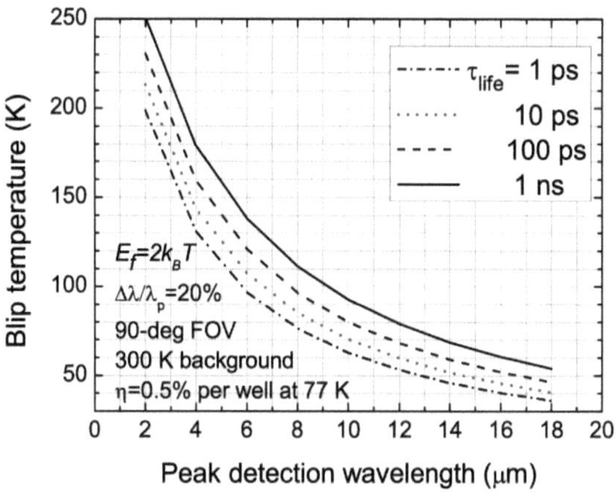

Fig. 4.16. Calculated background-limited infrared performance (*blip*) temperature vs. peak detection wavelength for $L_P = 50$ and different carrier lifetimes (τ_{life}) or capture time (τ_c). For GaAs QWIPs, $\tau_c \approx 5\,\text{ps}$

or (using (4.13)):

$$\eta^{(1)}\tau_{\text{scatt}}\phi_{\text{B,ph}} = \frac{m}{\pi\hbar^2}k_{\text{B}}T\exp(-hc/\lambda_c k_{\text{B}}T + E_f/k_{\text{B}}T), \tag{4.45}$$

where $\eta^{(1)}$ is the peak absorption efficiency for one quantum well. Solving the above transcendental equation (4.44) for T gives T_{blip}. Figure 4.16 shows the calculated T_{blip} vs. QWIP peak detection wavelength (λ_p) for different values of τ_c. Typical values have been used: response bandwidth $\Delta\lambda/\lambda_p = 20\%$, $\eta^{(1)} = 0.5\%$, 90° full cone field of view (FOV), and 300 K background temperature. Note that the cutoff wavelength is 10% larger than the peak in this calculated example, i.e., $\lambda_c = 1.10 \times \lambda_p$. Another shortcoming of QWIPs is clearly displayed in (4.44) and (4.45). The short capture time (or high "capture velocity" L_p/τ_c of typically $10^4\,\text{ms}^{-1}$) results in a lower T_{blip} and the effect of a finite E_f on the exponential also leads to a lower T_{blip}.

From (4.44) and (4.45) for T_{blip}, the most sensitive parameter is λ_c, being in the exponent. The high capture velocity (short scattering time), although giving rise to a fast intrinsic response speed, is the cause for the low T_{blip}. It is interesting to note that T_{blip} depends on the *one*-well absorption, not the total absorption, and that improving $\eta^{(1)}$ has the *same* effect as improving τ_c/L_p or τ_{scatt}. Detectors made of HgCdTe or InSb have a performance similar to or slightly better than the solid line ($\tau_c = 1\,\text{ns}$) of Fig. 4.16. The practical values of τ_c for QWIPs fall in the range of 1–10 ps. It then follows that if $\eta^{(1)}$ can be enhanced by about 100 times, QWIP performance would be comparable to HgCdTe or InSb.

From (4.44), an optimum condition for the Fermi energy and hence the doping density can be found. Given λ_c, τ_c, T, and $\phi_{\text{B,ph}}$, (4.44) can be re-written as

$$E_{\mathrm{f}}/k_{\mathrm{B}}T \exp(-E_{\mathrm{f}}/k_{\mathrm{B}}T) = (\mathrm{Constant}) \times \exp(-hc/\lambda_{\mathrm{c}}k_{\mathrm{B}}T), \tag{4.46}$$

noting $\eta^{(1)} \propto N_{\mathrm{D}} \propto E_{\mathrm{f}}$. One can adjust E_{f} to maximize the left-hand side of the equation, which maximizes T_{blip}. The optimum condition is $E_{\mathrm{f}} = k_{\mathrm{B}}T$, which is different from the optimum condition for maximizing the detector limited detectivity by a factor of two (i.e., $E_{\mathrm{f}} = 2k_{\mathrm{B}}T$).

For completeness, the ideal (blackbody) background photon flux is given by

$$\phi_{\mathrm{B,ph}} = \int d\lambda \left(\pi \sin^2 \frac{\theta}{2}\right) \eta(\lambda)\, L_{\mathrm{B}}(\lambda), \tag{4.47}$$

where θ is the FOV full cone angle, the photon irradiance is given by

$$L_{\mathrm{B}}(\lambda) = \frac{2c}{\lambda^4} \frac{1}{\mathrm{e}^{hc/\lambda k_{\mathrm{B}} T_{\mathrm{B}}} - 1}, \tag{4.48}$$

T_{B} is the background temperature, and the spectral lineshape of a QWIP is modeled by

$$\eta(\lambda) = \frac{1}{1 + (\frac{\Delta\lambda}{2\lambda} - \frac{\Delta\lambda}{2\lambda_{\mathrm{p}}})^2}. \tag{4.49}$$

4.4 Design of an Optimized Detector

In this section, we summarize the guidelines in designing an optimum photoconductive QWIP, which involves choosing the following parameters: well width L_{w}, barrier height (Al fraction x, in the case of GaAs well and AlGaAs barrier), barrier width L_{b}, well doping density N_{D}, and number of wells N. We use the simplest structure made of GaAs/AlGaAs square quantum wells. The well region is GaAs, and the barrier is $\mathrm{Al}_x\mathrm{Ga}_{1-x}\mathrm{As}$ so that its height is controlled by Al fraction x.

As discussed before, the optimum well shape is the one having the first excited state in resonance with the top of the barrier. Given this design rule, the well width and barrier height are fixed once a desired detection wavelength (peak wavelength λ_{p}) is chosen. Figure 4.17 shows these parameters for a range of λ_{p} for a GaAs/AlGaAs quantum well. The peak detection wavelength λ_{p} corresponds to the energy difference between the first excited and the ground states. The calculation is a simple one-band effective mass model calculation. The difference in effective mass values between AlGaAs and GaAs is included. The barrier height V_{b} relates to Al fraction by $V_{\mathrm{b}} = 0.87 \times x\,\mathrm{eV}$. All higher order effects have been neglected, such as, band nonparabolicity, Coulomb interaction between ionized donors and electrons (Hartree correction), exchange-correlation effect, and depolarization-exciton effect. For a structure appropriate for $\lambda_{\mathrm{p}} = 10\,\mu\mathrm{m}$, these effects all lead to a modification of the transition energy in the few-percent range (discussed in Sect. 3.4).

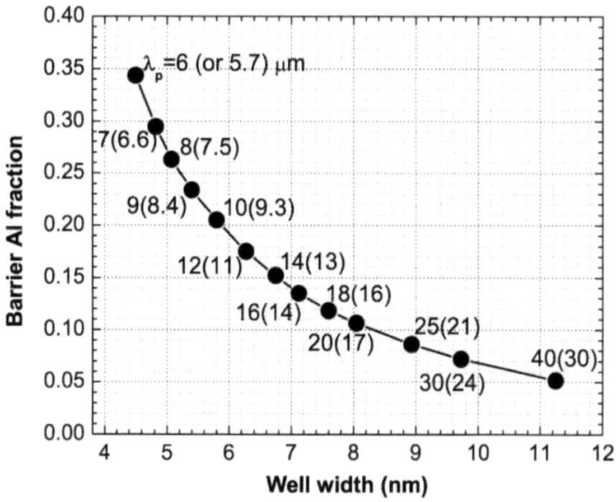

Fig. 4.17. Calculated parameters of barrier Al fraction and well width for a given peak detection wavelength (λ_p). The values in parentheses give λ_p including many-body effects for a doping density of $5 \times 10^{11}\,\mathrm{cm}^{-2}$ in the quantum well

As an example, the main effect of adding the exchange-correlation and depolarization-exciton effects is to shift the transition energy to a higher value. The amount is about 10 meV for a 2D electron (or doping) density of $N_D = 5 \times 10^{11}\,\mathrm{cm}^{-2}$. Taking this into account, the renormalized peak wavelengths are shown in Fig. 4.17 in parentheses. For this range of electron densities (optimum for detector sensitivity as discussed below), the effect results in only a correction for short wavelength structures, but is very important for long wavelengths. In general, the quantitative values of detection wavelength and doping density determine whether Hartree, exchange-correlation, and depolarization-exciton effects are just minor corrections or must be included to obtain a good fit to experiments. For $\lambda < 7\,\mu\mathrm{m}$, the nonparabolicity effect becomes important. Figure 4.17 stops at $x = 0.35$, and a high x value (e.g., > 0.4) is not advisable since such a QWIP results in a low responsivity [159] due to the poor transport properties in AlGaAs with a high x value. A better approach in this case is to use InGaAs wells [18].

The next parameter is the well doping. As discussed before, to maximize the detector-limited detectivity, the doping density should be such that the Fermi energy is $E_f = 2k_BT$, where T is the desired operating temperature. On the other hand, to maximize the blip temperature, one should have $E_f = k_BT$. As before, the doping density relates to Fermi energy by $N_D = (m/\pi\hbar^2)E_f$. Figure 4.18 shows these two (trivial) relations for QWIPs with GaAs wells.

The barrier width parameter L_b has so far been chosen intuitively. Only very limited experiments have been reported on the investigation of barriers. In general, the barrier width should be wide enough so that the interwell

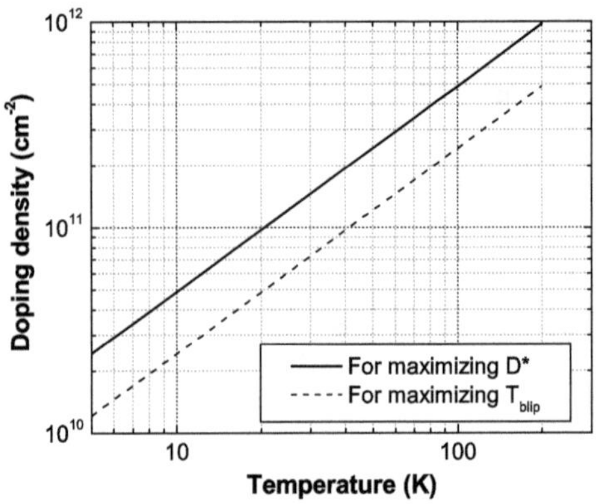

Fig. 4.18. Optimum doping density vs. device operating temperature. The well material is GaAs

tunneling current is suppressed; this means that one should use thick barriers. There is however a critical value beyond which any further increase in L_b does not lead to a lowering in dark current for the same applied field. Moreover, practical concerns for MBE growth mean that the thicker the total epilayer, the lower the material quality. Furthermore, if the barrier width can be reduced, one can grow more repeats for higher absorption for a given total epilayer thickness. In the literature [15], a barrier width in the range of 30–50 nm seems to be sufficient for QWIPs operating in the mid-infrared (MIR) region (3–12 µm).

An ideal estimate of the interwell direct tunneling can be given. Similar to the approximation used in Sect. 4.2.2 for the tunneling escape time, we calculate an interwell tunneling time by considering an "attempt frequency" $v_1/2L_w$ and the transmission probability $\mathcal{T}$, given by $\tau_{tunnel} \approx (2L_w/v_1)\mathcal{T}^{-1}$, where $v_1 = \sqrt{2E_1/m}$, E_1 is the ground state eigenenergy, and $\mathcal{T}$ at energy E_1 is calculated by a WKB approximation as in Sect. 4.1.1. The product of the tunneling time and the 2D well charge density yields the interwell tunneling current. Figure 4.19 shows the estimated results of the tunneling current vs. barrier width for several values of barrier Al fractions. The choice of barrier width must ensure that the interwell tunneling current is completely negligible in comparison with the background photocurrent. Taking a usual 300 K blackbody background temperature and common values of QWIP responsivity, the background photocurrent is in the range of 10^{-4}–10^{-5} A cm^{-2}. This was proven successful in recent work on THz QWIPs [54,55] (see Sect. 4.5). One can make an additional "hand-waving" argument in estimating the critical barrier thickness. For example, if the operating field is below about 10 kV cm^{-1}, to ensure that the tunneling of electrons near the top of the barrier (e.g., within 20 meV) directly into the next well is suppressed, one needs a barrier width of more than about 20 nm. A study of QWIPs for one particular wavelength (~9 µm) was carried out by Sim et al. [160], who found a criti-

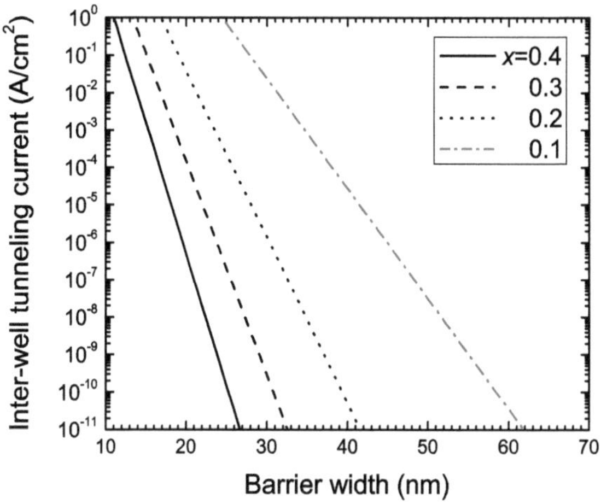

Fig. 4.19. Estimated interwell tunneling current vs. barrier width. The well is GaAs and the barrier is $Al_xGa_{1-x}As$. The well 2D electron density is set to $4 \times 10^{11}\,cm^{-2}$. Well widths are chosen according to Fig. 4.17

cal barrier value in the range of 20–30 nm, which depends on the operating temperature and the FOV.

Lastly, for the number of wells or repeats (N), only a general guideline can be given since the absorption depends strongly on the device geometry and optical coupling scheme: one should maximize the absorption with a minimum number of wells.

Note that these design guidelines are based on our current understanding and on our and others' published experimental results. As yet there have not been systematic studies to reenforce or confirm these design "rules."

4.5 THz QWIPs

Terahertz (THz) science and technology have attracted much recent attention [161,162]. The THz region is loosely defined as the range of frequencies from 0.1 to 10 THz (wavelengths from 30 to 3,000 µm or energies from 0.41 to 41 meV). THz frequencies fall partly into the FIR spectrum and partly into the millimeter-submillimeter wave band. Although many proof of concept experiments, demonstrating potential applications [161–163], have been carried out, the lack of compact and convenient THz generation and detection devices needs to be addressed before widespread applications become a reality. The recent development on THz QCL could fulfill the void in THz generation devices [164,165]. This section addresses THz detection and presents the design and experimental results on three test devices, all working at photon energies below and around the GaAs optical phonons. Blip operations were observed for all three devices designed for different wavelengths.

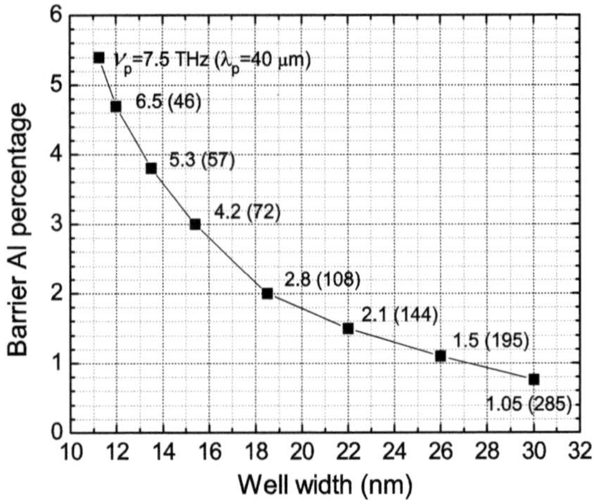

Fig. 4.20. Calculated parameters of barrier Al percentage and well width for a given peak detection frequency ν_p (wavelength λ_p)

4.5.1 Design Considerations

Using the same design rules discussed earlier in this chapter, QWIPs have been extended to the THz frequencies [54,55]. Figure 4.20 shows calculated quantum well parameters corresponding to THz QWIPs. The high absorption by the optical phonons of GaAs makes the region of 34–36 meV (34–36 μm or 8–9 THz) inaccessible. From Fig. 4.20, it is predicted that the THz frequency range of 1–8 THz is covered by QWIPs with low aluminum fractions from 0.8 to 5.4%. For completeness, the calculated blip temperatures are shown in Fig. 4.21. For achieving the highest dark-current-limited detectivity, if one follows the $E_f = 2k_BT$ rule, the density would become very low for low temperatures ($\ll$ 80 K), making the absorption also low. We therefore considered two cases in Fig. 4.21. The upper curve uses the $E_f = 2k_BT$ condition, while the lower curve starts with a higher absorption (using higher doping) $\eta^{(1)} = 0.5\%$ at 10 THz, and is reduced linearly down to 0.05% at 1 THz. The trade-off here is that if the operating temperature is desired to be as high as possible, the upper curve should be followed. However, if a high absorption is needed one should use the lower curve, which will mean a somewhat lower operation temperature. Since for these very FIR devices the doping densities are usually low, many-body effects result in small energy corrections. However, since the transition energies are also small, many-body effects need to be considered. The exact values depend on doping densities, but could be a substantial fraction of the transition energy as in the test devices discussed in Sect. 4.5.2. The detection frequency given in Fig. 4.20 and the horizontal axis in Fig. 4.21 should therefore be shifted to higher values by about 30% if these effects are included.

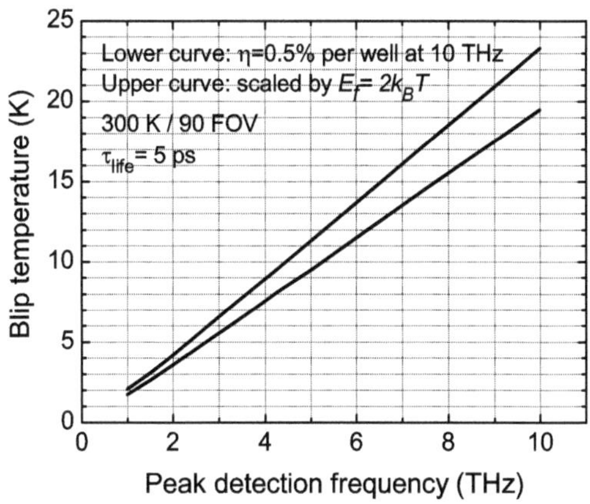

Fig. 4.21. Calculated background-limited infrared performance (blip) temperature vs. peak detection frequency

In general, to qualify as a good detector there must be a sufficiently high absorption. On one hand, a high doping is desirable for achieving high absorption; but on the other hand, high doping leads to a high dark current and low operating temperature. A trade-off must therefore be made for a given application. For most applications, it is desirable to operate the detector under the *blip* condition for detecting weak signals. In some applications involving a strong source such as a THz QCL, the requirement is different. Here as long as the dark current is lower than the signal photocurrent, photon noise-limited detection is achieved. In such a case, the detector operating temperature can be raised.

So far, the above analysis assumes that the tunneling contribution (such as, direct interwell, scattering assisted, and hopping-like via deep impurity levels in the barriers) is negligible. Some of the mechanisms are highly sample dependent, such as those related to impurity and defect. From the design point of view, as before, we must ensure that the barrier thickness is sufficient to suppress the interwell tunneling contribution. Figure 4.22 shows the estimated tunneling current vs. barrier thickness for three values of aluminum fraction x.

4.5.2 Experimental and Discussion

Under the guidance of the above design considerations, three test samples were fabricated. The sample parameters are listed in Table 4.3. The barrier thickness was chosen according to Fig. 4.22 so that the interwell tunneling currents are below $10^{-5}\,\mathrm{A\,cm^{-2}}$ which corresponds to the estimated background current. The center 10 nm of each well was doped with Si. The doping values were chosen to be close to the condition for optimizing detectivity. These values give rise to an estimated absorption in the range of a fraction of a

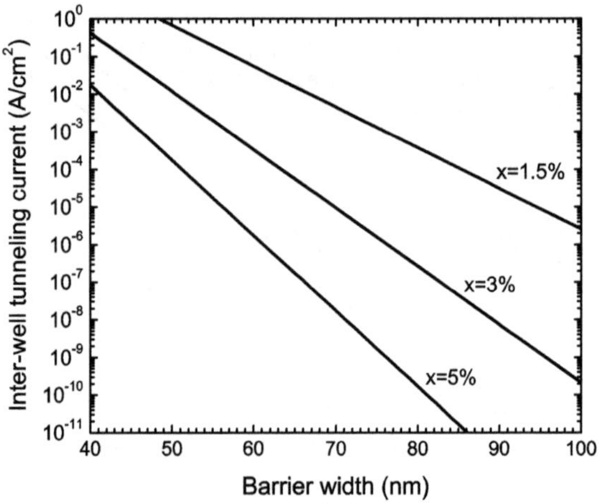

Fig. 4.22. Estimated direct interwell tunneling current vs. barrier thickness for three cases of barrier aluminum fractions of $x = 1.5$, 3, and 5%. Other parameters used for the estimate are $E_1 = 4.3$, 8.6, and 14 meV, $n_{2D} = 0.3$, 0.6, and $1.0 \times 10^{11}\,\text{cm}^{-2}$, and $L_w = 22.0$, 15.4, and 12.0 nm, for the three x values, respectively

percent per well. The top 400-nm and bottom 800-nm GaAs contact layers were doped with Si to $10^{17}\,\text{cm}^{-3}$. The relatively low contact doping value was used to reduce the contact layer free carrier absorption and plasma reflection in the THz region. The numbers of wells were varied among the three samples to keep the total epilayer thickness at no more than about 3.5 µm. In Table 4.3, the calculated transition frequencies include exchange and depolarization corrections. Using formulae given in Sect. 3.4.2, the values are $E_{ex} = 7.5$, 5.8, and 4.1, and $E_{depol} = 2.2$, 1.6, and 1.0 meV, for the three samples, respectively.

Mesa devices of different sizes were fabricated using standard GaAs processing techniques. Test devices were angle polished to give rise to a 45° facet and packaged in the standard double-pass backside illumination geom-

Table 4.3. Structure parameters for the THz QWIPs

Sample	L_w (nm)	L_b (nm)	[Al] (%)	N_d (cm^{-3})	N	ν (THz)
v265	11.9	55.2	5	1×10^{17}	40	9.3
v266	15.5	70.2	3	6×10^{16}	30	6.0
v267	22.1	95.1	1.5	3×10^{16}	23	3.3

L_w is the well width, L_b is the barrier width, [Al] is the barrier aluminum fraction, N_d is the doping value in the center 10 nm of each quantum well, N is the number of quantum wells, and ν is the frequency corresponding to the calculated intersubband transition energy (including exchange and depolarization corrections). The GaAs/AlGaAs MQWs are sandwiched between 400-nm top and 800-nm bottom GaAs contact layers doped with Si to $10^{17}\,\text{cm}^{-3}$.

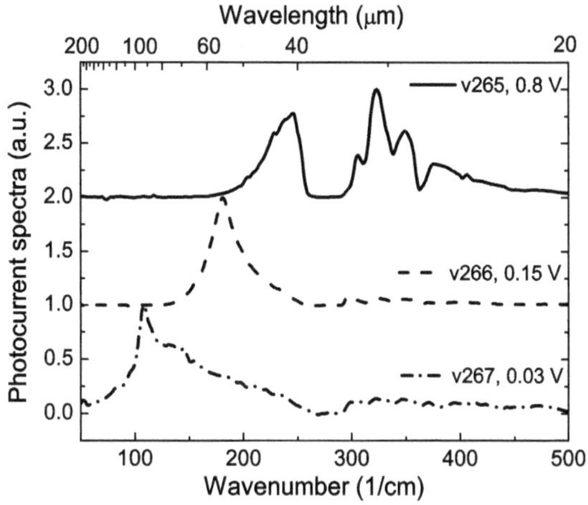

Fig. 4.23. Normalized photocurrent spectra at 8 K

etry with a 45° internal incident angle. The photocurrent spectra at 8 K for the three samples are shown in Fig. 4.23. Photocurrent was observed only under p-polarization, confirming its origin from (ISBT). The response peaks were measured at 322 cm^{-1} (31 μm, 9.7 THz), 180 cm^{-1} (56 μm, 5.4 THz), and 108 cm^{-1} (93 μm, 3.2 THz) for sample v265, v266, and v267, respectively. These agree reasonably well with the design value, considering the uncertainty in the growth and the untested range of very low aluminum fractions. The main dark region from 265 to 297 cm^{-1} (33–37 meV) is due to GaAs optical phonon absorption. Other smaller features are also related to phonons: The dip at about 360 cm^{-1} is caused by the AlAs-like phonon, and the two at about 310 and 340 cm^{-1} are due to zone-edge two-phonon processes [59]. Because of the strong phonon absorption, the as-observed peak at 322 cm^{-1} for sample v265 has a large uncertainty. The photoresponse spectra for all three samples are quite broad, indicating that the ISBTs are of bound-to-continuum origin. The doping value in the GaAs contact layer (10^{17} cm^{-3}) gives rise to a plasma edge frequency of 101 cm^{-1}, which may contribute to the relatively sharp decline at the longer wavelength side around 100 cm^{-1} for sample v267. For all three samples, the spectral shapes are nearly independent of bias voltage up to a certain voltage value. Beyond the voltage limits, the dark currents increase rapidly and the photocurrent spectra cannot be taken reliably. The voltage limits are ±0.8, ±0.2, and ±0.03 V for sample v265, v266, and v267, respectively.

The calibrated peak responsivity vs. bias voltage at different temperatures is shown in Fig. 4.24. All three curves display very similar trends. The responsivity increases nearly linearly with bias from the origin, another indication that the transition is of bound-to-continuum origin. The responsivity–voltage curves are largely independent of temperature up to a certain value. Higher than this temperature, the responsivity curves show a marked increase. This

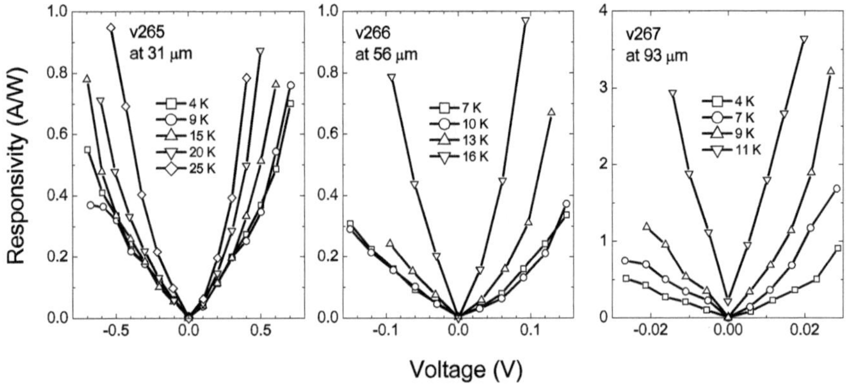

Fig. 4.24. Peak responsivity vs. bias voltage at different temperatures

upward trend behavior may be related to the increase in mobility with temperature in the impurity scattering-limited regime. The responsivity values at low temperature are in the range of 0.4–1.0 A W^{-1}, comparable to those from MIR QWIPs. This result implies that the photoconductive gain is similar to those of MIR QWIPs and excited carrier lifetimes are also similar.

Blip temperatures were measured by comparing current–voltage curves under dark condition and under a 300 K background with a 90° FOV at different temperatures. The measured current–voltage curves are shown in Fig. 4.25. R_{blip} is taken as the temperature at which the background photocurrent equals the dark current. The measured blip temperatures are 17, 13, and 12 K for sample v265, v266, and v267, respectively. The results agree reasonably well with the calculated values for v265 and v266 in Fig. 4.21. For sample v267, the measured value of 12 K is much higher than the calculated 5 K. The reason for this is not known presently. It is interesting to compare dark current of sample v265 with a previous device [54], which had a similar quantum well structure but a narrower barrier width (40 nm). V265 has much

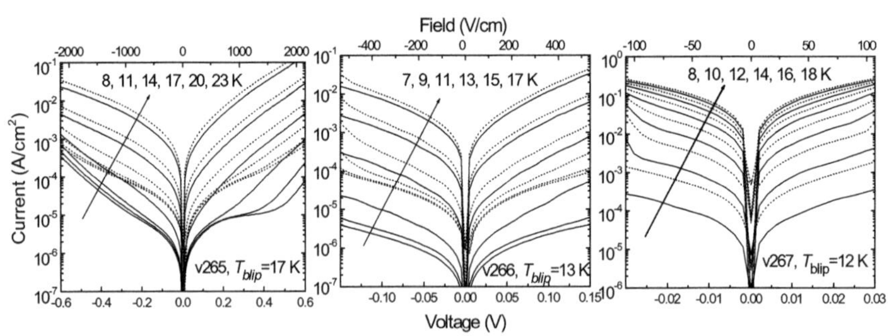

Fig. 4.25. Current–voltage curves under dark condition (*solid*) and under a 90° FOV 300-K background (dash)

lower dark current, about three orders lower at temperatures below 17 K. We attribute the substantial decrease in dark current to the reduction of interwell tunneling by employing wider barriers. When the temperature dropped below a certain point, the dark current stopped decreasing and remained the same. This temperature is 14 and 9 K for samples v265 and v266, respectively. This bottom-out behavior is due to the remaining interwell tunneling or other mechanisms. Since these tunneling currents are much lower than the background photocurrent, they do not affect device performance under normal background-limited operation. For sample v267, because thermionic emission and field-assisted tunneling were much stronger than interwell tunneling within thle experiment temperature range, we did not observe any current bottom-out behavior.

To end this section, we have discussed the design parameters and projected characteristics, and presented the experimental results on a set of test devices. We have demonstrated blip operation of three QWIPs with different detection wavelengths in the terahertz region. Increasing barrier thickness is found to be an efficient method to reduce the dark current and to enable blip operation. blip temperatures of 17, 13, and 12 K were achieved for peak detection wavelengths of 31, 56, and 94 µm, respectively. Although substantial follow-up work is needed, we are optimistic that we will further improve the device performance, and we project that optimized THz quantum devices will be useful for the booming field of THz science and technology and create new and unique applications.

5

Photovoltaic QWIP

5.1 General Concept

Most QWIP studies have so far concentrated on the photoconductive detection mechanism discussed in the previous chapter, where photoexcited carriers are swept out of the quantum well (QW) region by an *externally* applied electric field. A different class of QWIP structures is formed by devices with a built-in inversion asymmetry acting as an *internal* field; the devices can in principle be operated without external bias voltage, i.e., in a photovoltaic mode (see Chap. 2). At first sight, this concept has the advantage of a vanishing dark current, such that no generation-recombination noise is present under dark conditions. However, the photocurrent of these photovoltaic QWIPs is associated with a much smaller gain by comparison to photoconductive QWIPs. As we will show below, the reduced photocurrent and the reduced noise floor give rise to similar detectivities for optimized photoconductive and photovoltaic QWIPs. Therefore, photoconductive QWIPs are preferrable for applications which require high responsivity, e.g., for sensors operating in the 3–5 μm regime. Photovoltaic QWIPs will be a good choice if the integration time of the FPA is limited by the storage capacity of the readout circuit. In this case, the benefits of the photovoltaic QWIP arise from the facts that, (1) the capacitor is not loaded by any dark current and (2) the noise associated with the collected photocharge is extremely small.

Early device concepts for photovoltaic QWIP-like structures are shown in Fig. 5.1a and b. The device reported by Kastalsky et al. [166] relies on interminiband absorption in a superlattice with an adjacent barrier layer. While electron motion in the lower miniband is blocked by this barrier, electrons excited into the upper miniband gain enough energy to traverse this barrier, thus giving rise to a photocurrent already without external bias voltage. Goossen et al. [11] reported a different approach, in which the capacitance of the device is influenced by carriers emitted into a depletion layer located between an n-type QW and a surface metallization (see Fig. 5.1b). This device structure

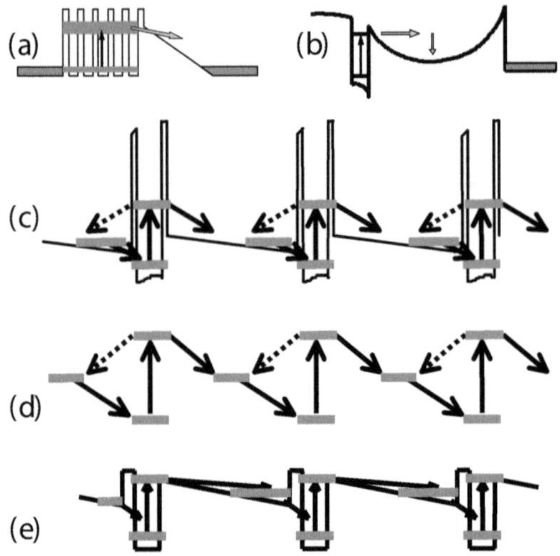

Fig. 5.1. Transport mechanism (*arrows*) of photovoltaic infrared detectors involving intersubband transitions: (**a**) superlattice with blocking barrier after Kastalsky et al. [166], (**b**) single quantum well with surface depletion layer after Goossen et al. [11], (**c**) asymmetrically doped double-barrier quantum well [167], (**d**) general subband configuration for photovoltaic detection, and (**e**) modulation-doped single-barrier quantum well [168]

was shown to exhibit an ac photovoltaic effect, i.e., periodically modulated illumination induced an alternating current without external bias voltage.

While these approaches were based on one single absorbing region, huge photovoltaic signals have been observed in *periodic* GaAs/AlAs/AlGaAs double barrier QW structures detecting radiation in the 3–5 µm regime [47]. The asymmetrically doped double-barrier QW structure depicted in Fig. 5.1c consists of 5 nm GaAs QWs, 2 nm AlAs barriers, and 25 nm wide $Al_{0.3}Ga_{0.7}As$ layers, and gives rise to an efficient transport mechanism [167]. After intersubband excitation from the lower to the upper subband of a QW (vertical arrows), there is a finite probability for the photoexcited electrons to be emitted from the QW across the AlAs barriers before intersubband relaxation occurs. Emission proceeds both to the right-hand and the left-hand sides of the QW. However, the space charge field across the $Al_{0.3}Ga_{0.7}As$ layers causes a net electron current towards the right, as the majority of the emitted carriers relax back into the QW located at the right-hand side of the $Al_{0.3}Ga_{0.7}As$ layers. In [167], the position of the dopant spike was varied systematically, resulting in a strong impact on the photovoltaic properties. It was shown that the "built-in voltage" of those detectors is about as large as the external bias necessary to compensate the built-in field across the $Al_{0.3}Ga_{0.7}As$ layers.

The general concept behind this transport mechanism is characterized by the configuration of subband levels shown in Fig. 5.1d, which forms the basis for the photovoltaic QWIP structures discussed below. As a first application of this concept, it is straightforward to conceive of a detector similar to Fig. 5.1c, but with reduced barrier heights. This idea leads to the single-barrier QW of Fig. 5.1e. In fact, model calculations show [169] that the latter approach should lead to similar detectivities to those observed for photoconductive QWIPs.

To demonstrate the detector of Fig. 5.1e experimentally, we have realized a device structure where photovoltaic operation was achieved by a tunnel barrier on one side of the QW and by using modulation doping, giving rise to an internal electric field [168,170]. The detector was found to have a reasonable detectivity which was about three times less as compared with an optimized photoconductive QWIP with the same cutoff wavelength of 10.5 μm. This reduction in detectivity of this initial photovoltaic QWIP is not unexpected due to the bidirectional emission of the photoexcited carriers at zero bias, the unoptimized internal electric field, and the possibility of thermionic reemission from the intermediate state across the triangular barrier.

Recently a configuration similar to Fig. 5.1e was reported [171] by generating the built-in field by a digitally graded superlattice barrier rather than modulation doping. The photovoltaic effect was successfully achieved. However, this "quantum cascade detector" (the band diagram actually resembles a quantum cascade laser [QCL]) showed limited performance due to dark currents caused by residual carrier leakage into resonant states of the barrier. In spite of some drastic improvement [172], this problem has not been completely resolved.

5.2 The Four-Zone QWIP

5.2.1 Transport Mechanism and Device Structure

A similar detetectivity as for photoconductive QWIPs has been achieved for photovoltaic detection by using the concept of the "four-zone" QWIP [173, 174]. Here each period of the active QWIP region is divided into four zones which can be optimized independently. The transport mechanism of this class of IR detectors is summarized in Fig. 5.2a. In the excitation zone 1, carriers are optically excited and emitted into the quasicontinuum above the bandedge of the drift zone 2. A photocurrent without external bias voltage is achieved, if the excited carriers relax into the capture zone 3. Finally, the carriers have to cross the tunneling zone 4 in order to populate the adjacent QW. This tunneling process has to be fast enough to prevent the captured carriers from being reemitted thermionically into the original well. The second task of the tunneling zone is to prevent the photoexcited carriers from being emitted towards the left-hand side of the excitation zone.

For efficient implementation of this transport mechanism, several requirements have to be met, as summarized in Fig. 5.2b, which depicts the transport behavior under a finite applied electric field. The tunnel barrier has to exhibit low probability for tunneling at high energies, such that the capture time into

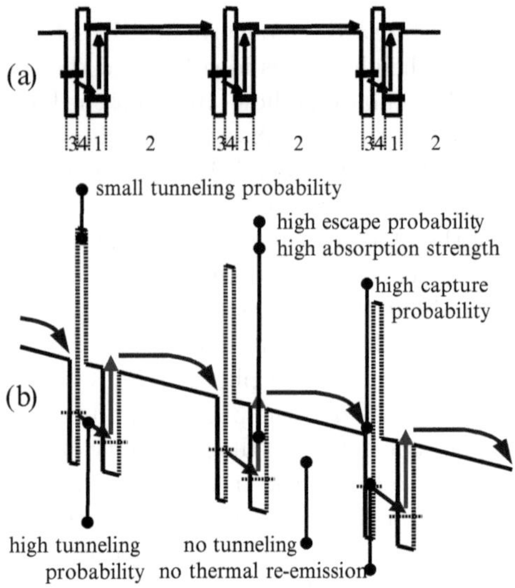

Fig. 5.2. (**a**) Schematics of the four-zone approach for photovoltaic intersubband photodetection; potential distribution (1: emission zone, 2: drift zone, 3: capture zone, and 4: tunneling zone) and basic operation (*arrows*). (**b**) Bandedge distribution of a four-zone QWIP in an electric field and considerations for optimization

the narrow QW is shorter than the tunneling escape time. Simultaneously, the time constant for tunneling has to be shorter than that for thermionic reemission from the narrow QW back into the wide QW. The requirements concerning high escape probability, high absorption strength, and suppression of tunneling across the wide barrier are analogous to the photoconductive QWIP. Another important detail is the steplike shape of the tunneling zone shown in Fig. 5.2b. This separation between the wide QW (emission zone) and the high-energy part of the tunneling zone is required to achieve a spectrally narrow absorption band and a high peak absorption.

For an initial demonstration of this approach [173], we have realized detector structures containing (in the growth direction) an active region with 20 periods of nominally 3.6 nm GaAs (the capture zone), 45 nm $Al_{0.24}Ga_{0.76}As$ (the drift zone), 4.8 nm GaAs (the excitation zone), and a sequence of 3.6 nm $Al_{0.24}Ga_{0.76}As$, 0.6 nm AlAs, 1.8 nm $Al_{0.24}Ga_{0.76}As$, and 0.6 nm AlAs (the tunneling zone). The 4.8 nm GaAs wells are n-doped to a sheet concentration of $4 \times 10^{11}\,cm^{-2}$ per well. The active region is sandwiched between n-type ($1.0 \times 10^{18}\,cm^{-3}$ Si) contact layers. While this detector structure will be referred to as sample 1, we have also produced a second structure (sample 2), in which the capture zone is replaced by a sequence of 3.0 nm GaAs, 1.8 nm $Al_{0.24}Ga_{0.76}As$, and 1.8 nm GaAs. The samples were grown by MBE on (100)-oriented, semi-insulating GaAs substrates. Mesa diodes with $120 \times 120\,\mu m^2$ area were fabricated using standard photolithography, wet chemical etching, and ohmic contact metallization.

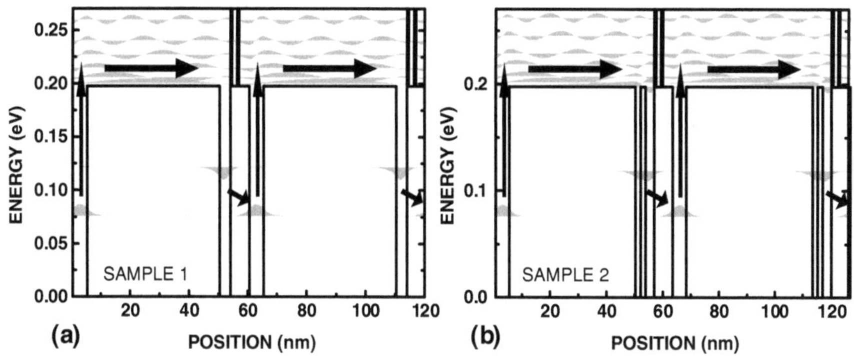

Fig. 5.3. Potential distribution and calculated periodic wavefunctions of (**a**) sample 1 and (**b**) sample 2 (after [173]). The transport mechanism is indicated by *arrows*.

Figure 5.3a indicates the conduction bandedge distribution and the calculated electronic wavefunctions for sample 1. Here the tunneling zone contains two AlAs spikes, which are introduced in order to suppress tunneling processes at energies above the $Al_{0.24}Ga_{0.76}As$ bandedge, and to induce emission across the drift zone. The energy of the subband located in the narrow well (capture zone) is as low as possible, with the restriction that this subband does not mix substantially with the ground state, and that only a negligible thermal population occurs at 77 K. By design, these two subbands are separated by a longitudinal optical phonon energy in order to ensure efficient phonon-assisted inelastic tunneling across the tunneling zone. This configuration also minimizes thermal reemission of the captured carriers, thus maximizing the probability for tunneling across the tunneling zone into the excitation zone of the subsequent period.

For this device structure, photoexcited electrons arriving at the capture zone can be reflected at the AlAs spikes without being captured by the narrow well, which will reduce the responsivity of the device, as the returning carriers can then be captured back into the excitation zone. It is therefore desirable to generate a resonance to increase the capture probability in the capture zone. An interesting concept to achieve this goal is the potential-inserted QW structure, which has been proposed by Akiyama et al. [175] for the case of bound-to-bound transitions. They introduced an AlAs spike at the center of a GaAs/AlGaAs QW, based on the idea that the second subband has a node at the spike position and the subband energy will not be affected. However, the spike will raise the groundstate energy, as its wavefunction has a maximum in the QW center.

The subband structure of our sample 2 is shown in Fig. 5.3b. Since the energy of the lowest subband in a potential-inserted QW depends critically on the thickness and the band parameters of the spike material, we used a strongly coupled double QW containing an $Al_{0.24}Ga_{0.76}As$ layer rather than an AlAs spike for the capture zone. In Fig. 5.3b, the signature of an above-barrier resonance manifests itself by a large amplitude of the higher-energy states. This resonantly enhanced amplitude is now present not only at the emitting QW, but also at the capture zone.

In addition to the four-zone QWIP structures described here, several modifications and parameter variations have been conducted, as described in more detail [176].

5.2.2 Responsivity and Dark Current

The optical properties of the detector structures discussed here are summarized in Fig. 5.4. The inset shows the normalized spectral dependence of the photocurrent, which is nearly identical for the two samples. Both samples show a peak wavelength of 8.4 μm and a 50% cutoff wavelength of 9.0 μm. At 0 V, samples 1 and 2 show peak responsivities of $3.5\,\mathrm{mA\,W^{-1}}$ and $10\,\mathrm{mA\,W^{-1}}$, respectively. At negative bias, the voltage dependence of both samples exhibits a plateau region from −2 to −3 V, more pronounced for sample 2, at a responsivity of about $37\,\mathrm{mA\,W^{-1}}$. We associate this plateau region with the field regime where complete emission of the photoexcited electrons occurs, while the capture and tunneling zones still reabsorb the emitted carriers efficiently. The subsequent increase at higher negative fields is due to the increasing probability for carrier transport across the tunneling zone without relaxation into the capture zone.

This interpretation is also consistent with the observation that the two responsivity curves cross each other at −2 V. At low fields, carrier relaxation

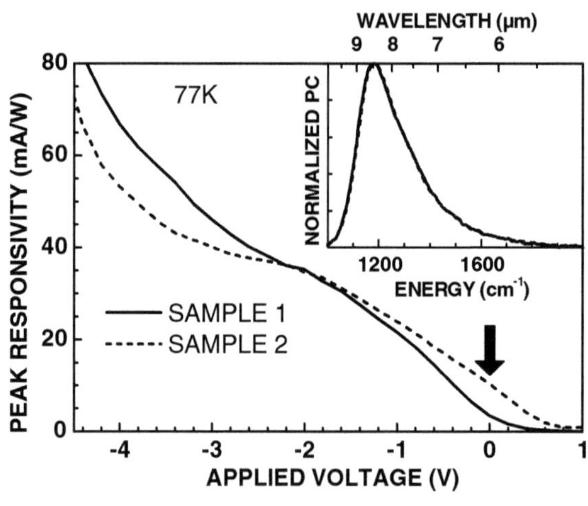

Fig. 5.4. Responsivity vs. applied voltage. The *arrow* indicates the responsivity at zero external bias. *Inset*: normalized photocurrent spectra measured at 0 V applied bias (after [173]).

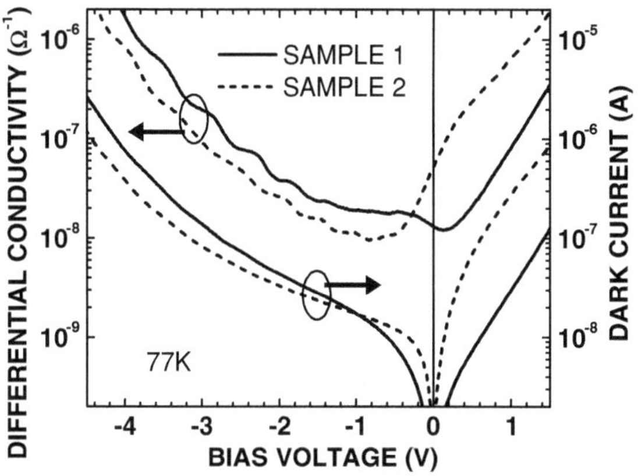

Fig. 5.5. Dark conductivity and dark current vs. applied voltage at a temperature of 77 K (after [173])

at the capture zone increases the responsivity; this is no longer the case at high negative fields, since the relaxation back into the initial well is already suppressed by the field. Instead, the probability for traversing the tunneling zone without relaxation into the capture zone is reduced upon optimizing the capture zone, as this also lowers the responsivity at high voltages.

We therefore relate the plateau region at a responsivity of $37\,\mathrm{mA\,W^{-1}}$ in Fig. 5.4 with a photocarrier mean free path of one period, i.e., to a photocurrent gain of $1/20 = 0.05$. These values are in fact consistent with the usual behavior of photoconductive QWIPs with similar doping densities, which typically exhibit mean free paths of 10 periods, and the responsivities are also about 10 times as large [15].

Figure 5.5 shows the differential conductivity σ and the dark current I_{dark}. At positive bias, σ and I_{dark} both show a steep increase for both samples. For an explanation of this behavior we note that the carrier distributions within the bound states located on both sides of the tunneling zone are in thermal equilibrium, since the tunneling probability is finite. With increasing positive bias, the capture zone is energetically lowered with respect to the excitation zone adjacent to the tunneling zone, thus giving rise to a considerable increase of the carrier density within the capture zone, and to a strong increase of the current. This effect is more pronounced for sample 2, as the width of the capture zone is larger than for sample 1. It is beneficial for the detector performance that a high photoconductivity and a low dark conductivity are observed at the same (negative) polarity, as this results in a high detectivity.

5.2.3 Noise

The appropriate noise model for the four-zone QWIP was first given by Beck [140]. His model explicitly considers the case of high capture probability $p_c \approx 1$

(or equivalently low noise gain g_{noise}). The model was obtained by stochastic considerations (for derivations we refer [140,177]) and take into account that a high capture probability is not necessarily connected with a low escape probability. This is exactly the case for the four-zone QWIP. The g-r noise current expression obtained in this model is given by

$$i_{\text{n}}^2 = 4eg_{\text{noise}}I_{\text{dark}}\left(1 - \frac{p_{\text{c}}}{2}\right)\Delta f. \tag{5.1}$$

where I_{dark} is the dark current and Δf the measurement bandwidth. For the case of $p_{\text{c}} \approx 1$, this expression equals the shot noise expression of N series-connected photodiodes.

The noise gain in this model is given by $g_{\text{noise}} = 1/(Np_{\text{c}})$, which, in combination with (5.1), yields

$$g_{\text{noise}} = \frac{i_n^2}{4eI_{\text{dark}}\Delta f} + \frac{1}{2N}. \tag{5.2}$$

Assuming a capture probability of unity, the lower limit for the gain is given by $1/N = 0.05$ for $N = 20$ periods.

For comparison, we apply the conventional noise model (4.30) of a photoconductor. This model is strictly applicable only to the case of a low capture probability and is often used to evaluate the noise gain of *conventional* QWIP structures. The noise gain obtained from noise and dark current measurements according to expression (4.30) is then given by

$$g_{\text{noise}} = \frac{i_{\text{n}}^2}{4eI_{\text{dark}}\Delta f}. \tag{5.3}$$

Noise measurements were performed using a spectrum analyzer in combination with a low noise current preamplifier, as described in detail by Rehm et al. [178]. The samples were mounted on a shielded sample holder, and directly immersed into liquid nitrogen. The spectra exhibited white noise up to several kHz. Measurements were conducted at 1430 Hz. The g–r noise is obtained from the measured noise current i_{n} by the equation $i_{\text{n,gr}}^2 = i_{\text{n}}^2 - i_{\text{amp}}^2 - i_{\text{J}}^2$, which includes small corrections to i_{n} due to the amplifier noise i_{amp} and the Johnson noise $i_{\text{n,J}}$ (see (4.26)). For this purpose, I_{dark} and $\text{d}I_{\text{dark}}/\text{d}V \equiv 1/R$ were measured simultaneously with the noise current.

The evaluation of the noise gain from the noise and dark current measurements on sample 2 according to (5.3) is indicated by the short dashed line in Fig. 5.6a. It is clearly seen that the conventional photoconductive model is not applicable to the four-zone QWIP since the result lies far beyond the theoretical limit $g_{\text{noise}} \geq 1/N = 5\%$ of these $N = 20$ period structures. The failure of the photoconductive noise model is attributed to the large capture probability of these four-zone QWIPs. In contrast, the model according to (5.1) (full line in Fig. 5.6a) yields exactly the expected result for moderate reverse bias ($V \geq -2\,\text{V}$).

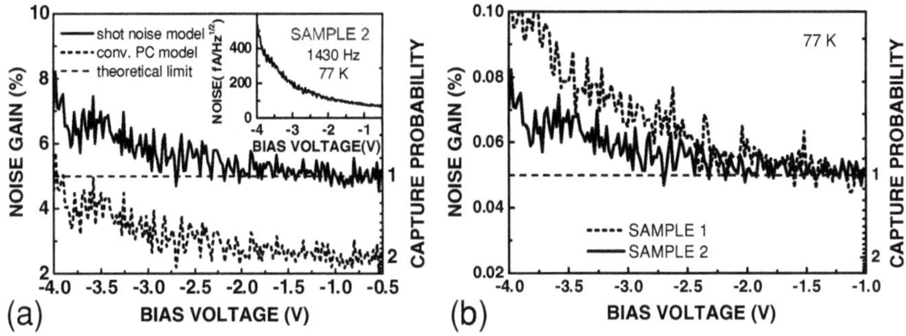

Fig. 5.6. Comparison between (**a**) the gain values for sample 2 evaluated according to the shot noise model (*solid line*) and the conventional photoconductive noise model (*dashed line*), and between (**b**) the gain values of both device structures according to Fig. 5.3 (after [143]). The *inset* of (**a**) shows the measured 77 K noise current at 1,430 Hz (0.23 mm^2 mesa size)

The noise gain of both device structures is compared in Fig 5.6b. Sample 1 exhibits a larger noise gain than sample 2 at high reverse fields, in addition to showing the same value of $g_{\text{noise}} = 5\%$ at moderate reverse bias. The larger gain is completely consistent with the observed behavior of the photo- and dark currents; it thus confirms the discussion in the Sect. 5.2.2.

Figure 5.7 relates the noise gain to the responsivity [143,178]. In the photoconductive QWIP of Fig. 5.7a, g is proportional to the responsivity, since the escape probability p_e has saturated to a value close to one, and the internal quantum efficiency of the detector does not depend on the bias (see Sect. 4.3). In the four-zone QWIPs of Fig. 5.7b, a gain of $1/N = 5\%$ is observed according to the transport mechanism discussed above. At high bias (above 2.5 V), g

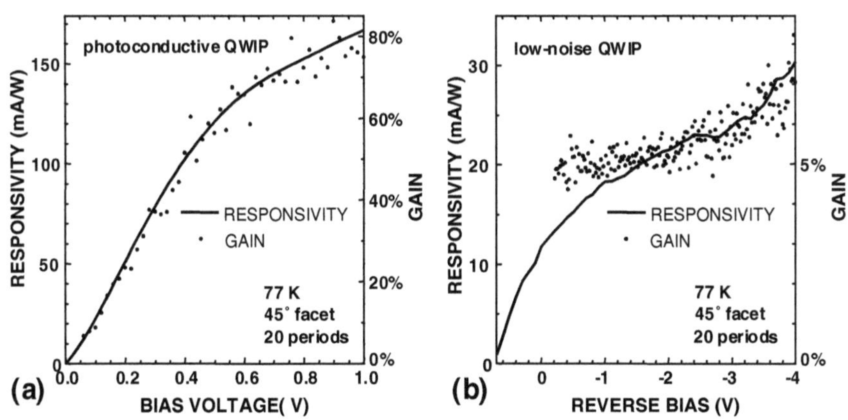

Fig. 5.7. Peak responsivity $\mathcal{R}$ and photoconductive gain g of (**a**) a photoconductive and (**b**) a four-zone QWIP, vs. bias voltage

increases in a similar way as the responsivity does, indicating that some electrons are penetrating the tunneling zone without relaxation into the QWs. Interestingly, g remains constant below 1 V, even though R decreases significantly. This deviation is attributed to the fact that a significant percentage of the photoexcited carriers are captured in the emission zone at these voltages, such that p_e decreases.

In particular, these experimental results, in combination with (5.1), clearly demonstrate the "$\sqrt{2}$-advantage" in the signal-to-noise ratio for four-zone QWIPs over photoconductive QWIPs, which is analogous to a similar advantage of photodiodes vs. photoconductors in the interband case (see (2.11) and (2.12)). In order to emphasize this improved noise behavior, photovoltaic QWIPs are referred to as "low-noise" QWIPs.

5.2.4 Detectivity

We have calculated the detectivity from the measured voltage dependence of responsivity and dark current. The 77 K detectivities obtained for samples 1 and 2, respectively, are 5.4×10^9 cm Hz$^{1/2}$ W^{-1} and 8.1×10^9 cm Hz$^{1/2}$ W^{-1} at 0 V. The highest detectivities are observed around −1.4 V, with a value of 1.2×10^{10} cm Hz$^{1/2}$ W^{-1} for both samples. Here the detectivity is already limited by g-r noise rather than Johnson noise, so the increase of the detectivity upon going from 0– −1.4 V is much less pronounced than the increase of the responsivity.

Figure 5.8 compares the bias dependencies of the dark D^* observed for low-noise QWIPs and photoconductive devices [176], both with 20 period active regions. With a somewhat different design of the active region (three rather than two AlAs spikes in the tunneling zone), this particular low-noise QWIP reaches its largest D^* at around −0.5 V, and about 70% of this maximum at 0 V (circle in Fig. 5.8). For both devices, the detectivity strongly

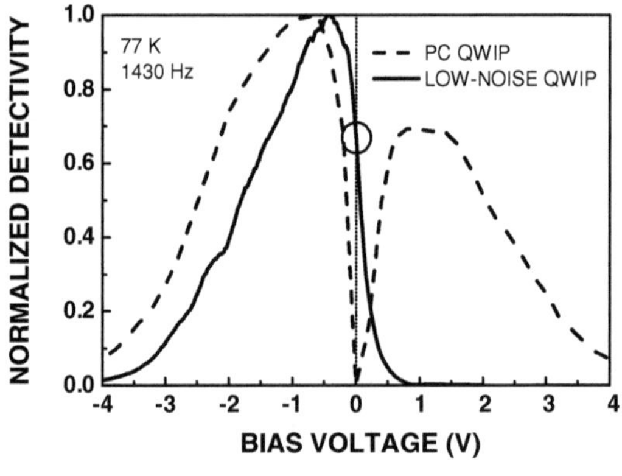

Fig. 5.8. Bias dependence of the detectivity of a four-zone QWIP as compared to a photoconductive QWIP (after [176])

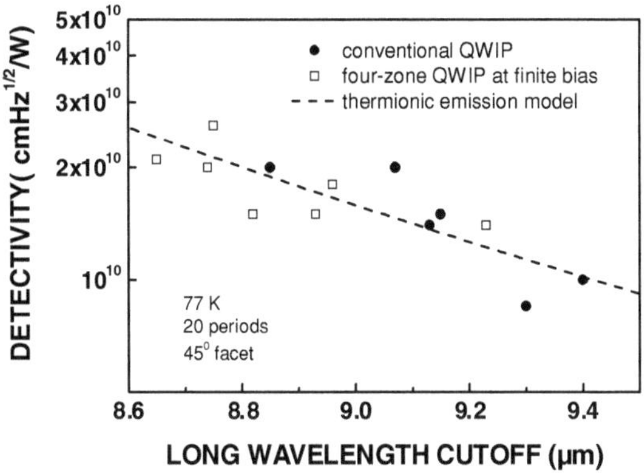

Fig. 5.9. Peak detectivity of low-noise QWIPs and photoconductive QWIPs at 77 K vs. cutoff wavelength (after [17])

decreases towards large bias for both polarities. This decrease is caused by an exponential-like bias depencence of the dark current at high electric fields (see Chap. 4).

Figure 5.9 shows peak detectivities of both conventional and low-noise QWIP structures as a function of the cutoff wavelength [17]. A good description of the functional dependence is obtained within a thermionic emission model (*dashed line*). The low-noise QWIPs show similar detectivities as the conventional ones. Since these detectivities are limited by the dark current and not by the background photocurrent, the improvement due to the suppression of the recombination noise is not prominent in Fig. 5.9. In fact, the low-noise QWIPs were grown with much higher carrier densities than the conventional QWIPs ($4 \times 10^{11}\,\mathrm{cm}^{-2}$ and $1 \times 10^{11}\,\mathrm{cm}^{-2}$ per QW, respectively), which resulted in a decrease of the activation energy. In addition, the spectral linewidth of the responsivity in low-noise QWIPs is typically 10–20% larger than for the conventional QWIPs, such that the peak detectivity is reduced correspondingly. However, in contrast to the dark detectivities of Fig. 5.9, the suppression of the recombination noise does have a significant influence on the background-limited detectivity which determines the temperature resolution of a thermal imager.

5.2.5 Time Dependence

Time-resolved measurements were performed on low-noise QWIPs in order to obtain further insight into the transport mechanism. The experiment was carried out using IR, pulses of <200 fs duration. These pulses were generated by difference frequency mixing of the signal and idler beams of an optical parametric oscillator, pumped by a mode-locked Ti:sapphire laser at a repetition rate of 76 MHz [179]. The intersubband photocurrent was measured with a sampling oscilloscope using the methods of [148].

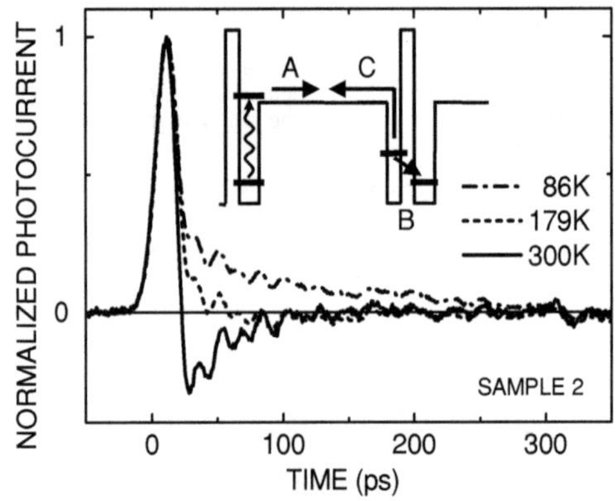

Fig. 5.10. Normalized photocurrent vs. time at different temperatures as indicated. The *inset* shows the schematics of the observed transport processes (after [174])

Figure 5.10 shows the time dependence of the photocurrent of sample 2 at different temperatures [173]. The curve obtained at 86 K represents the behavior observed at temperatures below 100 K. Here the photocurrent consists of two components. The fast component (about 18 ps full width at half maximum) is limited by the time resolution of the experiment. This component is attributed to carrier capture of the photoemitted carriers into the capture zone, as indicated in the inset (process A). The slow component, which corresponds to a decay time of about 0.11 ns, arises from tunneling from the capture zone across the tunneling zone (process B). In fact, the time constant can be explained quantitatively by inelastic tunneling assisted by longitudinal optical phonons [92,93].

At higher temperatures, we observed an additional component due to thermionic emission of carriers from the capture zone back into the emission zone (process C in Fig. 5.10). This process becomes important as soon as the associated thermionic emission time is comparable to or shorter than the time constant of the tunneling process. Since thermionic back emission gives rise to the opposite transport direction, the slow component is strongly reduced at 179 K and becomes negative at room temperature.

5.2.6 Theoretical Performance of Low-Noise QWIPs

From a theoretical point of view, the comparable performance of the detectivity as experimentally observed is not unexpected since the dark current mechanism of the low-noise QWIP under bias is closely analogous to the photoconductive structure. In fact, the thermionic emission current is determined by the emission and drift zones, which play essentially the same roles as the QW and thermionic barrier, respectively, of the photoconductive QWIP.

As long as electron reemission from the capture zone is negligible with respect to tunneling across the tunneling zone (which is the case according to Sect. 5.2.5), the main effect of these two zones is to increase the capture probability p_c and to reduce the gain.

For these reasons, the dark current mechanism essentially agrees with the case of the photoconductive QWIP discussed in Chap. 4. However, we should keep in mind that the condition $p_c \approx 1 - p_e$ does *not* hold for low-noise QWIPs. Similarly, intersubband absorption follows the same physical laws and is therefore comparable for both devices.

Since, in spite of the large capture probability p_c, the probability p_e for escape of photoexcited carriers from the emission zone can be high, the quantum efficiency for detection also remains high. In this way, the signal-to-noise ratio in low-noise QWIPs remains high, even though the signal and dark currents are strongly reduced as compared to the photoconductive device.

We conclude this chapter by listing a few trade-offs which are different for low-noise and photoconductive QWIPs.

1. As already mentioned, the probabilities for emission and capture in low-noise QWIPs obey $p_c \approx 1$ and $p_e \approx 1$, while $p_c \approx 1 - p_e$ for photoconductive QWIPs.
2. In order to reach a certain NETD, a photovoltaic QWIP requires only half as much photocharge as a photoconductive QWIP with the same gain (see Chap. 9). The "$\sqrt{2}$-advantage" of the photovoltaic device is thus equivalent to a twofold increase in quantum efficiency. Presently, this advantage only comes into play for large photo charges (i.e., long integration times and/or large apertures) where detector noise becomes comparable to readout noise.
3. The tunneling zone has some influence on the spectral bandwidth of the detector. The bandwidth can be tuned by a spacer layer between the tunnel barrier and the emission zone. Since this spacer layer must not be too large, due to its influence on the tunneling, the spectral bandwidth of low-noise QWIPs is usually somewhat larger than for photoconductive QWIPs.
4. Low-noise QWIPs for thermal imaging applications usually use higher carrier densities as high internal quantum efficiencies can be realized without saturating the readout electronics, and as system noise is more critical than for photovoltaic QWIPs. The trade-off between quantum efficiency and BLIP temperature is similar for both devices.
5. The reduced gain of low-noise QWIPs is expected to lead to smaller space charge effects (see Sect. 10.1.1).

6

Optical Coupling

The ISBT selection rule requires a nonzero polarization component in the QW direction (the epitaxial growth direction), as discussed in Chap. 3. In the simplest experimental geometry of shining the incident light normal to the as-grown wafer, a negligible ISBT absorption should be observed [44]; so other experimental geometries must be used. This chapter discusses the various optical coupling schemes for ISBTs and QWIPs.

6.1 Simple Experimental Geometries

Other than shining the light normal to the wafer, the next simplest is to have the light incident at an angle. In this case, only p-polarized light will have a component in the correct direction. P-polarized light is completely transmitted at the Brewster angle, so the geometry shown in Fig. 6.1 is commonly used to get a quick measure of the ISBT absorption spectrum after wafer growth. The complete transmission automatically eliminates the interference fringes often seen in IR transmission spectra of flat samples. For GaAs materials, the refractive index is about 3.3 in the MIR, leading to a Brewster angle of about 73°. An example of measured spectra is given in Fig. 4.12 of Chap. 4. Although the Brewster angle geometry is very convenient, it is not efficient in coupling to ISBT. Because of the large value of the refractive index, the internal incident angle onto the MQW is relatively small (17° in this case).

Another common geometry is the 45° zigzag waveguide formed by polishing two parallel facets into the substrate (schematically shown in Fig. 6.2). The length of the waveguide determines how many double passes the light will interact with the sample. If the length is made exactly two times the substrate thickness, only one double pass is obtained, enabling the measurement of the absorption equivalent to the 45° facet detector.

Alternatively, one can use a longer sample and etch away the QWIP layers except in a stripe twice as wide as the substrate thickness. An advantage of this geometry is that one can use s-polarized light as a reference, eliminating

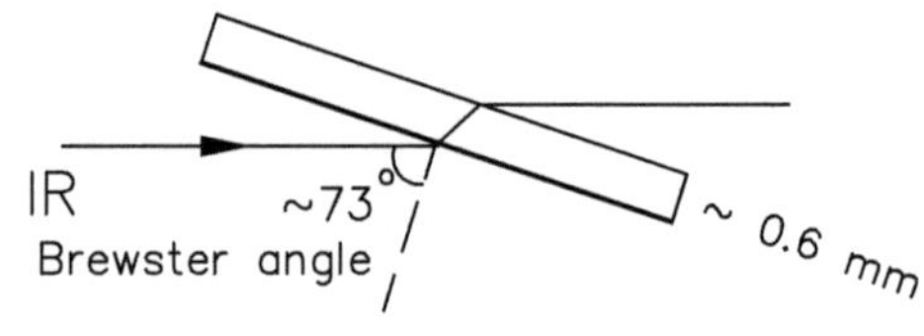

Fig. 6.1. Brewster angle incident measurement geometry for a typical GaAs substrate with a refractive index of 3.3. The backside of the substrate needs to be polished

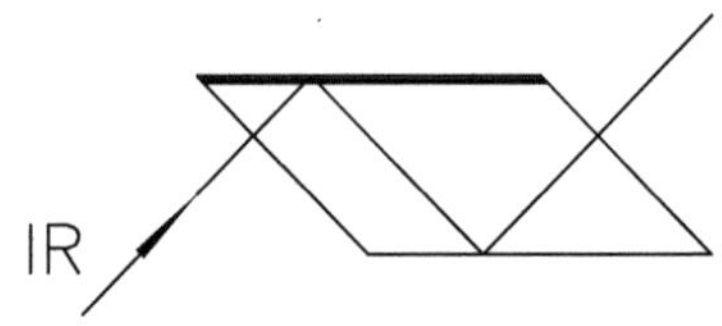

Fig. 6.2. Two parallel polished 45° facets forming a zigzag waveguide. The backside of the substrate needs to be polished

the need for a separate reference sample and hence the error associated with inequivalent optical beam paths. Measured transmission spectra on a set of samples for different wavelengths are given in Fig. 8.1 of Chap. 8.

Using this geometry, interference between the incident and reflected radiation gives rise to standing wave effects [180]. In order to determine the spatial distribution of the optical excitation density along the growth axis theoretically, we assume that a light beam is incident on a metal surface with a wave vector $\mathbf{k}$ at an angle α with respect to the growth direction, with the electric field $\mathbf{F}$ being polarized within the plane of incidence (see Fig. 6.3a). The polarization normal (y-direction) to the plane of incidence (x- and z-directions) is not relevant here since it does not satisfy the polarization rules for intersubband excitation. The perfectly conducting metal surface is located at $z = 0$. Introducing $k_z = |\mathbf{k}| \cos\alpha$, $k_x = |\mathbf{k}| \sin\alpha$, the electric field is then given by

$$\mathbf{F} = \begin{pmatrix} F_x \\ F_y \\ F_z \end{pmatrix} = 2F \exp(\mathrm{i}k_x x - \mathrm{i}\omega t \begin{pmatrix} \mathrm{i}\cos\alpha \sin(k_z z) \\ 0 \\ \sin\alpha \cos(k_z z) \end{pmatrix} . \tag{6.1}$$

This means, in particular, that the power density associated with F_z is proportional to $\cos^2(k_z z)$. The interference between the incident and the reflected beams thus has a periodicity of $\lambda/2n_\mathrm{r} \cos\alpha$. Both the photon fluxes polarized along the x- and z-directions show strong spatial variations due to this interference, even though the total power density for unpolarized light is constant at an angle of $\alpha = 45°$.

A surface which is not covered with metal, however, gives rise to a dielectric reflection. If α is far beyond the critical angle for total internal reflection (which is the case at $\alpha = 45°$), the electric field distribution is given by

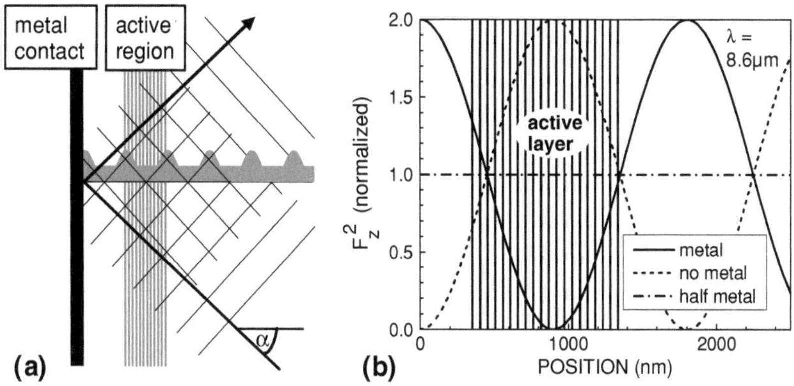

Fig. 6.3. (**a**) Optical interference of radiation incident at an angle α with respect to the direction normal to the metal surface (not to scale). Vertical lines indicate the quantum wells. The wave fronts of the incident and reflected beams are also shown. Interference gives rise to an intensity distribution as depicted by the shaded area. (**b**) Spatial dependence of F_x^2, normalized to its mean value, in the case of metallic reflection (*full line*), dielectric reflection (*dashed*), and 50% metal coverage (*dash-dotted*), for a 45° reflection geometry with 100% reflection at $z = 0$. Vertical lines indicate the actual positions of the quantum wells of an experimental QWIP structure (after [180])

$$\mathbf{F} = 2F\exp(\mathrm{i}k_x x - \mathrm{i}\omega t)\begin{pmatrix}\cos\alpha\cos(k_z z)\\ 0\\ \mathrm{i}\sin\alpha\sin(k_z z)\end{pmatrix}. \tag{6.2}$$

We use here the approximation to neglect the extra phase shift (which amounts to $\pi/33$ at $\alpha = 45°$) induced by the leakage of the light field into the vacuum [181]. Total internal reflection thus results in a phase shift of $\pi/2$ as compared to a metal reflector.

Figure 6.3b shows the spatial distribution of F_z^2, as predicted by (6.1) and (6.2), together with the actual positions of the QWs of a practical 20-period QWIP structure. In the case of a perfectly reflecting metal contact, the local power density has a minimum within the active region of the QWIP and the average power density is reduced by destructive interference. Dielectric reflection, however, gives rise to an increase of the power density due to an appropriate constructive interference.

The interference effect has several consequences on the detection properties of a QWIP structure. First, the total absorbed power depends on the precise location of the active QWIP region. In the example shown in Fig. 6.3b, the absorbed power is reduced almost by a factor of two due to a destructive interference. In fact, the active region will be located around a maximum of F_z^2 when inserting an additional 1 μm spacer between the metal contact and the active region. This influence of the active layer position has been clearly observed experimentally [180]. Also, variations of the absorbed power density along the growth direction of the QWIP induce inhomogeneous field distributions and nonlinear detection behavior. We will come back to this point in Chap. 7.

An approach to suppress the role of this interference on the detection properties of the QWIP consists of the use of mesa structures with a 50% metal coverage of the top surface. Within our simple calculation, the power density (averaged over the total mesa area) is constant along the whole active region. In fact, the in-plane conductivity in each quantum well is much higher than the conductivity across the barrier layers, such that each QW exhibits a well-defined potential along the whole area [182]. Therefore, the potential distribution is determined by the lateral average of the local power density with respect to the area of the QWIP, and the 50% metallized device is expected to behave in the same way as a device without any interference. In reality, some small deviations might still be present due to optical absorption by the carriers in the quantum wells [157] and contact layers, reflection losses at the metal surface, and distortions of the field distribution close to the side facets of the mesa devices. For most QWIP structures, the carrier-induced absorption is small (<10% typically), such that neglecting its influence on the phase of the reflected radiation is justifiable.

6.2 Gratings for Focal Plane Arrays

The majority of existing applications for IR detectors are related to imaging using 2D focal plane arrays (FPAs). A 2D array needs light incident normal to the wafer, so one has to use other means to bend the light than just discussed. Gratings have been exploited by Heitmann et al. [183] to excite ISBT in Si inversion layers, and their use was proposed in the very beginning of QWIP research [10]. 2D gratings have been modeled and tested systematically in a number of studies yielding quite high absorptions in large area devices [139, 184–187]. The gratings that have been mostly used so far have been made by etching them into the extra layer grown after the top contact layer. Gratings are in the form of either etched pits or trenches leaving unetched bumps, and gold metal is then evaporated for near perfect reflection. In a standard FPA, the illumination is through the substrate backside. Metal gratings (which do not need the extra layer) have also been studied [188]. In addition, biperiodic gratings have been studied [189,190], perhaps suited for dual-band or two-color QWIPs.

For the case of etched gratings, with the help of Fig. 6.4, simple design guidelines are as follows. The grating period d should be approximately the wavelength inside the material, i.e., $d = \lambda/n_r$, where λ is the wavelength to be detected and n_r is the refractive index. This will maximize the diffraction angle and lead to a high absorption. In practice λ should be chosen close to the cutoff wavelength. The etch depth h should be about one fourth of

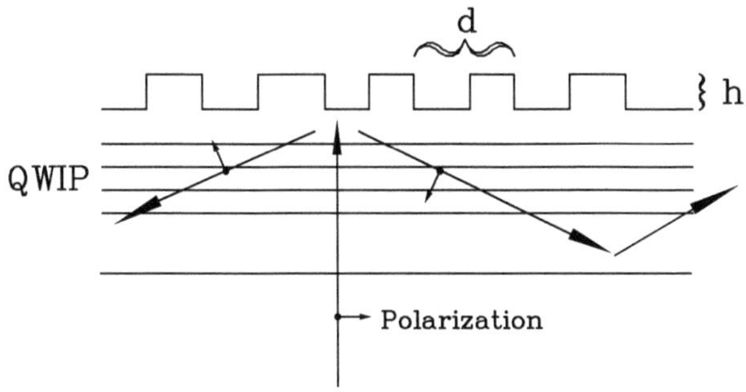

Fig. 6.4. Schematic of a grating fabricated on the top of a detector with infrared incident from the backside

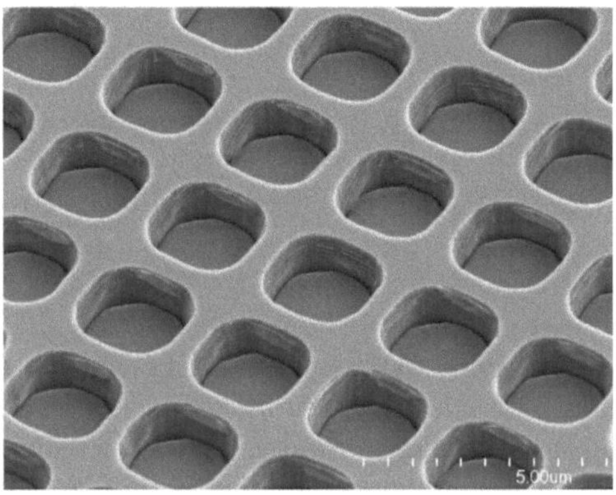

Fig. 6.5. Scanning electron microscope picture of a grating fabricated into GaAs (courtesy of M. Buchanan)

the inside wavelength, i.e., $h = \lambda/4n_r$. The areas of the etched and unetched regions should be equal. The last two conditions, under the simplest consideration, will eliminate the direct reflection (zero-order diffraction) due to the destructive interference. The grating features should be chosen to maximize the transverse-magnetic (TM) mode diffraction, as shown in Fig. 6.4. Finally, when the substrate is removed or a low index cladding layer is grown under the QWIP, an additional pass through the MQW is possible due to the reflection as shown in Fig. 6.4. An example of a pit grating is shown in Fig. 6.5. Details from FPAs [18] comprising gratings optimized to peak wavelengths of 4.8 and 8.5 μm are shown in Fig. 6.6.

Andersson and Lundqvist [187] reported a systematic study on gratings etched in the form of trenches. Their results are in reasonable agreement with the above general guidelines. They found that a square shaped bump grating tends to favor the TM diffraction. They also found that a slightly deeper etch seemed to enhance the performance. The range of values in their calculations corresponds to $h/(\lambda/n_r) \approx 0.25$–$0.28$; this represents a 10% etch depth

Fig. 6.6. Scanning electron microscope micrographs of 640×512 QWIP arrays with 24 µm pitch, comprising 2D diffraction gratings for the LWIR (*left*) with 2.95 µm period, and for the MWIR (*right*) with 1.65 µm period (from [18])

variation (a comfortable value in fabrication tolerance). Furthermore, they predicted that the edge length d_{etch} of the etched squares should fall in the range of $d_{\text{etch}}/d \sim 0.55$–$0.63$. Note that for equal etched and unetched areas (50% fill factor), one expects $d_{\text{etch}}/d = 1/\sqrt{2} = 0.707$. In an experimental test on the effect of fill factor for pit gratings (as in Fig. 6.5), we found quite a good agreement with the $d_{\text{etch}}/d = 1/\sqrt{2}$ condition [191]. A summary of this test is shown in Fig. 6.7. With a systematic change of the etch dimension from smaller than the $d/\sqrt{2}$ condition to larger, we observed a gradual change in the resulting QWIP responsivity. The exact value of $d/\sqrt{2}$ for these data is $2.8/\sqrt{2} = 1.980$ µm. We found that within ± 0.1 µm, the resulting

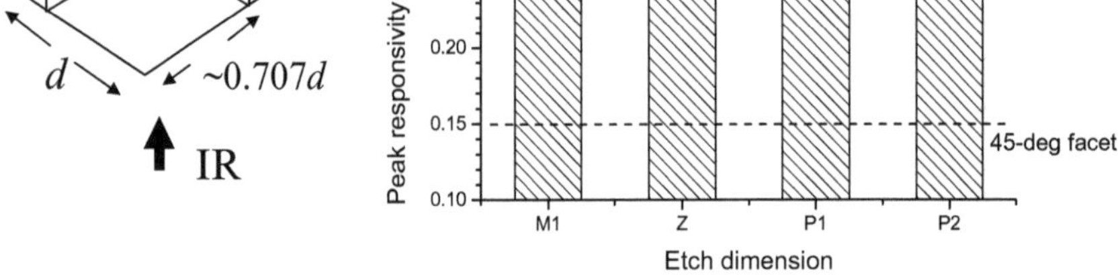

Fig. 6.7. (*Left*) Schematic of one period of etched pit grating and (*right*) peak responsivity vs. fill factor. The device is a standard 9-µm peaked QWIP. The grating constant d is fixed at 2.8 µm. The target etch size is 2 µm (labeled by Z), and slightly smaller and larger sizes are included: M1 – 1.9 µm, P1 – 2.1 µm, and P2 – 2.2 µm. The *dashed line* shows the 45° facet value

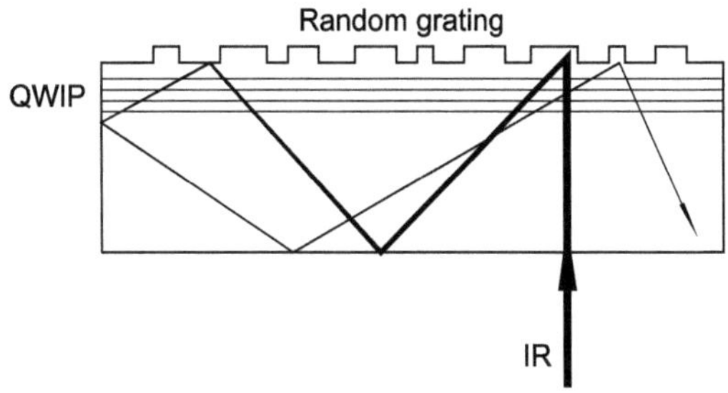

Fig. 6.8. Schematic illustration of the idea of a random grating

performance is not substantially affected. Again this gives a comfortable fabrication tolerance.

We emphasize that the grating problem is a very complicated one and many higher order effects play an important role, in some cases the dominant role. One such example is the near field effect [192]. This effect has been put to a good use in two cases. Bois et al. [192] maximized the near field absorption in a two-stack QWIP structure and used the upper stack as the active detector and the lower as a reference for subtracting the dark current. This serves as an effective "skimming" of the dark current in an FPA. In another case, gratings (pits/holes) are etched directly into the active QWIP layers, making the coupling completely near field [193].

Even 1D gratings are not simple in the presence of ISBT and free carrier absorption in both the MQWs and the contact layers. Dupont [194] modeled this case and found a highly nontrivial dependence of ISBT absorption on grating parameters. For example, an etch depth close to one eighth of the inside wavelength, rather than one fourth, gave the highest absorption.

The idea of random "gratings" [195–199] is simple and elegant as shown in Fig. 6.8. The incident light is scattered by the grating into a random direction, in most cases, with an angle larger than the total internal reflection, so that the light is trapped in the detector pixel. On each pass through the MQW a fraction of the IR gets absorbed, eventually leading to a high or near 100% absorption.

As a general comment, although they are successfully incorporated in QWIP FPAs which are commercially available, we believe that the gratings, or more generally the optical couplers, can be further improved. The difficulty, or the uncertainty, comes from two areas. (1) To accurately model a grating-coupled QWIP is a highly nontrivial problem [196,200,201], and presently involves heavy, sometimes unreliable, computation. The effects that must be included are the dielectric function change due to ISBT, free carrier absorption, multilayer structures, finite mesa sizes, realistic grating shapes and metal coatings on them, and others. Perhaps techniques from other fields, such as computer-generated holograms [202] and photonic crystals [203], could be explored for a more efficient and accurate computation. (2) The second factor

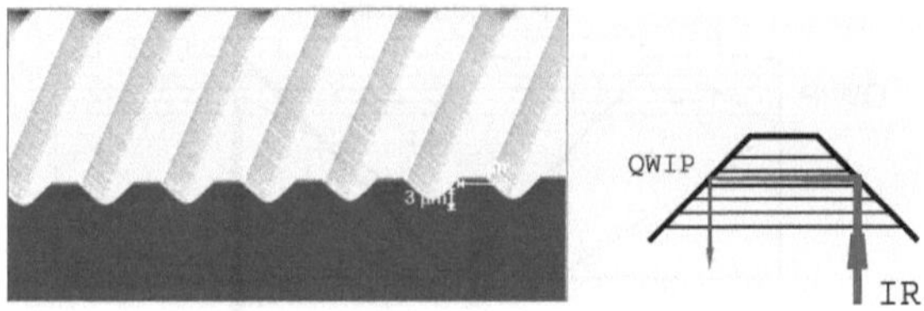

Fig. 6.9. *Left*: scanning electron microscope picture of V-grooves fabricated into GaAs (courtesy of M. Buchanan), and *right*: schematic optical coupling scheme

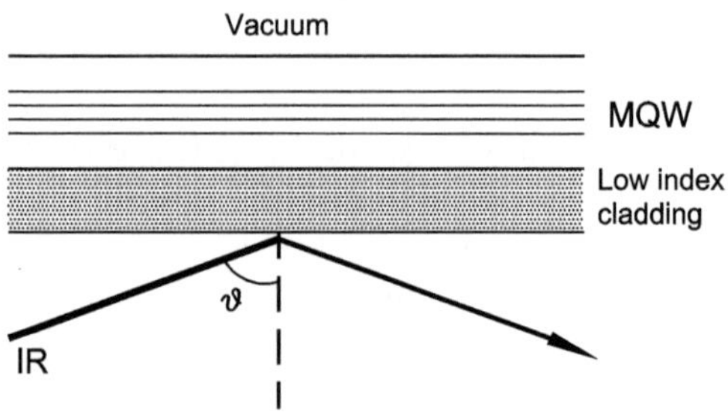

Fig. 6.10. Schematic layers of a dielectric waveguide

involves the microfabrication technology. The minimum feature size required for gratings is often in the submicron region. With the common optical lithography tools (not the state of the art achieved in Si technology), this is close to the limit. The fabricated features then often deviate from the design.

Finally, an alternative technique has been explored by Choi et al. [27], involving fabricating V-grooves with nearly 45° slopes. The process relies on the anisotropic wet chemical etch rates in GaAs. An example of fabricated V-grooves and the idea of the V-groove optical coupler are shown in Fig. 6.9.

6.3 Strong Coupling in Waveguides, Polaritons, and Vacuum-Field Rabi Splitting

If an MQW is embedded in a waveguide, e.g., by incorporating a sufficiently thick low index layer under the MQW, a strong coupling between the ISBT and the waveguide mode can occur. This leads to the formation of polaritons. By measuring the transmission or reflection spectra, a vacuum-field-like Rabi splitting is observed [204,205]. A simple structure, like the one in [204], is schematically shown in Fig. 6.10. The MQW is the core of the waveguide, with the QW made of GaAs and low Al fraction AlGaAs. The cladding layer is high Al fraction AlGaAs, or even pure AlAs as in [204]. The experimental results are reproduced in Fig. 6.11, showing a clear anticrossing behavior by varying the angle of incidence. The left inset shows the two polariton positions.

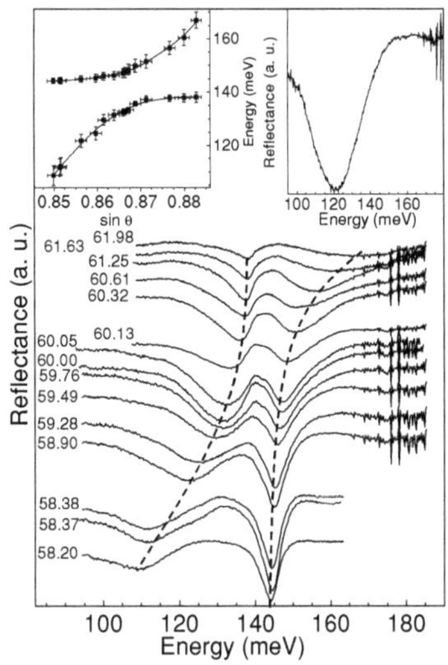

Fig. 6.11. Observed TM or p-polarized reflection spectra (*left inset*) polariton resonance positions, and (*right inset*) TE or s-polarized waveguide mode [204] (courtesy of A. Tredicucci)

Under transverse-electric (TE) or s-polarized light, the ISBT is "turned off," and a pure waveguide mode is seen in the spectrum due to the residual free carrier absorption. Of course, the TE mode position is not exactly the same as that for TM polarization, but slightly red shifted.

A potential use of the strong coupling is to yield a QWIP with very low doping density, and hence low dark current, and yet high absorption [205]. This may lead to high temperature operation.

7

Miscellaneous Effects

7.1 Intersubband Absorption Saturation

So far the excitation infrared intensity has been assumed to be low so that the absorption is far from saturated. Absorption saturation was studied in the early years of ISBT physics research [206,207]. The power level at which saturation occurs can be easily estimated by the following consideration. Suppose an intense infrared radiation resonantly couples the ground state and the first excited state of a quantum well structure. The 2D electron density of the excited state (n_2) is determined by

$$\frac{\mathrm{d}n_2}{\mathrm{d}t} = \phi_{\mathrm{ph}}\sigma(n_1 - n_2) - n_2/\tau_{\mathrm{relax}}, \tag{7.1}$$

where ϕ_{ph} is the photon number flux, σ is the absorption cross section, n_1 is the ground state 2D electron density, and τ_{relax} is the intersubband relaxation time. In steady state $\mathrm{d}n_2/\mathrm{d}t = 0$, one finds

$$n_2 = \frac{n_1}{1 + \frac{1}{\phi_{\mathrm{ph}}\sigma\tau_{\mathrm{relax}}}} = \frac{n_1}{1 + \phi_{\mathrm{ph,sat}}/\phi_{\mathrm{ph}}}, \tag{7.2}$$

where the saturation photon flux is $\phi_{\mathrm{ph,sat}} = 1/\sigma\tau_{\mathrm{relax}}$. As the intensity increases and far exceeds saturation, the two densities become equal ($n_2 \to n_1$ as $\phi_{\mathrm{ph}} \to \infty$). The photon energy flux ϕ is a more practical unit, which is given by $\phi = h\nu\phi_{\mathrm{ph}}$. The absorption efficiency as a function of ϕ is $\eta(\phi) = \sigma(n_1 - n_2)$. Using (7.2), we get the standard expression for absorption saturation:

$$\eta(\phi) = \frac{\eta(0)}{1 + \phi/\phi_{\mathrm{sat}}}, \tag{7.3}$$

where the saturation flux is

$$\phi_{\mathrm{sat}} = h\nu/\sigma\tau_{\mathrm{relax}}. \tag{7.4}$$

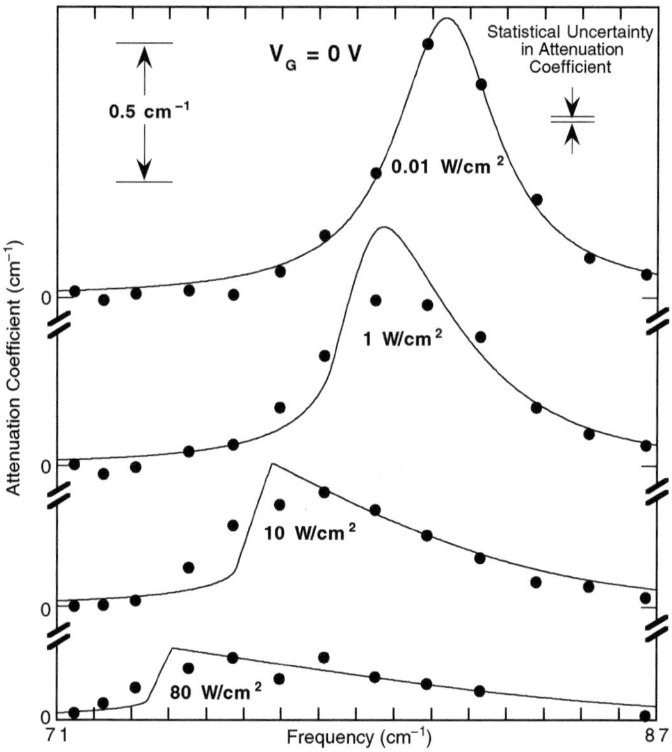

Fig. 7.1. Far-infrared absorption vs. frequency at various intensities for gate voltage $V_G = 0$ with an electron density of $1.1 \times 10^{11}\,\mathrm{cm}^{-2}$. The dots are experimental data, and the solid curves are calculated (courtesy of Mark Sherwin). See additional details [210]

We can easily estimate the order of magnitude of the saturation intensity. Referring to (3.7) and the discussion thereafter in Chap. 3 for a typical case of a simple GaAs/AlGaAs quantum well, the low power absorption efficiency per quantum well is $\eta \approx 0.5\%$ for polarized light at a 45° angle-of-incidence. The electron density was $n_{2D} = 5 \times 10^{11}\,\mathrm{cm}^{-2}$. The absorption efficiency and crosssection are related by $\eta = n_{2D}\sigma$. We then have $\sigma \approx 10^{-14}\,\mathrm{cm}^2$ or $\sigma \approx (1\,\mathrm{nm})^2$, and for $\tau_{\mathrm{relax}} \sim 1\,\mathrm{ps}$ and 10-μm light, $\phi_{\mathrm{sat}} \sim 2\,\mathrm{MW\,cm}^{-2}$.

A more in-depth discussion and experimental results are found [208]. Including many-body effects, the absorption spectrum displays interesting physics shown theoretically by Załużny [209] and experimentally [210]. Figure 7.1 reproduces the results of [210]. The experiments were performed on a MQW with 40-nm GaAs wells and modulation doped $Al_{0.3}Ga_{0.7}As$ barriers with the UCSB free electron laser. Gates were fabricated on the sample to provide tuning of the electron density. The low-intensity absorption spectral peak is depolarization-shifted substantially from the intersubband separation.

At higher intensities, the absorption peak shifts to lower frequencies, in effect "undresses" the collective effect. It is interesting to note that because of the smaller photon energy and the longer relaxation time the saturation intensity is much lower than that for 10-μm light case.

Using optical pulses of ≤1 ps duration, $\phi_{\text{sat}} \sim 2\,\text{MW}\,\text{cm}^{-2}$ corresponds to ≈2 μJcm^{-2} or to ≈200 nJ for a (100 μm)^2 spot. Even much higher pulse energies are readily achieved by table-top laser systems incorporating an optical amplifier. The coherent nonlinear response of modulation-doped GaAs QWs has been studied recently by Luo et al. [211] using intense 200 fs long pulses at 12.4 μm wavelength. Phase-resolved detection of the transmitted pulses provided clear evidence of Rabi oscillations.

7.2 Nonlinear Transport and Optical Effects

As a QWIP is a photoconductive and intersubband dipole transition based detector, the nonlinear behavior in QWIPs is caused by both the (extrinsic) transport and (intrinsic) optical processes. Extrinsic nonlinearity leads to a degradation of QWIP performance at high incident power or low operating temperatures. In this case, as the responsivity (the ratio between the photocurrent and the incident power) is measured as a function of increasing power, the observed value decreases from a constant at lower power. In contrast, some intrinsic nonlinear QWIP properties are useful in applications. An example is in autocorrelation of short pulses by two-photon absorption, where the photocurrent is proportional to the square of the incident power. The general area of QWIP nonlinear properties [212] has not been extensively investigated: only limited work has been carried out [118,213,214]. To study nonlinear effects a strong infrared excitation is required. Hence laser is commonly used. The study of QWIPs under strong illumination is relevant to a number of applications such as heterodyne detection [215], infrared pulse characterization [216,217], and free space optical communication [30,218]. In view of the high intrinsic speed [219], and the demonstration of near ideal absorption efficiency [220,221], QWIPs are well suited for these more exploratory applications.

7.2.1 Extrinsic (Photoconductive) Nonlinearity

An ideal photoconductor has a responsivity independent of the illumination power if the power is low; i.e., the photocurrent is linearly proportional to the power. The dependence of responsivity on the fundamental QWIP parameters is given by the standard expression:

$$\mathcal{R}(F) = \frac{e}{h\nu}\eta g(F) = \frac{e}{h\nu}\eta p_{\text{e}}(F)\frac{\tau_{\text{c}}(F)}{\tau_{\text{trans,tot}}(F)} \tag{7.5}$$

where $h\nu$ is the photon energy, η is the total absorption quantum efficiency, g is the photocurrent gain, p_e is the photoexcited carrier escape probability, τ_c is the photoexcited carrier lifetime, and $\tau_{trans,tot}$ is the transit time across the QWIP ($\tau_{trans,tot} = L/v(F)$, where L is the QWIP length, and v is the drift velocity). In an idealized case, the electric field F across the QWIP is constant and equal to $F_0 = V/L$, where V is the applied voltage. This condition is satisfied in QWIPs with a large number of equally doped QWs at low excitation power. The condition that the electric field is constant and independent of the illumination power can be violated at high incident power due to several reasons [212].

Contact Effects

Under uniform QW excitation (including both thermo- and photoexcitation) the electric field in QWIPs with multiple QWs is constant in the bulk of the detector [99,116]. However, the balance between the injection from the emitter and the bulk current in the steady state leads to the creation of a high electric field domain near the injection contact [99]. QWs in the bulk of a QWIP are electrically neutral (i.e., the QW electron density equals the doping), while a few (2–5) QWs near the emitter are partially depleted to provide the high contact electric field required for injection. At high illumination power, the total current and therefore the injection current may become so high that the voltage drop across the high field domain becomes comparable to the total applied voltage. This leads to a reduction of the electric field in the bulk of the QWIP (since the total voltage is kept constant), and hence to a decrease of the escape probability and an increase of the transit time (due to the decreased drift velocity). As a result, the responsivity is decreased at high power and the photocurrent may saturate with power. This behavior, investigated in detail in [118], is illustrated in Fig. 7.2. The experiments were conducted on standard 9 μm QWIPs at 77 K (nonbackground limited) using a CO_2 laser. The experimental results were explained by self-consistent numerical simulations. For these experiments, the nonlinearity was observed at a power level greater than about 1 W cm^{-2}, and the power was increased up to about 1,000 W cm^{-2}. With a CW CO_2 laser we did not observe any damage to the device. This property of laser "hardness" is quite unique to QWIPs, especially in view of the low damage threshold (∼1 W cm^{-2}) of the standard infrared detectors made of HgCdTe. This hardness in resisting high power illumination is an advantageous property for heterodyne detection.

Figure 7.3 shows experimental results from a different study [222] which was conducted at 65 K (a typical operation temperature for QWIP cameras, where the dark current is at least ten times lower than typical background photocurrents) and low excitation powers. In Fig. 7.3a, symbols indicate measured responsivities of a 4-period QWIP which fall off significantly well below 0.1 W cm^{-2}. The lines are the result of a simple phenomenological model, in which we approximate the field distribution inside the QWIP by electric

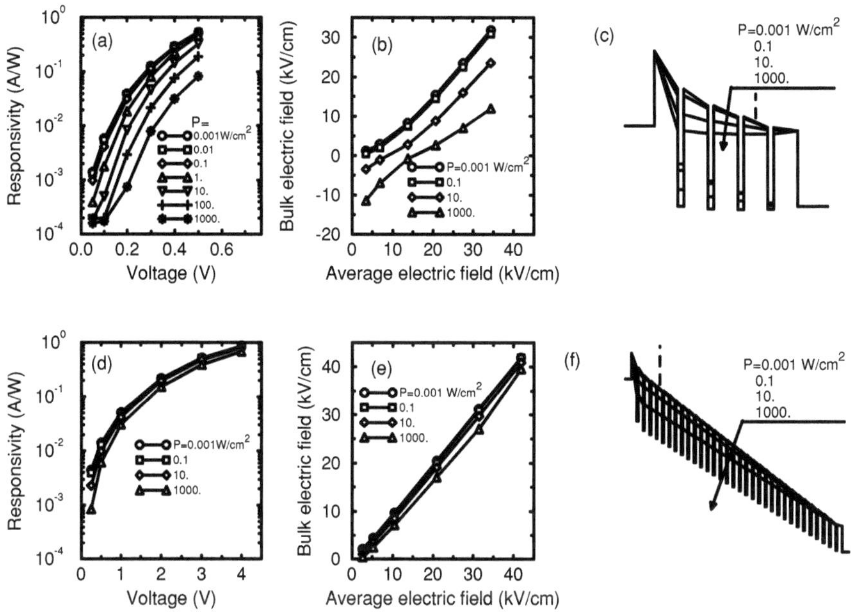

Fig. 7.2. Simulation results for (**a**)–(**c**) 4-well and (**d**)–(**f**) 32-well QWIPs under different infrared illumination power densities: (**a**) and (**d**) responsivity vs. voltage, (**b**) and (**e**) electric field in the bulk vs. average field, and (**c**) and (**f**) conduction bandedge profile at an average field of $15\,\mathrm{kV\,cm}^{-1}$ (courtesy of M. Ershov). Values of the bulk field were taken at positions shown by the dashed lines in (**c**) and (**f**)

fields F_1 at the emitter barrier and F_2 which is common to the remaining barriers (see also Fig. 7.2c and f for the validity of this approximation). Within this approximation, the photocurrent nonlinearity is readily obtained from the measured dark current and low-power responsivity curves without any free parameter. Figure 7.3b displays the corresponding result for a 50-period QWIP, which already exhibits the nonlinearity at the thermal background.

When a QWIP is operated at low temperatures and low backgrounds, so that the total current is low, the nonlinearity can thus be observed under a very low IR illumination [222,223]. The key factor is the relative magnitude of the total current and photocurrent. If the photocurrent is a large or dominant fraction of the total current, the response becomes nonlinear. As nonlinearity is an important issue for the correctability of focal plane arrays, we point out that the observed nonlinearity is still very weak and that it can be reduced further by using appropriate operating conditions (e.g., suitable bias voltages).

With the demonstration of near 100% absorption efficiencies [220,221], QWIPs are ideally suited for heterodyne detection [215]. Thus, the first mechanism of high power QWIP nonlinearity is related to the contact effects and modulation (reduction) of the bulk electric field in QWIPs. This mechanism is shown much more strongly in QWIPs with a small number of QWs, where the

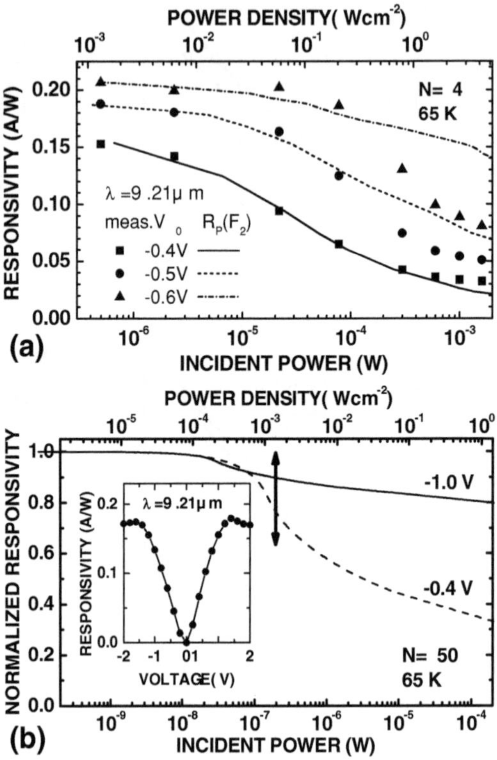

Fig. 7.3. Responsivity vs. incident power of a QWIP with (**a**) 4 wells and (**b**) 50 wells at different voltages as indicated. Symbols indicate the measured results, *lines* denote the values derived from the experimental dark current and low-power responsivity. In (**b**), the inset shows the measured low-power responsivity vs. applied voltage, and the vertical double arrow indicates the equivalent power due to thermal background radiation (after [222])

role of contact effects is especially important [99]. Contact related nonlinearity appears to be relatively unimportant in QWIPs with many QWs (>10) [98], where the voltage drop in the high field domain is small compared to the total applied voltage. It should be noted that in this mode of nonlinearity (7.5) and other classic formulae for the ideal photoconductor are still applicable, but the electric field F entering these formulae becomes lower than the average electric field $F_0 = V/L$.

Nonuniform Optical Field Distribution

Another mechanism of nonlinearity is due to a nonuniform optical field distribution in QWIPs [224]. The nonuniformity may be caused by the attenuation of the optical power due to absorption [156]. This effect may be pronounced in

QWIPs optimized for heterodyne operation, because they have much higher QW doping [220,221] and hence higher absorption than QWIPs designed for low-temperature and low-power applications. Optical power nonuniformity may also be caused by IR radiation reflection from the top metal contact, from sidewalls, and by interference. The monochromatic light could lead to strong interference patterns in a given experimental geometry. For example, with a polished 45° facet-coupled QWIP [180], the standing wave pattern leads to a strongly nonuniform illumination. This nonuniform distribution of the optical power leads to a coordinate-dependent photoexcitation rate from the QWs and therefore to a nonuniform concentration of the photoexcited carriers (see Chap. 6). If the nonuniform photoexcitation rate exceeds the thermal or background excitation, current continuity can no longer be provided by a constant electric field across the QWIP. The electric field readjusts itself so that the field is increased in regions with low optical power, and decreased where the optical intensity is high (this effect is illustrated in Fig. 7.4). The electric field

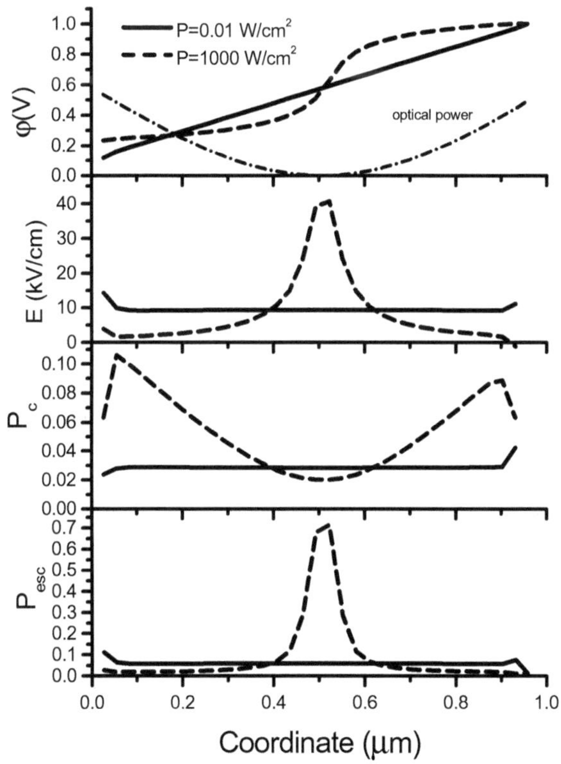

Fig. 7.4. Coordinate dependence of (**a**) potential, (**b**) electric field, (**c**) QW capture probability, and (**d**) photoexcited electron escape probability for low-power (*solid line*) and high-power (*dashed line*) densities, all for a 32-well QWIP at temperature $T = 77\,\mathrm{K}$ and applied voltage $V = 1\,\mathrm{V}$ [224]. The *dash-dotted* line shows the distribution of optical power

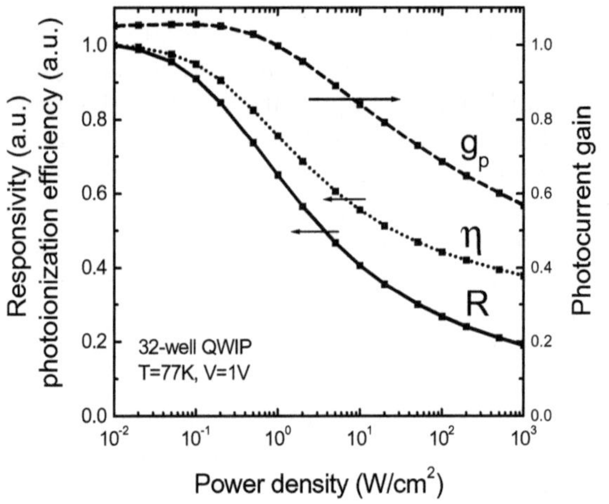

Fig. 7.5. Responsivity $\mathcal{R}$, photocurrent gain g, and photoemission efficiency $\eta' = \eta p_e$ vs. incident infrared power [224]

distribution becomes nonuniform, making invalid the simple standard formulae for responsivity, noise, etc. The nonuniformity of the electric field is supported by a slight recharging of the QWs. Theoretical calculations of detector characteristics in this case require self-consistent modelling. Photoconductive noise and detector responsivity decrease, while the noise gain and noise power are increased, with respect to the uniform electric field case (see Fig. 7.5). QWIP nonlinearity due to nonuniform optical power distribution takes place when the photocurrent exceeds the dark current or the background current.

Quantum Well Depletion

In the steady-state regime considered above, the variation of the QW electron density with incident power is not significant (unless the number of QWs is very small), so the depletion of the QWs by electrons causes a negligible reduction in the absorption efficiency. The situation however may be different in the nonequilibrium regime. If a high-power short infrared pulse is incident on the QWIP, a significant fraction of electrons are excited from the QWs [225]. On the short time scale, the absorption and the photocurrent may saturate because the electron density in the QWs responsible for the absorption is decreased. Experimental studies of these effects permitted the estimation of the photoexcited electron lifetime [213,214] and the study of the carrier dynamics [225]. The saturation of the absorption and photoconductivity happens at a very high power–over 10–100 MW cm^{-2}. However, this type of nonlinearity can only exist on a very short time scale (the lifetime or transit time). This is because the photoexcited carriers quickly relax back to the QWs, or exit to the collector (in which case they are replenished by an extra electron injection from the emitter).

Finally, extrinsic or photoconductive nonlinearity of QWIPs can be caused by a number of physical phenomena discussed above. These effects are parasitic, i.e., they lead to a deterioration of the detector characteristics (responsivity, noise power, etc.). In addition, detector nonlinearity complicates the calibration procedure. Extra care must be taken to avoid these effects, whenever possible. The standard formulae describing QWIP characteristics may lead to erroneous results, and the basic detector parameters (photocurrent gain, noise gain, etc.) lose their straightforward physical meanings due to the nonuniformity of the electric field.

7.2.2 Negative Differential Photoconductivity and Electric Field Domains

In most III–V bulk semiconductors, negative differential conductivity (NDC) is observed in n-type material at sufficiently high electric fields (e.g., at $F > 5\,\mathrm{kV\,cm^{-1}}$ in GaAs). This NDC is induced by scattering of the carriers from the Γ-minimum into the L- and X-minima of the conduction band. These intervalley scattering processes give rise to a reduced conductivity, since due to the higher effective mass, the electron drift velocity at an indirect minimum is smaller than at the Γ-point. In GaAs, the negative differential dependence of the average drift velocity v as a function of the electric field F is well described by the expression [226,227]

$$v(F) = \frac{\mu_1 F + v_\mathrm{v}(F/F_\mathrm{c})^4}{1 + (F/F_\mathrm{c})^4}, \tag{7.6}$$

where μ_1 is the mobility at the Γ-point, v_v is the asymptotic drift velocity for $F \to \infty$, and F_c is the characteristic field for intervalley scattering. The NDC gives rise to the formation of stable electric field domains at small carrier densities (typically $<10^{14}\,\mathrm{cm^{-3}}$). Propagating domains are observed at higher carrier densities. In the latter case, pronounced oscillations of the current are observed [226]. These Gunn-oscillations have found widespread use in microwave oscillators and amplifiers.

Negative Differential Photoconductive Gain in QWIPs

QWIPs are typically operated at electric fields of the order of $10\,\mathrm{kV\,cm^{-1}}$. This field value is expected to be high enough to induce intervalley scattering. In order to obtain more information about the transport properties and precise values of the gain, we have carefully investigated the noise gain g_noise associated with the dark current using the methods described in Sect. 5.2.3. Since $p_\mathrm{e} \approx 1$, we do not distinguish here between noise gain and photoconductive gain, as $g \approx g_\mathrm{noise} \approx g_\mathrm{photo}$.

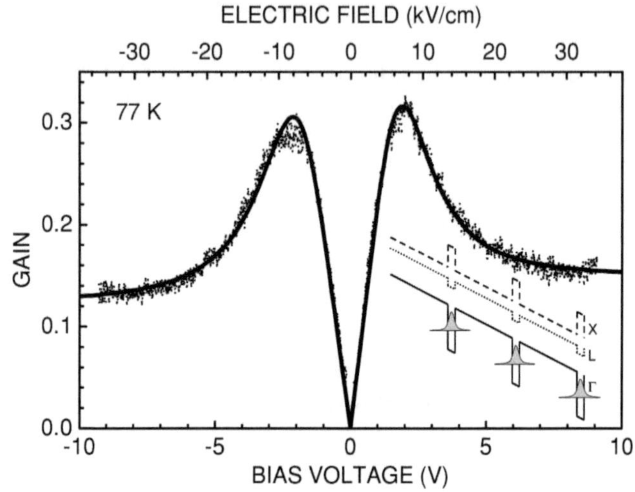

Fig. 7.6. Measured gain g (*dots*) of a 50-period QWIP structure and theoretical fit (*solid line*) vs. applied voltage and electric field. The inset shows the potential distribution of the Γ, L, and X minima in a GaAs/$Al_{0.26}Ga_{0.74}$As QWIP (after [228])

Figure 7.6 shows the field dependence of g as obtained from a noise measurement with the QWIP covered by a cold shield [228,229]. In the low-field regime, we observe a strictly linear dependence of g on F up to about $\pm 5\,\mathrm{kV\,cm^{-1}}$. The linear regime is followed by a strong saturation and, at fields above $\pm 8\,\mathrm{kV\,cm^{-1}}$, by a pronounced negative differential behavior. For a constant carrier density in the continuum, this will induce NDC.

Using (4.23), we now relate g with v. Assuming a constant capture time τ_c, the gain g should be proportional to v. The observed negative differential field dependence of g obviously contradicts the usual form (4.3). Therefore, we have used (7.6) to obtain the theoretical fit shown in Fig. 7.6, which is in excellent agreement with the experimental data. For positive (negative) polarity, the fit yields $F_c = 8.0\,\mathrm{kV\,cm^{-1}}$ $(9.3\,\mathrm{kV\,cm^{-1}})$, $v_v\tau_c = 0.40\,\mu\mathrm{m}$ $(0.32\,\mu\mathrm{m})$, and $\mu_1\tau_c = 1.57\,\mu\mathrm{m}^2\,\mathrm{V}^{-1}$ $(1.35\,\mu\mathrm{m}^2\,\mathrm{V}^{-1})$. Since the value obtained for F_c is only twice as large as in bulk GaAs [226], it is plausible that the observed decrease of g is indeed caused by intervalley scattering.

Assuming a value of 5 ps for the capture time τ_c (see the discussions in Sects. 4.2, 10.1.2, and 10.2.2), the fit yields $v_v = 7.9 \times 10^6\,\mathrm{cm\,s^{-1}}$ $(6.5 \times 10^6\,\mathrm{cm\,s^{-1}}$ for negative bias) and $\mu_1 = 3{,}100\,\mathrm{cm^2\,V^{-1}s}$ $(2{,}700\,\mathrm{cm^2\,Vs^{-1}})$. v_v has a similar value as typically observed in bulk GaAs, while μ_1 is about 3–4 times smaller. The lower value of μ_1 (as compared to bulk GaAs) is in good correspondence with the observed higher value of F_c. The noise gain in Fig. 7.6 shows a slight asymmetry with respect to the polarity of the bias, which we attribute to segregation of the dopant atoms (see Sect. 7.3).

The potential distributions for the Γ, L, and X-minima are indicated in the inset of Fig. 7.6. Both the L- and X-minima are expected to contribute

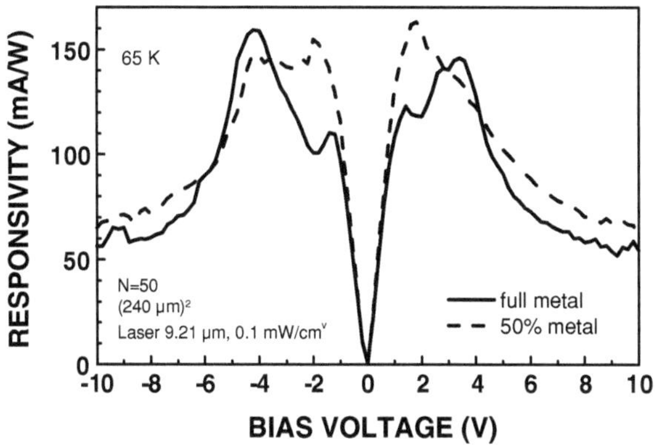

Fig. 7.7. Responsivity vs. applied voltage of a 50-period QWIP with completely metallized (*solid line*) and with 50% metallized (*dashed line*) top contact (after [229])

to the inter-valley transfer since the L-minimum of $Al_{0.26}Ga_{0.74}As$ lies only slightly below the X-minimum, and since the X-minimum contains the largest density of states. However, the experimentally observed reduction of g at high fields suggests that $\Gamma-L$-transfer is more efficient than Γ–X transfer, since the GaAs layers induce shallow wells at the L-point and potential barriers at the X-point [230]. Therefore, the lowest subbands at the L-point have a much stronger spatial overlap with the GaAs wells than the lowest X-subbands have, such that an efficient recapture into the Γ-point is possible. This interpretation is further supported by the observation of extremely high gain values in a GaAs/$Al_{0.55}Ga_{0.45}As$ QWIP, where carrier transport occurs mainly at the X-minimum [15].

In order to demonstrate that a similar negative differential dependence is also experienced by the photocurrent, Fig. 7.7 shows the bias-dependent responsivity $\mathcal{R}$ of the same 50 period QWIP structure at 9.21 μm excitation wavelength. The measurement was performed with a lock-in technique at a small excitation density of 0.1 mW cm^{-2}, which is more than an order of magnitude lower than the intensity of the thermal 300 K background radiation. In fact, $\mathcal{R}$ shows a negative differential dependence at high bias similar to the one observed in Fig. 7.6.

$\mathcal{R}$ shows somewhat different voltage dependencies for mesa devices with 100% and 50% metal coverage of the top surface. In fact, the fully metallized device has its highest responsivities at −4.2 V and +3.4 V, while the 50% metallized structure has its peak values at −2 V and +1.8 V. This difference is attributed to different electric field distributions. For 100% metallization, optical interference induces a strong spatial dependence of the optical excitation (see Chap. 6), thus giving rise to a "dark zone" where most of the applied voltage is dropped, and to a reduced field in the illuminated region. Since the interference is approximately opposite for metallic and dielectric reflections, the total absorbed power becomes homogeneous along the structure in the case of a 50% metallized mesa; and since the in-plane conductivity in the wells is much higher than the conductivity across the barriers, each well exhibits, in fact, a well-defined potential along the whole area of the detector [182]. Therefore, the 50% metallized QWIP is expected to behave like a device without any interference, such that the highest responsivity is reached at lower bias than for 100% metallization. In spite of the strongly different bias dependencies, the overall responsivity values are similar for the two devices. In fact, the length of the active region has a similar value to π/k_z, leading to about the same average excitation density for both devices.

Comparing Fig. 7.7 with Fig. 7.6, we see that the behavior of the responsivity of the 50% metallized detector is closer to the noise data than that of the completely covered one, which is in turn explained by the associated field distributions.

Electric Field Domains

More information about the transport behavior can be obtained at higher excitation densities where the nonlinear transport effects become more prominent. Figure 7.8 shows I–V curves measured at different power densities. We have used here half-metallized mesa structures in order to suppress features induced by the interference effect. The data reveal a prominent plateau behavior, which is limited by the mobility regime at low bias (up to about $\pm$ 1.5 V) and by the increase of the dark current at high bias.

Before discussing the plateau formation, let us first focus on the behavior of the dark current. As can be seen in Fig. 7.8, the dark current increases monotonically with applied voltage. Using (4.1) and (4.23), we have $J_{\text{dark}} = eN_{\text{3D}}\,NL_{\text{p}}g/\tau_{\text{c}}$. The monotonic increase of J_{dark} means that, with increasing field F, the carrier density in the continuum $n_{\text{3D}}(F)$ grows faster than the gain $g(F)$ shrinks, such that the differential conductivity is always positive. The

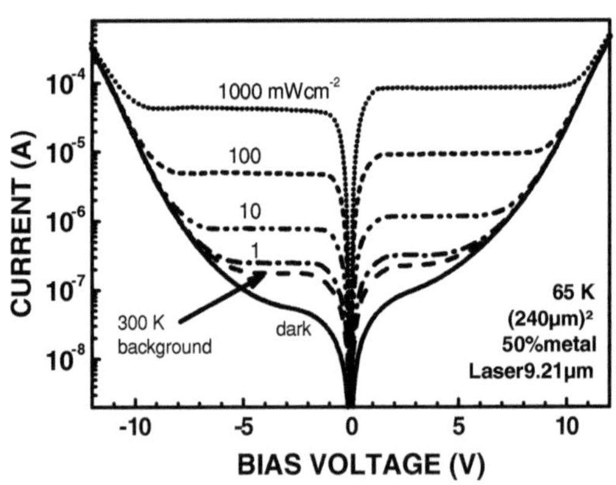

Fig. 7.8. Total current vs. applied voltage at different incident power densities as indicated (after [229])

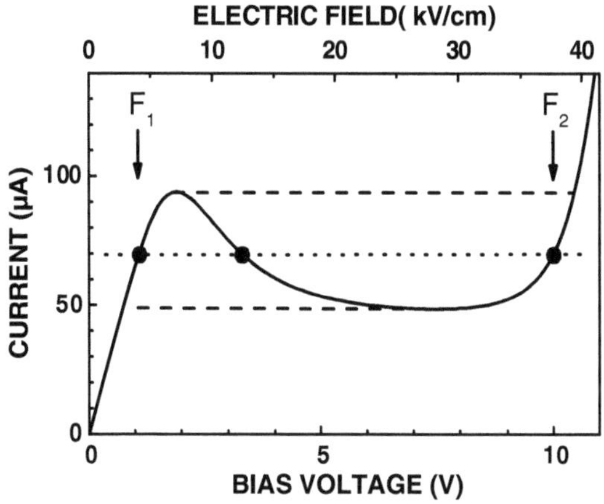

Fig. 7.9. Total current vs. applied voltage as expected in the case of a constant electric field across the QWIP for $P = 1\,\mathrm{W\,cm^{-2}}$. Full circles indicate the possible local field values (two stable ones and an instable one in between) for a given current indicated by the dotted line. Dashed lines show the regime where multiple fields are possible for a fixed current (after [229])

fast increase of $n_{3\mathrm{D}}(F)$ is due to the field-induced reduction of the effective barrier height, which reduces the activation energy of the dark current [15]. $I_\mathrm{D}(F)$ thus exhibits an exponential-like behavior at higher fields. From the absence of NDC, we then expect an approximately homogeneous electric field distribution in the dark.

The situation changes drastically on illuminating the device. As a function of the electric field F, the total current can be expressed as

$$I = e\left(n_{3\mathrm{D}}(F)\frac{NL_\mathrm{p}}{\tau_\mathrm{c}} + \frac{\eta P_\mathrm{E}}{h\nu}\right) g(F). \tag{7.7}$$

NDC thus becomes prominent as soon as the optical excitation rate significantly exceeds the thermal excitation rate. In Fig. 7.9, we have plotted the I–V curve as expected for a homogeneous field. Here the photocurrent has been obtained from the measured gain (see Fig. 7.6), assuming a power density of $P = 1\,\mathrm{W\,cm^{-2}}$ and a quantum efficiency of $\eta = 7\%$, and has been added to the experimental dark current.

The origin of the plateau formation is as follows. The homogeneous field distribution becomes unstable if the applied field is in the the NDC regime. However, two stable field values, F_1 and F_2, are possible if the total current is between the peak current and the valley current (between the dashed lines in Fig. 7.9). Therefore, the field distribution splits up into low-field and high-field domains, characterized by F_1 and F_2, respectively. Since the current (which is already determined by the local field and illumination density within such a domain) is not affected by the domain sizes (as long as the domain

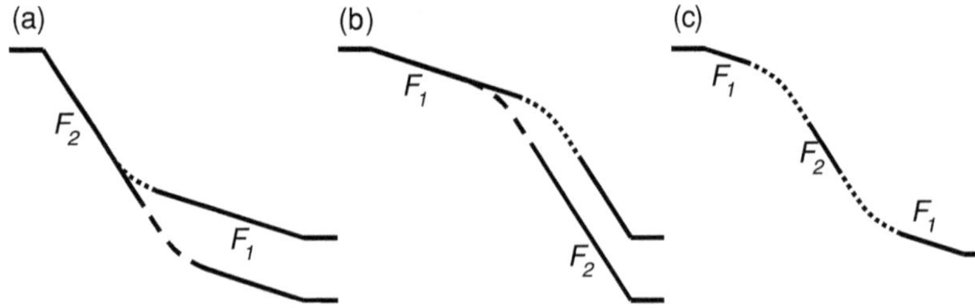

Fig. 7.10. (**a**) Schematics of stable electric field domains for positive space charge. Upon changing the bias voltage, the transition region (indicated by the *dashed* and *dotted lines*, respectively, for two different applied voltages) is shifted while the field values F_1 and F_2 remain constant. (**b**) Same for a negatively charged transition region. (**c**) Domain configuration where the high-field domain is located inside the active region

field remains constant), this property automatically gives rise to the plateau formation as observed in Fig. 7.8.

The exact location of the respective electric field domains and of the precise value of the plateau current (within the allowed window as indicated by the dashed lines in Fig. 7.9) is expected to depend on the actual device structure, its structural inhomogeneities, the spatial distribution of the excitation density, and the details of the transport mechanism.

A few possible domain configurations are summarized in Fig. 7.10. As a common signature of these distributions, the high-field domain increases at the expense of the low-field domain for increasing applied bias, while the values of F_1 and F_2 remain constant. In between two domains there must exist a transition region in which F_1 changes to F_2 or vice versa. This transition region is expected to contain one or several quantum wells carrying a positive or negative space charge. In Fig. 7.10a, the high-field domain is located adjacent to the emitter contact. This configuration is associated with a positive space charge (i.e., fewer electrons than donor atoms) at the transition region. The other case, with the low-field domain close to the emitter contact and a negative space charge, is shown in Fig. 7.10b. Finally, a potential distribution with three domains, comprising both positive and negative space charges, is shown in Fig. 7.10c. Configurations with three or more domains are expected to become increasingly probable for very thick active regions.

Besides n-GaAs and QWIPs, domain formation also occurs in weakly coupled superlattices [231,232], where NDC is induced by interwell tunneling rather than intervalley scattering: a recent overview can be found in [233]. Of interest is also the actual domain configuration present in the device, and the role of inhomogeneities. Domain formation in GaAs/AlGaAs superlattices containing a *gradient* in the well widths has been investigated by Han et al. [234]. They found that the high-field domain is always formed on the side with the wider wells, such that the location of the high-field domain does not depend on the polarity of the applied bias. A similar conclusion has been drawn by Yamashita et al. [235], who studied domain formation in n-type GaAs and attributed the observed location of the high-field domain to an inhomogeneous local conductivity.

Inhomogeneities are also expected to play a key role for electric field domains in QWIPs. In this context, inhomogeneities associated with the emitter and collector barriers, as well as inhomogeneities arising from the interference of the IR illumination (including residual interference in the case of partial metal coverage) are expected to be relevant to the nucleation of high-field domains, rather than structural inhomogeneities within the active region. These effects can also lead to changes of the domain configuration upon reversing the polarity of the applied bias. In fact, the configurations in Figs. 7.10a and b have been identified in a 50-period QWIP structure at opposite polarities [236]. The experimental signature for different domain configurations is obtained from the fact that the configuration of Fig. 7.10a is associated with the peak current, whereas Fig. 7.10b is associated with the valley current.

In conclusion, the photoconductivity in GaAs/AlGaAs QWIPs exhibits a negative differential behavior, which is attributed to intervalley scattering. In spite of this NDC, the dark current behaves normally since the gain reduction is overcompensated by the field-induced increase of the thermal excitation rate. The field distribution in an illuminated QWIP is strongly influenced by NDC, giving rise to high- and low-field domains. This behavior, which can readily occur under background-limited operation, is very typical for GaAs/AlGaAs-based QWIPs, in particular for bound-to-continuum ones. A nice feature of the effect from the viewpoint of detector applications is its role in stabilizing the current with respect to fluctuations of the bias voltage – which is also the reason why the phenomenon is easily overlooked.

7.2.3 Intrinsic Nonlinearity

With increasing power, the standard nonlinear optical processes become important. For the dipole intersubband transition itself, the full range of nonlinear optical processes should be observable. So far, absorption saturation [213,214], hole burning [237], harmonic (including sum frequency) and difference frequency generation [238–240], and two-photon and multiphoton absorption [239,241] have been reported. A review of intersubband nonlinear optics in quantum well structures is given [239]. Here, specific to QWIPs, we describe the two-photon absorption process occurring in QWIPs; and its potential applications.

Using the simplest picture of 1-dimensionally confined quantum wells, the calculation of two-photon absorption is straightforward. For a square quantum well, three configurations are schematically shown in Fig. 7.11. For a given photon energy, the three configurations correspond to photo-ionization from the ground state to the continuum via an intermediate resonant real state (1), off-resonant state, (2), and virtual state (3). The first case is expected to

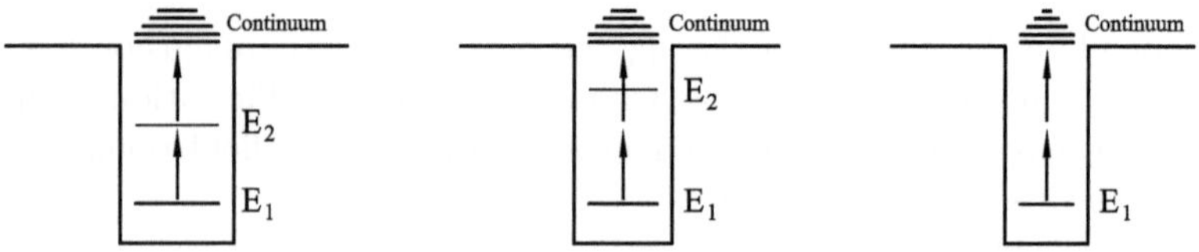

Fig. 7.11. Schematic two-photon absorption from ground (bound) state to the continuum via an intermediate (*left*) real resonant state, (*center*) off-resonant state, and (*right*) virtual state

have the highest two-photon absorption efficiency – close to that for a true double-resonance case. A high efficiency is the advantage of the resonant case, but the intrinsic time scale would be limited by the intersubband relaxation of the order of 1 ps. With a sufficient detuning (off-resonant) or for the virtual state case, the intrinsic speed of the device could be much faster than 1 ps. A simple calculation [241] in Fig. 7.12 shows the efficiency vs. photon energy for the three cases. As an estimate of the maximum possible efficiency, we can evaluate a double-resonance case analytically. For example, for an equally spaced three-level structure and a photon energy exactly on resonance ($\hbar\omega = E_2 - E_1 = E_3 - E_2$), the efficiency (for one QW) is

$$\eta^{(2)} = \left(\frac{e^2 h}{4\epsilon_0 n_r mc}\right)^2 \frac{\sin^4\theta}{\cos\theta} n_{2D} f_{23} f_{12} \frac{P}{2\pi^2 \omega \gamma^3}, \quad (7.8)$$

where n_r is the refractive index, m is the effective mass, θ is the angle of incidence, n_{2D} is the 2D electron density, f_{mn} is the oscillator strength between m and n states, P is the infrared power flux, and γ is the (half width) linewidth parameter. For the same $\gamma = 10$ meV as used in Fig. 7.12 and assuming $f_{12} \sim 1$

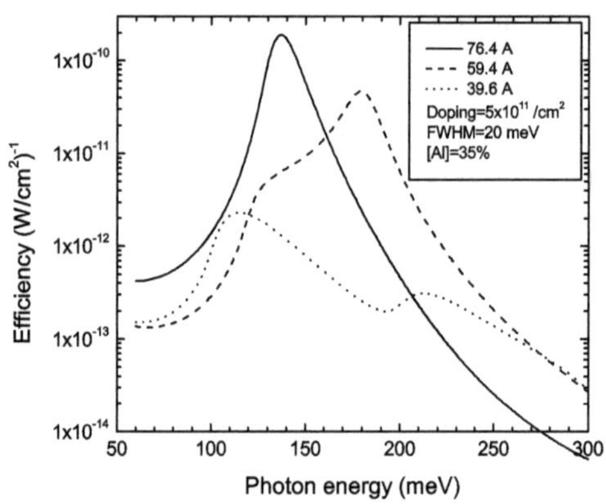

Fig. 7.12. Calculated two-photon absorption efficiency for three quantum well structures. The GaAs/AlGaAs quantum well parameters are given in the inset. The linewidth parameter is set to be 20 meV. The schematic shown in Fig. 7.11 corresponds to a photon energy of 136 meV

and $f_{23} \sim 2$, the two-photon absorption efficiency for the double-resonance case is $\eta^{(2)}/P \sim 3\times10^{-10}$ (W cm^{-2})$^{-1}$. The peak value in Fig. 7.12 for the solid curve is close to this value, and hence the case of bound–bound continuum resonant configuration is close to the highest efficiency that we expect to achieve.

The first experimental study on two-photon absorption and photocurrent was reported [216,241]. They used a standard 8-μm QWIP operated at 77 K, the situation of the center part of Fig. 7.12. The device was excited by CO_2 laser pulses at 10.6 μm with a flat-top shape and 3-ps duration, generated by semiconductor optical switching [216]. As an experimental proof of a two-photon detection, the QWIP signal was shown to be quadratically proportional to the linear monitor signal.

Of particular interest is the use of a two-photon QWIP to characterize the pulses produced by mode-locked quantum cascade lasers [217,242]. This however may require a higher efficiency than the substantially detuned quantum well. This goal has been achieved by resonant two-photon QWIPs (Fig. 7.11, left panel), as described later in Sect. 10.3.

7.3 Asymmetry Caused by Dopant Segregation

In the measured QWIP I–V curves, an asymmetry is often seen, as in results reported in [105]. The main cause of the asymmetry is attributed to the segregation of the dopants during growth [114,115,167]. For QWIPs with intended doping in the center of the well, the segregation leads to an asymmetric distribution of the dopants and therefore the positive background charge seen by the electrons. Experiments on a set of samples with purposely shifted δ-doping positions away from the center led to the above conclusion [114]. From Fig. 7.13, it is seen as the shift increases the I–Vs become more and more symmetric. From this study, the "true" shape of the δ-doping profile for Si at a growth substrate temperature of $T_{\text{sub}} = 605°$C is broadened asymmetrically with a width of 27 Å. This inferred true profile is shown in the right panel of Fig. 7.14. The left panel shows the substantial asymmetry in the potential if the δ-doping is shifted (and also remained sharp). The center panel shows the case of a smaller segregation which still leads to an asymmetry. The amount of segregation is sensitive to the growth substrate temperature: as T_{sub} is reduced, the segregation decreases, as seen in the curve from a $T_{\text{sub}} = 550°$C sample in Fig. 7.13.

Furthermore, using the measured asymmetric I–Vs together with modeling, we can infer the segregation length [115]. A set of samples grown at different growth temperatures have been measured and modeled and the results are summarized in Fig. 7.15. Finally, the As overpressure during growth is shown to suppress the Si segregation for both As_2 and As_4 sources [115].

Similar results have been obtained by Larkins et al. [243], who studied Si incorporation in QWIP structures based on GaAs/AlAs/AlGaAs double-

Line	Sample	T_{sub} (°C)	δ shift (Å)	Well width (Å)	Barrier width (Å)
— —	1300	605	0	59	350
– – –	1301	605	6	59	350
- - - - - -	1302	605	12	59	350
———	1327	**550**	**0**	55	326
-·-·-·-	1328	605	11	55	326
··············	1329	605	22	55	326

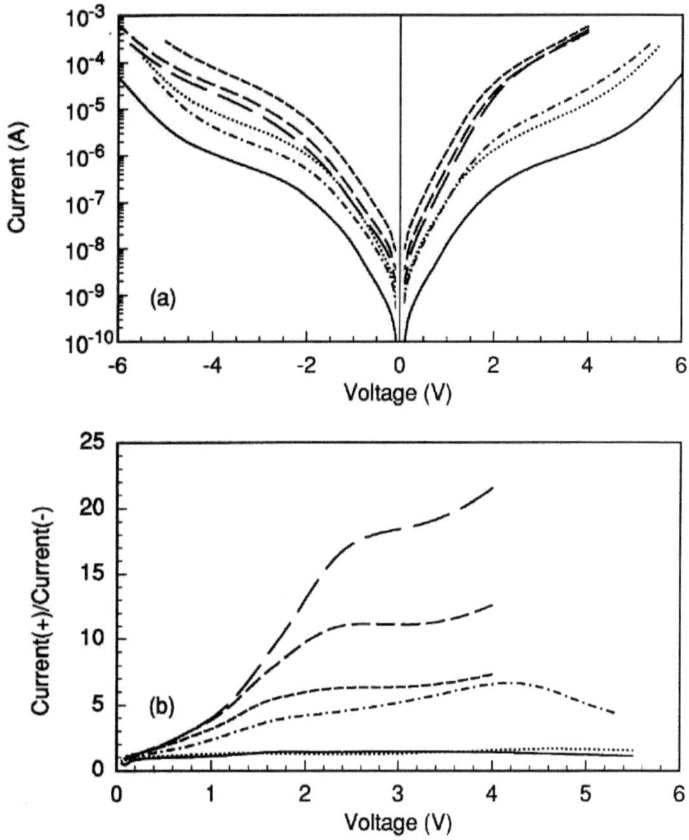

Fig. 7.13. Sample parameters (*top table*), measured current-voltage characteristics (**a**) and the ratio of the forward and reverse currents (**b**) for six samples at 77 K

barrier quantum wells. These investigations were triggered by experimental observations of a significant photovoltaic behavior in these double-barrier QWIPs [167,244]. Figure 7.16 shows simulated Si doping profiles obtained by two different models. The observed photovoltaic effect was successfully explained in terms of the resulting dipole fields between ionized impurities and the electrons in the quantum wells.

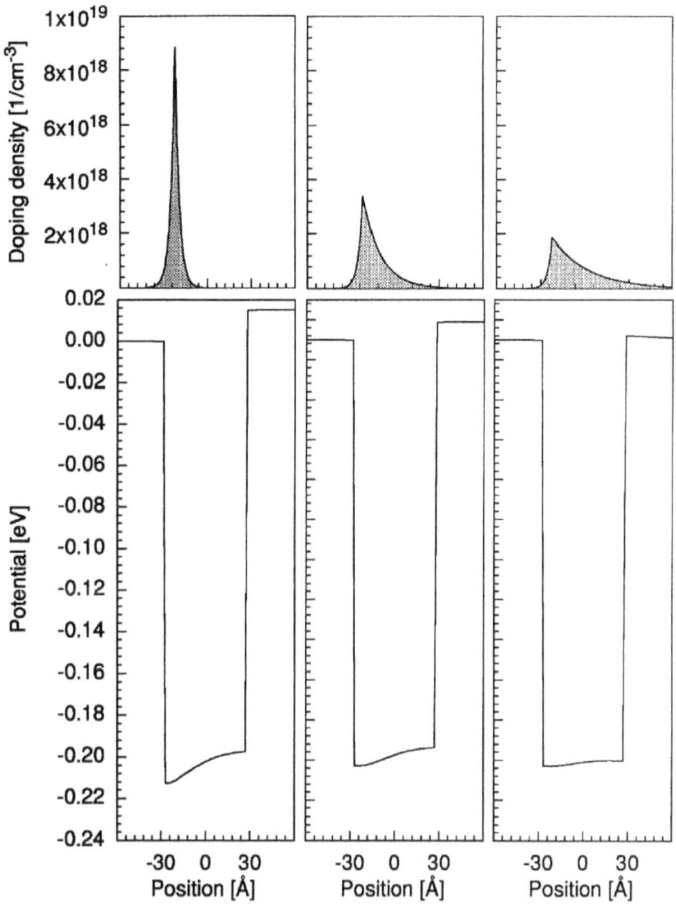

Fig. 7.14. Model simulations of doping profiles and quantum well potentials for a nominally δ-doping shifted from the center by 22 Å. The left part is for the case of no segregation and the δ-doping is modeled by exponentials with a width of a monolayer. The center and the right parts are for segregation of 11 and 22 Å, respectively

7.4 Coherent Photocurrent

7.4.1 Coherent Control by Optical Fields

An interesting experiment demonstrating the coherent nature of the electron waves was carried out by Dupont et al. [245]. Using a bound-to-bound transition QWIP and excitations of both one- and two-photon transitions, the interference photocurrent polarity or direction was shown to be controlled by the relative phase between the two coherently related IR beams. The second harmonic frequency light generated in a nonlinear crystal from a pulsed CO_2

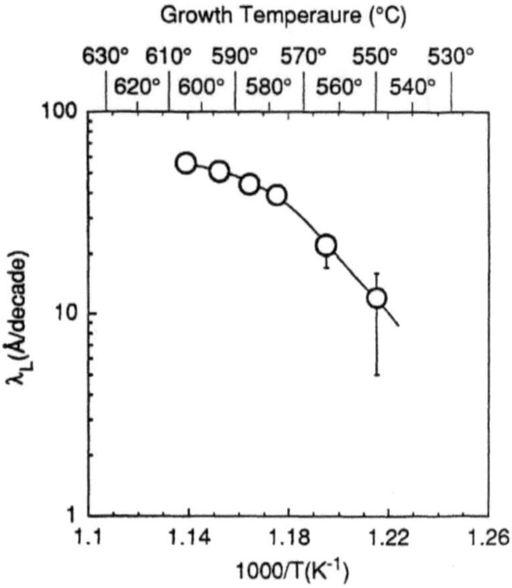

Fig. 7.15. Si segregation length λ_L vs. inverse temperature

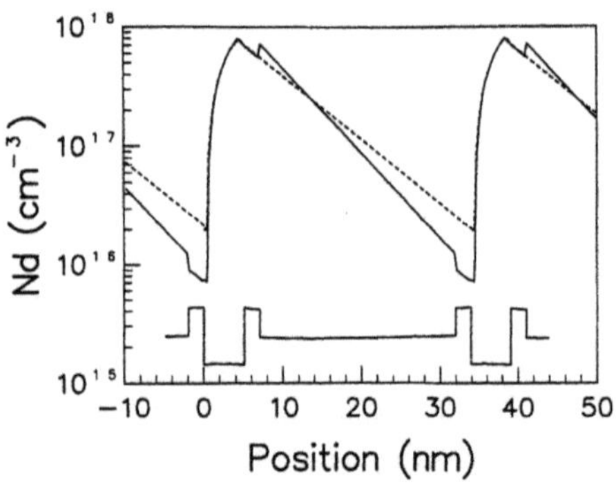

Fig. 7.16. Comparison of the Si doping profiles obtained using kinetically (growth rate) limited Si incorporation (*solid curve*) and equilibrium Si incorporation (*dashed curve*). The growth direction is toward the right. A profile of the conduction band minima is shown below these curves for visual reference [243]

laser was used for one-photon excitation, and a part of the CO_2 light for two-photon excitation. Neither the one-photon nor the two-photon transition were in resonance with a final or intermediate bound state. The two-photon transition corresponded to the center part of Fig. 7.11. In a symmetric quantum well, the final state of the one-photon transition was parity asymmetric, and that for two-photon transition was symmetric. This is schematically shown in Fig. 7.17. When the two degenerate final states are superimposed, the resulting wavefunction could resemble a left or a right-going planewave, depending

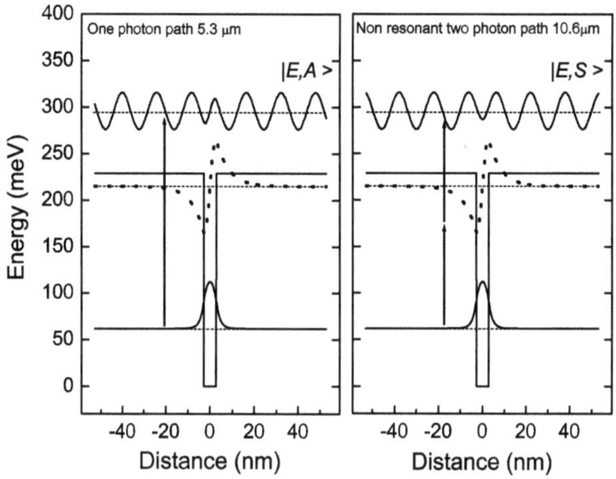

Fig. 7.17. Intersubband excitations by one-photon (*left*) and two-photon (*right*) transitions. The excited final state wavefunctions are asymmetric (*left*) $|E, A>$ and symmetric (*right*) $|E, S>$. See more details in Dupont et al. [245]

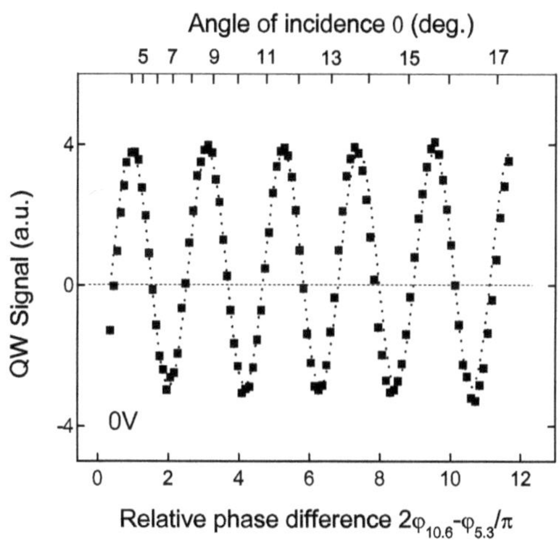

Fig. 7.18. Observed photocurrent vs. phase. The phase change is done by rotating a piece of NaCl crystal using its dispersion between 5.3 and 10.6 μm wavelengths. The rotating angle is plotted on the top axis. The relevant phase in expected interference photocurrent is plotted on the lower axis. See more details in Dupont et al. [245]

on the relative phase. The observed photocurrent at zero external bias should then oscillate with phase. This quantum interference phenomenon is experimentally shown in Fig. 7.18.

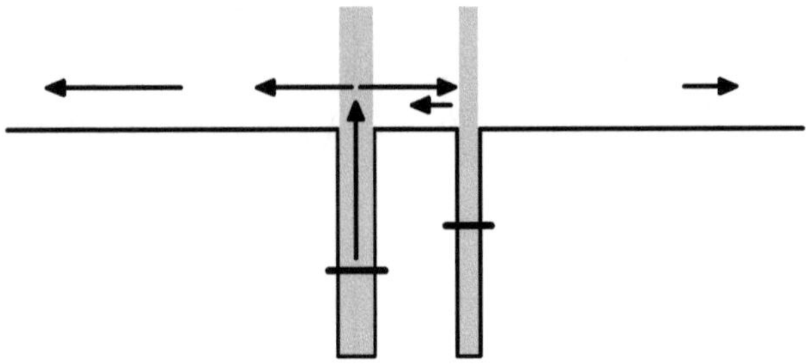

Fig. 7.19. Coherent transmission and reflection processes occurring in an asymmetric double quantum well structure upon optical carrier excitation

7.4.2 Coherent Control Through Potential Offsets

Coherent control of the photocurrent polarity can also be achieved by appropriate heterostructures. Schönbein et al. [246] have investigated the photocurrent in asymmetric GaAs/AlGaAs double quantum well structures. The samples under study contained 20 asymmetric double quantum wells separated by 45 nm thick $Al_{0.24}Ga_{0.76}As$ barrier layers. Each double quantum well consists of nominally 4.8 nm n-type GaAs, 7.8 nm $Al_{0.24}Ga_{0.76}As$, and 3 nm GaAs (sample 1). In sample 2, the 7.8 nm wide $Al_{0.24}Ga_{0.76}As$ layer is replaced by 11 nm $Al_{0.24}Ga_{0.76}As$.

Semiclassically, the situation upon optical carrier excitation in these structures is indicated by the horizontal arrows in Fig. 7.19. Due to the presence of the narrow GaAs quantum well, the inversion symmetry of the continuum states is broken. Carriers, which are initially emitted from the wide doped quantum well with equal probability toward both sides, undergo reflections when passing the second, narrow quantum well. The reflected waves can interfere constructively or destructively with the incoming wave [247] according to their de Broglie wavelengths. The occurrence of standing waves between the wide and the narrow quantum well is expected in the case when the spacing between the two quantum wells is within the *coherent mean free path* l_c of the excited carriers.

Photocurrent spectra were taken at 77 K using a Fourier spectrometer equipped with a glowbar source. In order to obtain the correct sign of the photocurrent at 0 V external bias, the phase of a spectrum at higher bias voltages (−2 V) where no current sign reversal occurs was stored and used for the zero bias measurement. A positive sign of the photocurrent is associated with carrier propagation in the growth direction.

Figure 7.20 shows photocurrent spectra of both samples at zero bias. Since the lowest subband of the narrow well is not populated under the experimental conditions, this well does not contribute to the total photocurrent. Due to the broken inversion symmetry, the structure has a slight photovoltaic behavior, favoring a current flow towards the substrate direction. The spectra

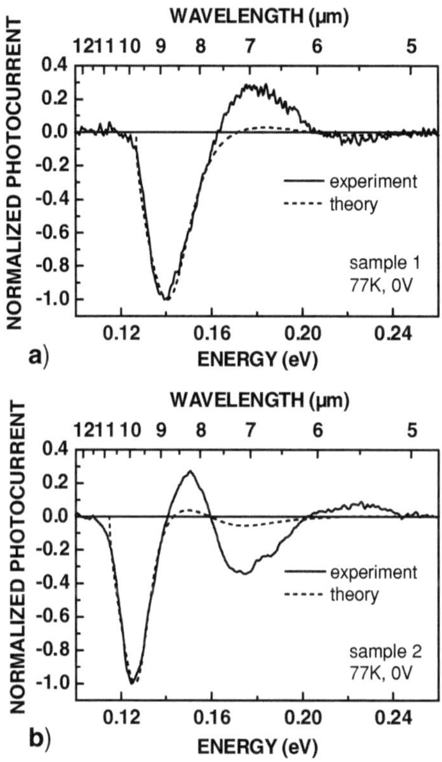

Fig. 7.20. Comparison of measurement and theory for sample 1 (**a**) and sample 2 (**b**). The measurements were conducted at 77 K without externally applied bias (after [246])

exhibit several sign reversals, which are only present at small fields and disappear above $+0.5\,\mathrm{kV\,cm^{-1}}$ ($+60\,\mathrm{mV}$ applied bias) and below $-0.83\,\mathrm{kV\,cm^{-1}}$ ($-100\,\mathrm{mV}$ applied bias) [246].

A semiclassical calculation of the current sign reversals has been given by calculating the transmission probability across wide and narrow quantum wells [247], which qualitatively explains the observed sign reversals.

In a quantum mechanical approach, the bound-to-continuum transition is described by the time dependent perturbation operator $H_1(z,t) = (\mathrm{e}^{\mathrm{i}\omega t} + \mathrm{e}^{-\mathrm{i}\omega t})H_1(z)$ in addition to the stationary one-electron Hamiltonian $H_0(z)$, where the growth direction is denoted by z. Nonparabolicities are taken into account within an empirical two-band model [87], where H_0 is given by

$$H_0(z) = \begin{pmatrix} V_\mathrm{c}(z) & -\mathrm{i}\gamma\frac{\partial}{\partial z} \\ -\mathrm{i}\gamma\frac{\partial}{\partial z} & V_\mathrm{v}(z) \end{pmatrix} . \tag{7.9}$$

Here, $V_\mathrm{c}(z)$ and $V_\mathrm{v}(z)$ are the piecewise constant potential distributions of the conduction and valence bandedges, respectively. γ is related to the effective

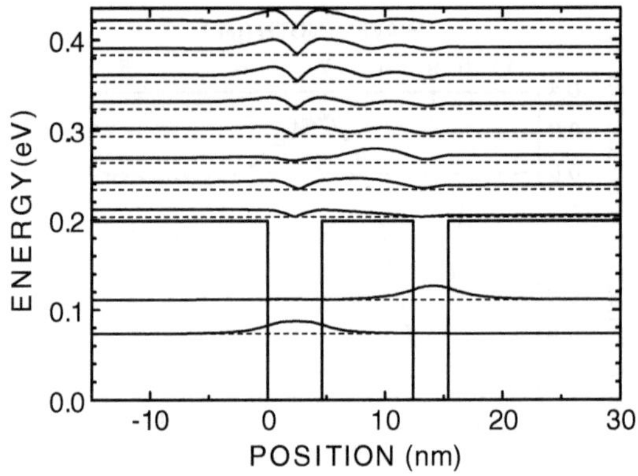

Fig. 7.21. Conduction bandedge distribution and probability densities of bound states as well as continuum states with asymptotic boundary conditions for the parameters of sample 1. The virtual states have been arbitrarily chosen at equidistant energies to demonstrate standing-wave effects between the two wells (after [246])

mass m by $\gamma = \sqrt{\hbar^2(V_c(z) - V_v(z))/(2m(z))}$, where $\gamma = 0.9294$ eVnm assures the correct reduced effective mass of $m^* = m/m_e = 0.067$ at the GaAs conduction bandedge. Neglecting the photon wave vector, the perturbation Hamiltonian H_1 is obtained by the substitution $\hat{p} \to \hat{p} - \frac{e}{c}\boldsymbol{A}$. The vector potential $\boldsymbol{A} = -(cF/\omega)\cos\omega t \boldsymbol{e}_z$, with F the amplitude of the radiation field and $\boldsymbol{e}_z$ the unit vector in z-direction, thus gives rise to the dipole Hamiltonian

$$H_1 = \frac{\gamma F}{2\hbar\omega}\begin{pmatrix} 0 & 1 \\ 1 & 0 \end{pmatrix}. \tag{7.10}$$

The total wave function $\psi(z,t) \equiv (\psi_c(z,t), \psi_v(z,t))^T$ in this model is given by

$$\psi(z,t) = \psi^i(z)e^{-(i/\hbar)Et} + \psi^f(z)e^{-(i/\hbar)(E+\hbar\omega)t}. \tag{7.11}$$

Here the ground state ψ^i is located at the first subband in the emitting quantum well, while the excited state ψ^f is located in the continuum above the AlGaAs conduction bandedge. Wave functions ψ^f at arbitrarily chosen equidistant energies are shown in Fig. 7.21. Substituting (7.11) into the time dependent Schrödinger equation with the Hamiltonian $H = H_0 + H_1$ and expanding to the first order in $e^{-i\omega t}$ yields

$$H_0\psi^i = E\psi^i \tag{7.12}$$

$$H_0\psi^f + H_1\psi^i = (E + \hbar\omega)\psi^f. \tag{7.13}$$

These equations correspond to the approach given by Sumetskii et al. [248], who model the quantum well structure by short range delta potentials.

The equations have been solved within a modified transfer matrix calculation developed for arbitrary, piecewise constant potentials. In the calculation, *outgoing boundary conditions* are assumed for the excited state wave function $\psi^f \equiv (\psi_c^f, \psi_v^f)^T$, characterized by $\psi_c^f(z \to \infty) = Ae^{ikz}$, and $\psi_c^f(z \to -\infty) = Be^{-ikz}$, with the 45 nm thick AlGaAs barrier layers replaced by infinitely extended barriers in the calculation. This is a good approximation in the case where this width is large compared to the coherent mean free path of the excited charge carriers. The coefficients A and B are obtained using the transfer matrix algorithm. The boundary conditions in the transfer matrix algorithm are obtained from the requirement that $\psi(z,t)$ (see (7.11)) is continuous. The current density j is obtained from the wave function ψ^f via the relation $j = \frac{\gamma}{\hbar}((\psi_c^f)^* \psi_v^f + (\psi_v^f)^* \psi_c^f)$. The expression for the total coherent current is then given by

$$j = \frac{2\gamma^2 k}{\hbar(E - V_v)}(|A|^2 - |B|^2). \tag{7.14}$$

The term associated with the coefficient $|A|^2$ can be viewed as an outgoing current to the right side, while the term associated with the coefficient $|B|^2$ represents an outgoing current to the left side. The "inhomogeneity" $H_1\psi^i$ in (7.13) acts as a source term for the outgoing currents.

Figure 7.20 also shows theoretical spectra as obtained from this model. For both samples, the spectral shape of the first peak is well reproduced. The model also correctly describes the spectral positions of the remaining maxima and minima. The strong influence of the quantum well spacing on the current sign reversals of the spectra can be clearly seen. The spacing between the energy values where the current drops to zero is narrower for the sample with the wider barrier.

Although the shape of the first peak and the sign reversals are nicely predicted by the calculation, the relative height of the first peak as compared to the following peaks is too large. A proper description of the amplitudes would thus require taking inelastic scattering into account, not contained in the present theory.

The interference effect discussed here also provides information on the coherent mean-free-path l_c of the excited charge carriers, as coherent effects are only observed if the width of the barrier layer separating the wide quantum well from the narrow quantum well is smaller than l_c. The absence of high-frequency oscillations indicates that the approximation of the 45 nm thick AlGaAs layers by infinitely extended layers in the simulation is justified, i.e., the width of these layers is larger than l_c.

7.5 Impact Ionization and Avalanche Multiplication

If carriers gain sufficiently high kinetic energy, there is a finite probability that this energy will be used to create new carriers through the ionization of impurities or the creation of electron-hole pairs. Since this process can occur a multiple times during the drift across the active region of a device, the phenomenon is similar to an avalanche. This avalanche multiplication is an important mechanism in a large number of semiconductor devices [34,249]. The avalanche process is generally described by a multiplication factor $\mathcal{M}$, which specifies the ratio between the total current and the primary current. In bipolar avalanche photodiodes, avalanche multiplication is usually accompanied by excess noise, which results in a reduction of the signal-to-noise ratio even though the signal itself is enhanced [34,249]. Only for a unipolar multiplication process, i.e., if only either electrons or holes cause impact ionization, avalanche multiplication without excess noise is possible, and enables "noiseless" amplification of the signal.

In addition to avalanche multiplication involving interband excitations, several theoretical and experimental studies have investigated avalanche processes based on intersubband transitions in MQW heterostructures [250–252]. The inset of Fig. 7.22 illustrates the avalanche process in a QWIP. Mobile electrons in continuum states of the conduction band are accelerated by the electric field during their drift in the barrier regions. These carriers may experience an energy relaxation during their drift due to scattering processes (indicated by the open arrows), resulting in a "ballistic" mean free path. Provided that the kinetic energy of the incident carriers exceeds the activation energy E_A, the confined ground-state electrons can be promoted to an excited state within the continuum [250]. Subsequently, both the primary and the secondary electron leave the quantum-well region and gather kinetic energy from the electric field across the following barrier layer to continue the avalanche process. The first evidence of avalanche multiplication in GaAs/AlGaAs QWIPs was obtained by Levine et al. [253] through a comparison between measured responsivity data and a theoretical model. However,

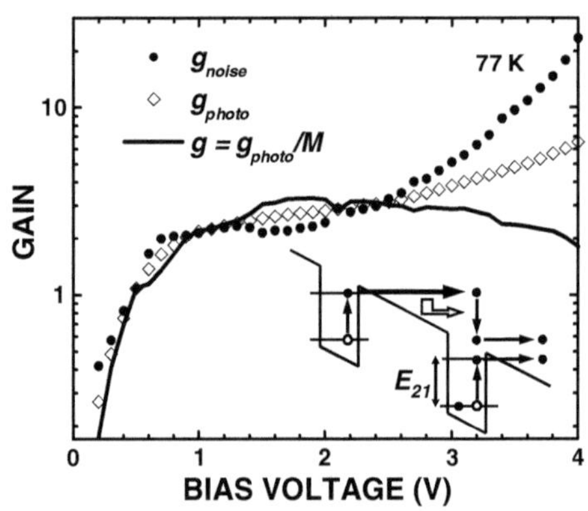

Fig. 7.22. Photoconductive gain g_{photo} and noise gain g_{noise} vs. bias voltage. The solid line displays the "net" gain g, which the investigated QWIP would posses in absence of avalanche processes. The *inset* shows the conduction band edge distribution in a photoconductive QWIP, and schematics of the avalanche multiplication process due to impact ionization (after [254])

their method did not allow them to derive the avalanche multiplication factor $\mathcal{M}$.

In QWIPs, the relation $g_{\text{photo}} \approx g_{\text{noise}} \approx g$ usually holds for $N \gg 1$ and $p_{\text{c}} \ll 1$, since avalanche processes are absent [255]. Taking avalanche multiplication into consideration, we express the photoconductive gain as $g_{\text{photo}} = g\mathcal{M}$, where the "net gain" g refers to the situation without avalanche multiplication. Thus, the responsivity is given by $\mathcal{R}_i = e\eta g\mathcal{M}/h\nu$. The presence of excess noise then causes an increase of the noise gain g_{noise} over the photoconductive gain g_{photo}. As experimental evidence for this phenomenon, Fig. 7.22 compares the values of g_{photo} and g_{noise} measured on a 20-period QWIP structure. To demonstrate avalanche multiplication, an InGaAs/GaAs-QWIP was used since alloy and intervalley scattering processes are less pronounced [256] than in conventional GaAs/AlGaAs-QWIPs. In fact, g_{noise} starts to increase more steeply than g_{photo} if the bias exceeds 2.4 V ($\approx$22 kV cm^{-1}).

For a quantitative description of this excess noise, we introduce the noise factor $\mathcal{F}(\mathcal{M})$, which defines the excess noise associated with the noise spectral density. The generalized g–r noise current associated with the current I within the bandwidth Δf can thus be written as

$$i_n^2 = 2eg_{\text{photo}}I\Delta f\mathcal{F}(\mathcal{M}). \tag{7.15}$$

In order to obtain a model for g–r noise in QWIPs, that includes avalanche multiplication, we have generalized the model developed by Beck [140] by introducing an additional term which takes avalanche processes into account. Re-expressing the result in terms of $\mathcal{M}$ and p_{c}, the calculation predicts [177]

$$\mathcal{F}(\mathcal{M}) = 2\mathcal{M} - p_{\text{c}}\left(2\mathcal{M} - 2 + \frac{1}{\mathcal{M}}\right). \tag{7.16}$$

Without impact ionization, i.e., at $\mathcal{M} = 1$, (7.15) and (7.16) reproduce Beck's result $g_{\text{noise}} = g_{\text{photo}} - 1/2N$. In particular (7.16) yields $\mathcal{F}(\mathcal{M}) = 1$ for photovoltaic QWIPs ($p_{\text{c}} = 1$) and $\mathcal{F}(\mathcal{M}) = 2$ for photoconductive QWIPs ($p_{\text{c}} \ll 1$). For $\mathcal{M} > 1$, and $p_{\text{c}} \ll 1$, (7.16) simplifies into $\mathcal{F}(\mathcal{M}) \approx 2\mathcal{M}(1 - p_{\text{c}})$. The excess noise factor is plotted in Fig. 7.23 for different values of p_{c}. The limiting cases $\mathcal{M} = 1$, $p_{\text{c}} \ll 1$, and $p_{\text{c}} = 1$ discussed above are clearly verified in the figure.

For the experiment of Fig. 7.22, we expect the condition $\mathcal{M}p_{\text{c}} \ll 1$ to be valid, such that $\mathcal{F}(\mathcal{M}) \approx 2\mathcal{M}$. In this case, g_{photo} and g_{noise} are connected by the simple relation

$$g_{\text{noise}}/g_{\text{photo}} = \mathcal{M}. \tag{7.17}$$

The extra noise contribution in the presence of impact ionization is due to the additional statistical indeterminacy inherent with the avalanche process. Equation (7.17) now allows us to determine the multiplication factor $\mathcal{M}$ which is plotted in Fig. 7.24. $\mathcal{M}$ is roughly equal to unity up to +2.4 V, and

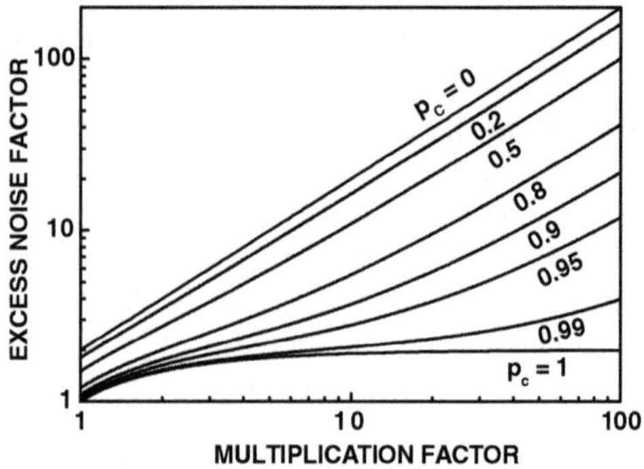

Fig. 7.23. Excess noise factor $\mathcal{F}$ vs. avalanche multiplication factor $\mathcal{M}$ for different capture probabilities p_c as indicated

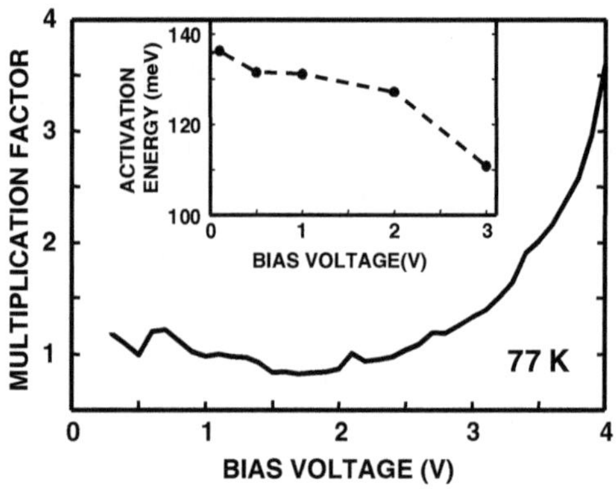

Fig. 7.24. Avalanche multiplication factor $\mathcal{M}$ vs. bias voltage. The inset shows the bias dependence of the activation energy E_{act} (after [254])

rises significantly above 1 at higher bias voltages, up to $\mathcal{M} = 3.6$ at $+4.0$ V (≈ 37 kV cm^{-1}).

In order to initiate avalanche multiplication processes, the kinetic energy E_{kin} accumulated by drifting carriers along their coherent (or ballistic) mean free path l_c must reach a threshold value before a ground state electron can be scattered out into the continuum. This threshold energy is expected to be comparable to the activation energy E_{act} of the dark current [250]. A rough estimation of l_c is thus obtained by dividing the experimental value $E_{act} = 121$ meV (at $+2.4$ V), as determined from the temperature dependence of the dark current (see inset of Fig. 7.24), by the threshold field F_{th} of ≈ 22 kV cm^{-1}; yielding $l_c = E_{act}/eF_{th} = 55$ nm. In fact, this value is consistent with results from independent experiments. Investigations of the coherent mean free path of electrons in the continuum of asymmetric GaAs double quantum-wells (see Sect. 7.4.2) suggest a value above 20 nm at 77 K and about 100 nm at liquid helium temperatures. Studies of GaAs hot-electron transistor structures also point to l_c-values of about 70 nm at low temperatures [257,258]. Knowing $\mathcal{M}$, we can now determine the "net" gain g of the investigated QWIP in the absence of avalanche processes using the relation $g = g_{\text{photo}}/\mathcal{M}$. The gain obtained in this way is shown as a function of the bias voltage by the solid line in Fig. 7.22. While $g_{\text{photo}} \approx g_{\text{noise}} \approx g$ in the regime below +2.4 V, g_{photo} and g_{noise} show a substantial increase when avalanche multiplication sets in. In contrast, g increases linearly below +1.0 V (9.2 kV cm^{-1}) and saturates at a value of about $g = 3$, followed by a slight decrease at high bias (above ≈3 V).

Similar avalanche multiplication effects have also been reported for InGaAs/InP devices by Aslan et al. [259]. The effect appears to be much less important in GaAs/AlGaAs QWIPs.

7.6 Radiation Hardness

QWIPs are suited to and are used for space applications [16,260]. It is well known that permanent degradation of semiconductor device performance can result from a prolonged operation in the space environment. GaAs based devices are generally regarded as more radiation resistant than others. In the space environment, there exists a wide range of light and heavy particles with different energies. A systematic study of QWIP radiation damage is therefore important for predicting their performance, and for qualifying them for space applications.

Measurements [261,262] have been carried out on the effect of high energy proton, α-particle, oxygen ion, and gold ion radiation on the performance of GaAs/AlGaAs QWIPs. In addition, Akhmetov et al. [263] have studied the effect due to 3.5 MeV electron radiation. In our experiments, the particle energies of protons, α-particles and oxygen ions ranged from 0.8 to 30 MeV, with the fluences from 10^{10} to 10^{16} cm^{-2}. The energy of Au ions was 1.5 GeV, and the fluence ranged from 10^6 to 10^9 cm^{-2}. The dark current and spectral response of irradiated devices were measured. A device operability, defined by the fractional reduction of detector responsivity, was used to evaluate the performance degradation. Device operability degrades with fluence for all particles. It also degrades with the mass of the ion and with the decrease in the energy of the particle.

In these studies, a standard QWIP was used, with peak detection wavelength at 10.7 µm. The epitaxial layers in growth sequence were: a 0.8 µm GaAs bottom contact layer Si-doped to 1.5×10^{18} cm^{-3}, a 32-repeat multiple quantum well structure with 36 nm wide $Al_{0.19}Ga_{0.81}As$ barriers and 7.5 nm wide GaAs wells, and a 0.4 µm GaAs top contact layer Si-doped to 1.5×10^{18} cm^{-3}. The center of all wells was δ-doped with Si to 5.0×10^{11} cm^{-2}. Mesa devices with 120×120 µm^2 active areas were made by standard wet chemical etching, and contacts were formed with alloyed NiGeAu. The processed wafer was cleaved into chips of about 4×4 mm^2 in size, each containing many devices. Each chip was attached to a metal block at room temperature and was irradi-

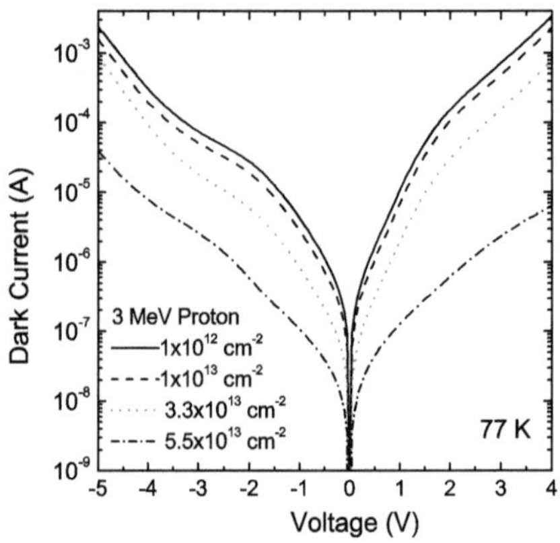

Fig. 7.25. Dark current vs. bias voltage characteristics at 77 K for 3 MeV proton irradiated QWIPs at different fluences

ated from the top in vacuum. The radiation beam current was sufficiently low, causing no appreciable change of the block temperature during irradiation.

The nuclear particles and energies used in the experiments were selected to provide a sample of radiation damage data at different energies. The range of these ions decreases with decreasing particle energy: the range of 0.8 MeV protons, the lowest energy protons used here, is about 8 μm in GaAs and AlGaAs, while the range of the lowest energy (2.4 MeV) α-particles is about 7 μm and that of 10 MeV oxygen ions is about 5 μm in these materials. The GeV energy Au particles used here have a range of about 50 μm. Since the QWIP active layer is thin (less than about 2 μm), the range of the irradiating particle in the device material is greater than the total thickness of the QWIP, which ensures a uniform damage in the device.

As an example, Figs. 7.25 and 7.26 show the dark current and the spectral responsivity curves. The I–V characteristic at the lowest fluence level is very similar to the unirradiated device. With increasing fluence, the dark current is reduced (Fig. 7.25). A dramatic decrease is seen over the range in fluence studied (approximately a 100-fold reduction). Further increases in fluence made the devices practically insulating. The responsivity measurements were performed using a liquid nitrogen cold-finger dewar with an estimated device temperature of 80 K. Spectra were taken over a range of applied voltages, and those for -2 V are shown. Comparing Figs. 7.25 and 7.26, the influence of radiation damage on the dark current is larger than that on the responsivity. Since the barrier is much thicker than the well, the defects created by the radiation are mainly in the barriers. Since the QWIP is a unipolar device, the effect of these defects is (a) to remove electrons from the wells and (b) to increase the effective barrier height because the defects trap electrons taken away from the

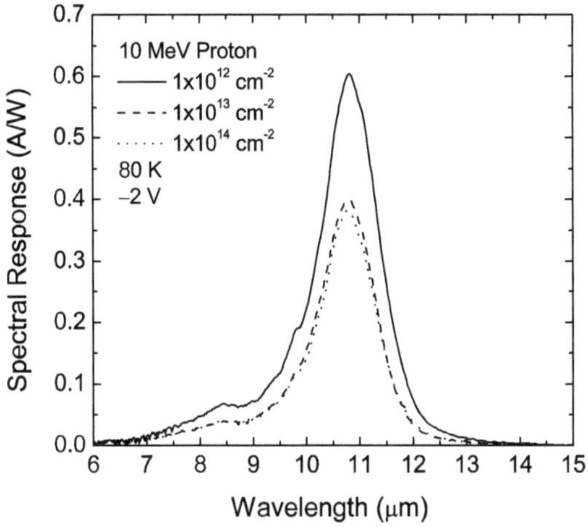

Fig. 7.26. Spectral responsivity vs. wavelength characteristics at 80 K and −2 V for 10 MeV proton irradiated QWIPs at different fluences

wells, resulting in a bandbending. For QWIPs, the dark current depends on electron density and barrier height exponentially; the absorption, however, is directly proportional to the electron density. Thus, the dark current is more sensitive to radiation damage than the detector responsivity. Moreover, defects and traps in the barriers could cause leakage paths leading to an increase in dark current, but this is not evidenced in the experiments.

We can compare the device responsivity dependence on radiation damage due to particles at different energies if we define a device operability by the

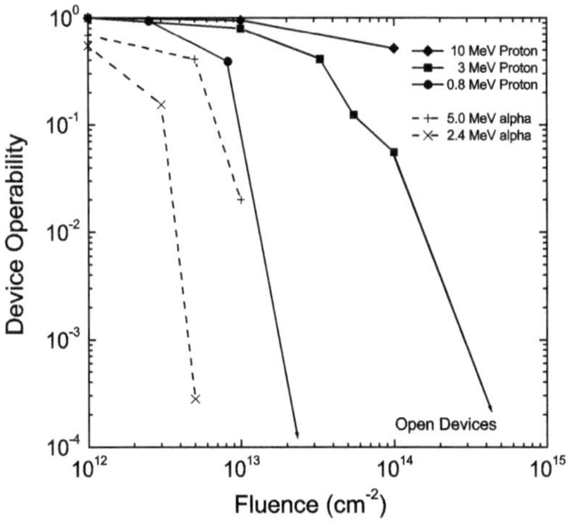

Fig. 7.27. Device operability (defined as the fractional reduction in responsivity) vs. fluence for QWIPs irradiated with different high energy proton and α-particles

reduction in responsivity: i.e., if the responsivity is reduced by a factor of ten, the operability is, by definition, 0.1 or 10%. Figure 7.27 gives the operability vs. proton and alpha fluence for different energies. It is clear that when the fluence is low, the device operability is nearly 100%, which means the effect of radiation damage is negligible. With increasing fluence, for all energy values, the device operability decreases monotonically. The devices became practically insulating at some fluence dependent on the radiating particle energy. Looking at the behavior at different energies, a clear trend is shown: at a fixed fluence, the device displays more damage for a lower energy proton than for a higher energy proton in the energy range tested here. Comparing the proton and alpha curves reveals that the degradation due to α-particles is much greater than that due to protons for the same particle energy.

An overall trend in the introduction of radiation damage can be stated after observing results from all particles [262]. For any particle at a fixed energy, there is a gradual and continuously increasing low level damage at low fluence levels. However, the rate of radiation damage changes at higher fluence levels, and this change in the rate as a function of fluence sets in at a lower fluence level with the same particle at a lower energy.

To compare damage due to different particles, we list a set of typical values. When the damage is relatively small, and not catastrophic, the device operability is 0.95 for a $10^{13}\,\mathrm{cm}^{-2}$ fluence of 10 MeV protons, while it is 0.43 for the same fluence of 10 MeV α-particles. For 10 MeV oxygen ions, the device operability is 0.34 at a $10^{11}\,\mathrm{cm}^{-2}$ fluence level. Thus, 10 MeV α-particles are about twice as damaging, and 10 MeV oxygen ions are about 300 times more damaging, than are protons at the same energy. In the same study, we found that the 1.5 GeV Au ions in the fluence range from 10^6 to $10^9\,\mathrm{cm}^{-2}$ did not cause measurable damage. Based on other experiments and simulations, a fluence of $10^{10}\,\mathrm{cm}^{-2}$ or greater is needed.

An interesting (somewhat unexpected) result was seen in the experiment of [262]: Low level irradiations slightly improved the detectivity. The behavior is explained as follows. Since the optimum well doping density for D^* corresponds to a Fermi energy $E_\mathrm{F} = 2k_\mathrm{B}T$, where T is the operating temperature (which is $T = 80\,\mathrm{K}$ here). The Fermi energy for this QWIP before irradiation is $E_\mathrm{F} = 18\,\mathrm{meV}$. For $T = 80\,\mathrm{K}$, $2k_\mathrm{B}T = 14\,\mathrm{meV}$. Since $E_\mathrm{F} > 2k_\mathrm{B}T$ before irradiation, the effect of removing electrons (which decreases E_F) brings the electron density closer to the optimum and hence results in an initial improvement in D^*. This observation also confirms the insensitivity of QWIPs (as majority unipolar devices) to irradiation.

8

Related Structures and Devices

This chapter discusses related and new/novel structures and devices for studying physical processes, which may lead to IR detection and imaging applications.

8.1 High Absorption QWIPs

When the signal to be detected is strong (e.g., from a laser) the signal induced photocurrent can dominate over the dark current. In this case, the detection sensitivity is only limited by the absorption, and hence a high absorption is most desirable. This is also true in the case of heterodyne detection (to be discussed in Sect. 10.2.2). In the simplest picture (Sect. 3.2), the absorption is proportional to the electron density and the number of wells. We demonstrate here that high absorption (~100%) can be achieved by simply changing these device parameters.

The device intrinsic absorption is best characterized by a direct measurement of transmission/absorption. In this section, we show experimental results to demonstrate the high absorption capability. We also show that these devices can work at elevated temperatures either within the thermoelectric cooling range (>200 K) or at room temperature.

8.1.1 Absorption Measurements

Starting from a standard QWIP structure, the simplest way to enhance absorption is to dope the wells more heavily and to grow more wells [220,221]. At 45° incidence with polarized light, the absorption per quantum well per pass is about 0.54% per well for a standard GaAs/AlGaAs QWIP with $5 \times 10^{11}\,\mathrm{cm}^{-2}$ doping (see Sect. 3.2). For a doping density of $1.5 \times 10^{12}\,\mathrm{cm}^{-2}$, the one-well/one-pass absorption is expected to be $\eta_1 \approx 1.6\%$. If a 90% QWIP absorption is desired, the number of wells needed is determined by:

Table 8.1. GaAs/AlGaAs QWIP sample parameters

Sample	λ_p (µm)	x	L_w (nm)	L_b (nm)
QWIP10	10.0	0.19	6.6	25
QWIP8	8.4	0.24	5.9	24
QWIP5	5.0	0.48	5.0	22

(The doping in the well is Si, δ-doped at the center to $1.5 \times 10^{12}\,\text{cm}^{-2}$. The number of periods is 100. The symbols are λ_p – peak detection wavelength, x – Al fraction, L_w – well width, and L_b – barrier thickness.)

$\exp(-2N\eta_1) = 10\%$, which gives $N = 72$. (The factor of 2 in the exponential accounts for the double passes in the 45° facet detector geometry.) $N = 100$ was chosen in the experimental verification, with a doping density of $1.5 \times 10^{12}\,\text{cm}^{-2}$. Other doping densities were also tried. The study of [220] indicated that a 100-well QWIP with 1–$1.5 \times 10^{12}\,\text{cm}^{-2}$ well doping is close to the optimum. Decreasing the doping substantially leads to a reduction in absorption, while increasing the doping seems to only result in a spectral broadening.

Here we show three examples with different detection wavelengths centered at about 10, 8, and 5 µm. The period of the 100-repeat multiple quantum well structure consists of a GaAs well and $Al_xGa_{1-x}As$ barriers. The GaAs well center region is doped with Si to give an equivalent 2D density of $1.5 \times 10^{12}\,\text{cm}^{-2}$. The top and bottom GaAs contact layers are 400 and 800 nm thick, doped with Si to $2\times10^{18}\,\text{cm}^{-3}$. Other device parameters are listed in Table 8.1. Mesa devices were fabricated by standard GaAs processing techniques. All devices were packaged into the 45° edge facet geometry.

With the high well doping densities used in these QWIPs, we expect a large increase in dark current. Using a set of samples having similar cutoff wavelengths, but different doping densities of 5×10^{11}, 1.0×10^{12}, 1.5×10^{12}, and $2.0 \times 10^{12}\,\text{cm}^{-2}$, measurement of current–voltage (I–V) characteristics at 77 K showed approximately a factor of 10 increase between consecutive samples. The increase is mainly caused by the increase in the well Fermi energy ΔE_f which raises the dark current by $\exp(\Delta E_f/k_BT)$. Measurements of blip temperature show that T_{blip} is degraded with increasing doping. $T_{blip} \approx 60$, 45, and 30 K for 1.0×10^{12}, 1.5×10^{12}, and $2.0 \times 10^{12}\,\text{cm}^{-2}$ doping samples, respectively, were found. By contrast, an optimized QWIP (with about 4×10^{11} cm^{-2} doping) covering the same wavelength range would have a T_{blip} about 77 K for a 2π-solid-angle field of view, so high doping devices are therefore far from optimum for the detection of weak signals.

To show the anticipated high absorption, Fig. 8.1 presents the measured double-pass polarized 45° incident transmission spectra for the three samples (see Fig. 6.2 for the measurement geometry). For samples QWIP10 and QWIP8, the high absorption is clearly demonstrated. We would have expected QWIP5 to have a similar absorption. Possibly carrier transfer to the L and X valleys may be responsible for the reduced absorption.

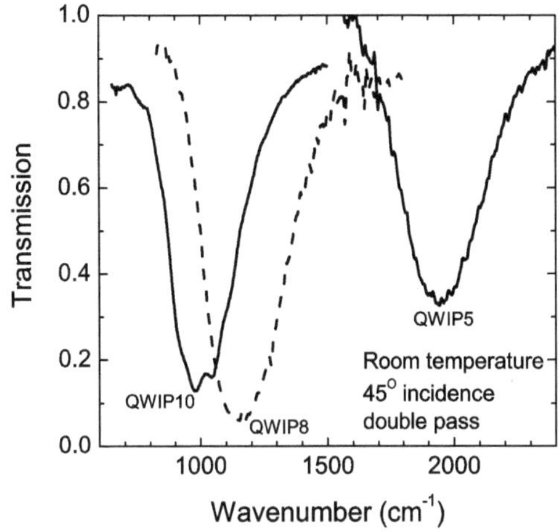

Fig. 8.1. Double-pass 45° incidence transmission spectra at room temperature and with polarized light

8.1.2 Detector Characteristics

Figure 8.2 shows the spectral response curves for the three samples, covering the designed wavelength regions. A further goal of these samples was to attempt operation at an elevated temperature. To maximize the dark current limited detectivity, it has been established (see Sects. 4.3 and 4.4) that the well doping density should be such that the Fermi energy is $E_f = 2k_BT$, where T

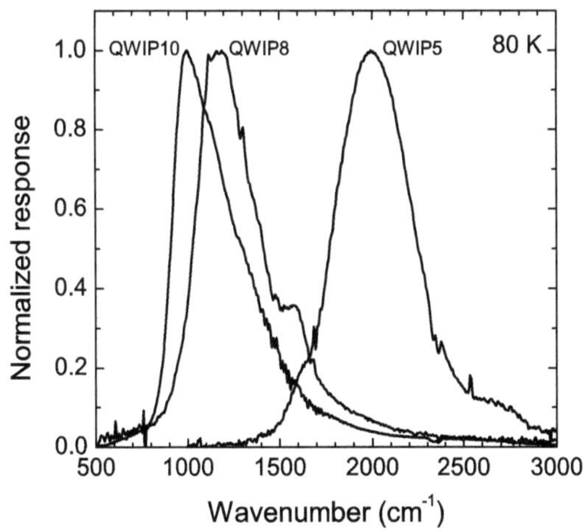

Fig. 8.2. Normalized spectral response curves at 80 K and 3 V. The device geometry is equivalent to that of Fig. 8.1 so that the absorption efficiency is represented by Fig. 8.1

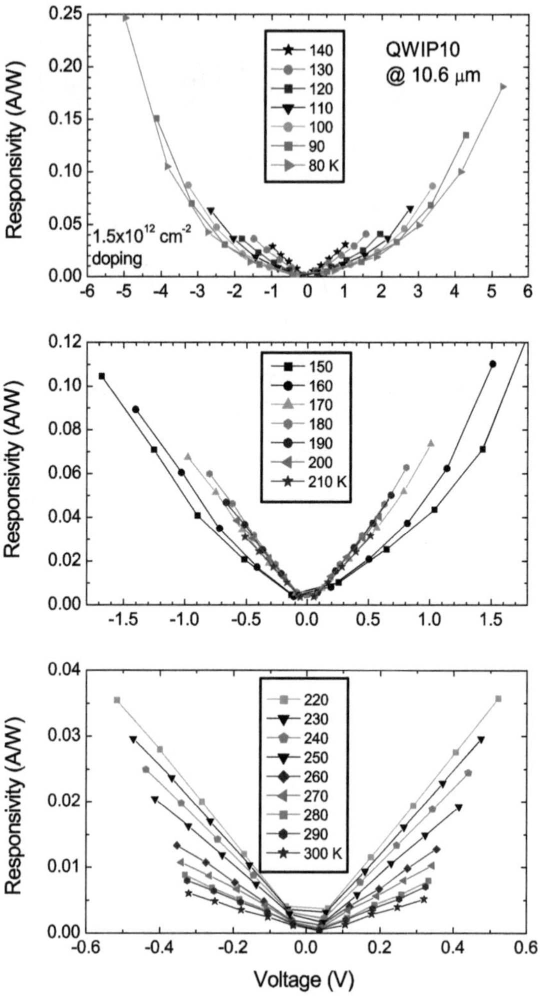

Fig. 8.3. Responsivity vs. applied voltage under a CO_2 laser (10.6 μm) illumination and at various temperatures for sample QWIP10

is the desired operating temperature. The 2D doping density is related to the Fermi energy by $N_D = (m/\pi\hbar^2)E_f$, where m is the well effective mass. For $T = 80\,K$, the required density is about $4 \times 10^{11}\,cm^{-2}$ for GaAs wells, and it is expected that QWIPs with doping in the 1–2 $\times$ $10^{12}\,cm^{-2}$ range may be operable near room temperature, albeit with a reduced sensitivity. Measured results at various temperatures using a CO_2 laser tuned to 10.6 μm are shown in Fig. 8.3 for sample QWIP10, and it is clear that the device does indeed work up to room temperature.

It is interesting to note that with increasing temperature the responsivity first increases and then decreases for a given bias voltage. This behavior is explained qualitatively as follows. Since the responsivity is proportional to the mobility, the behavior is attributed, in a large part, to the temperature dependence of the mobility. It is well known that the impurity scattering limited mobility increases with temperature, whereas the phonon scattering limited mobility decreases. The two scattering effects result in a mobility vs. temperature dependence similar to the observed responsivity vs. temperature behavior [264]. For elevated temperature operation, where the QWIP resistance is low, a high responsivity is highly desirable to overcome the large noise associated with the low resistance and the 50 Ω termination. There is a difficulty in accurately measuring the responsivity when the device resistance is low and comparable to contact and/or series resistance. The low resistance causes a "short-circuit" for the photocurrent, leading to an apparent low responsivity. The values in Fig. 8.3 for high temperatures (>200 K) could be somewhat lower than reality.

Given the measured *I–V*s, responsivity, and absorption, the detectivity $D^\star$ can be evaluated. Using the relations $\mathcal{R}_i = (e/h\nu)g\eta$ and $D^\star = \mathcal{R}_i\sqrt{A}/\sqrt{4egI}$, where $\mathcal{R}_i$ is the responsivity, ν is the photon frequency, g is the photoconductive gain, A is the device area, and I is the device current, the detectivity at 10.6 μm and for polarized light is calculated and plotted in Fig. 8.4 for various temperatures. A state-of-the-art QWIP for the same wavelength would have a $D^\star$ value of about 10^{10} cm Hz$^{1/2}$ W^{-1}. The device here is therefore far from optimum for low signal and low temperature use. Comparing with a room temperature thermal detector having a $D^\star$ value of about 10^8 cm Hz$^{1/2}$ W^{-1}, which is commonly used in MIR spectrometers, this device needs to be cooled to about 150 K to achieve the same $D^\star$. Note that temperatures about 200 K and higher are attainable by thermoelectric cooling. Operation at elevated temperatures implies high dark currents. In order to overcome the dark current noise at room temperature, a laser power of about 10 mW is required

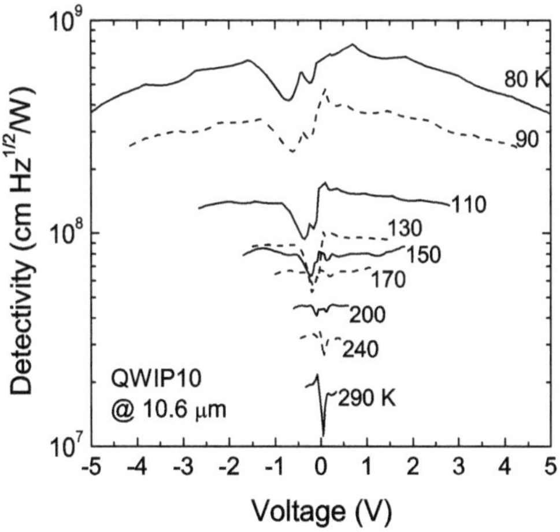

Fig. 8.4. Detectivity at 10.6 μm for sample QWIP10 and at various temperatures

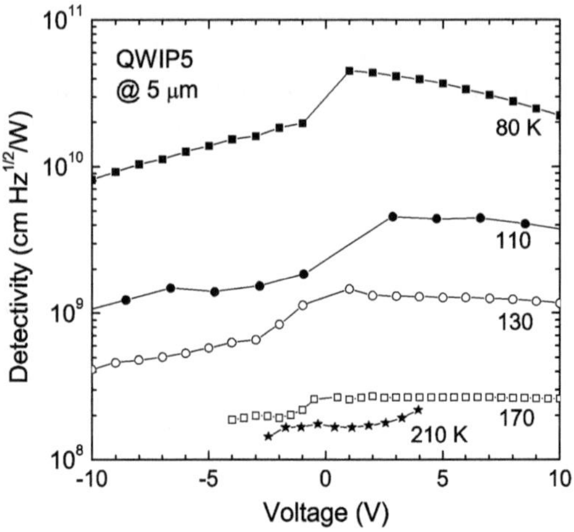

Fig. 8.5. Detectivity at 5 μm for sample QWIP5 and at various temperatures

for a $10 \times 10\,\mu m^2$ active-area device; and similarly at 200 K a power of about 5 mW is needed. High speed operation requires small device capacitances. The approach employing a large number of quantum wells makes the device thicker than a typical QWIP and hence has lower capacitance. For these devices having a thickness of about 3 μm, an active area of about $50 \times 50\,\mu m^2$ or smaller is sufficient for operation at 30 GHz or higher. Similarly, the detectivity for sample QWIP5 is shown in Fig. 8.5. For this shorter wavelength device, a higher detectivity is obtained, making this device able to operate with a thermoelectric cooler.

8.2 Multicolor QWIPs

Being based on thin multilayers grown by epitaxial techniques, the design of QWIPs is very flexible. This enables various implementations of multicolor and multispectral detectors. The word "color" used here is only by analogy to the visible spectrum, and is meant for different wavelengths in the IR spectrum. The approaches to multicolor QWIPs are divided into three basic categories: (1) multiple leads, (2) voltage switched, and (3) voltage tuned. The three cases are schematically shown in Fig. 8.6. The division here is somewhat arbitrary. In general, a multicolor or multispectral detector is a device having its spectral response varied with parameters like applied bias voltage or any other parameters such as temperature, pressure, magnetic field, filter position, etc. The latter cases are not as desirable in practical applications and are not discussed here.

Approach (1) is a direct one which involves contacting each intermediate conducting layer separating one-color QWIPs grown in a multistack. This re-

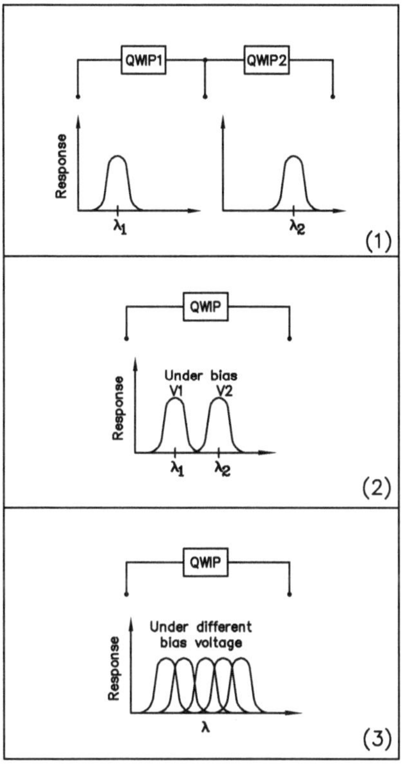

Fig. 8.6. Three different multicolor QWIP approaches. **(1)** Shows two QWIPs with different response wavelengths grown in a multistack. Intermediate leads are provided, so that each QWIP functions independently. **(2)** The QWIP has a response peaked at λ_1 at voltage V_1 and at λ_2 at voltage V_2. **(3)** The QWIP has a response that is continuously tuned by voltage

sults in a separately readable and addressable multicolor QWIP with multiple-electrical leads, and a two-color version has been demonstrated [265]. The advantage of this approach is its simplicity in design and its negligible electrical cross-talk between colors. Moreover, each QWIP region can be optimized independently for a desired detection wavelength. However, the additional electrodes require additional detector processing and special/complicated readout electronics. A further drawback is the difficulty in fabricating a many color version because of the many separate leads required in contacting each intermediate layer. Bois et al. [266,267] have developed processing technologies for implementing two-color imaging arrays. Other direct approaches include the use of different parts or rows for different colors. Gunapala et al. [268] demonstrated a four-color imager by separating a large array into four stripes, each responding to a different color. The technology developed at IAF (Fraunhofer Institute for Applied Solid State Physics) is described in Sect. 9.3.

8.2.1 Voltage Switched Multicolor QWIP

The second approach (middle part in Fig. 8.6) is to have a QWIP with a switchable response, e.g., for an applied voltage V_1 the response is at λ_1 and for V_2 at λ_2. One such example is realized by stacking one-color QWIPs separated by thin conducting layers [269,270]. The device operation relies on the highly nonlinear and exponential nature of the device dark I–V characteristics. This implies that an applied voltage across the entire multistack would be distributed among the one-color QWIPs according to their values of dc resistances. Thus, when the applied voltage is increased from zero, most of the voltage will be dropped across the one-color QWIP with the highest resistance (which will be the shortest wavelength). As the voltage is further increased, an increasing fraction of the voltage will be dropped across the next highest resistance one-color QWIP, and so on. Since the detector responsivity of a one-color QWIP gradually turns on with applied voltage, we therefore can achieve a multicolor QWIP with spectral response peaks that turn on sequentially with applied voltage.

The bandedge profiles of a three-color version are schematically shown in Fig. 8.7. To quantitatively predict the magnitude of the photocurrent, we need to consider what is being measured. The use of a photoconductive detector usually involves applying a constant dc bias across the detector in series with

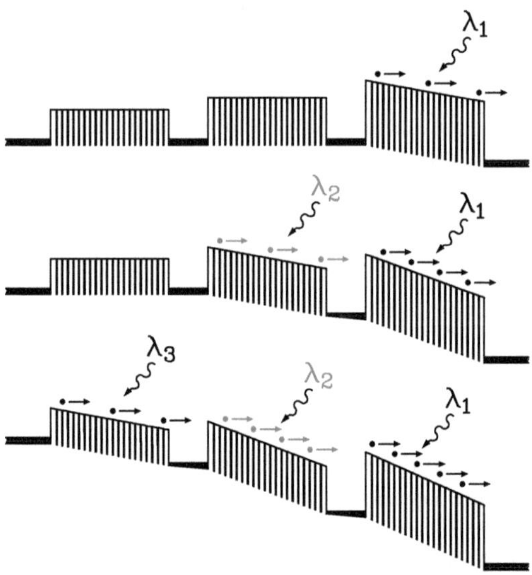

Fig. 8.7. Bandedge profile of a three-color QWIP at different bias voltages. The top part is for a small voltage where only the highest resistance one-color QWIP (at the shortest wavelength λ_1) is turned on, the middle part is appropriate for the situation where two of the three one-color QWIPs (at λ_1 and λ_2) contribute to the photocurrent, and in the lower part the applied voltage is high enough so that all three one-color QWIPs (at λ_1, λ_2, and λ_3) have nonnegligible electrical fields

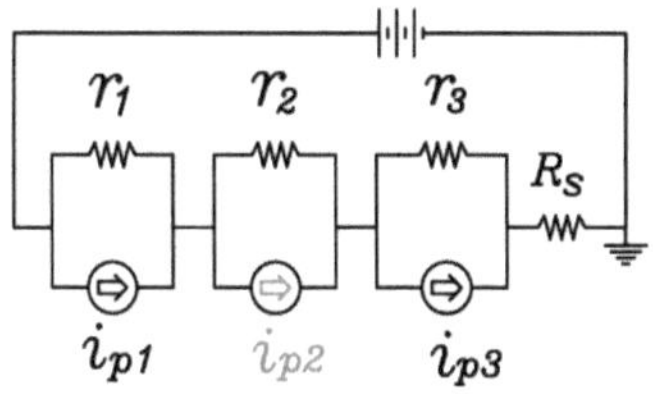

Fig. 8.8. Model equivalent circuit of a three-color QWIP biased through a series load resistor R_s. Symbols i_p and r are the photocurrent source and the internal resistance

a load resistor R_s. The equivalent circuit of this three-color detector involves a network of photocurrent sources (i_{p1}, i_{p2}, and i_{p3}) and dynamic device resistances (r_1, r_2, and r_3), as shown in Fig. 8.8. Under small signal condition, the measured photoresponse current is

$$I_{\text{photo}} = \frac{i_{p1}r_1 + i_{p2}r_2 + i_{p3}r_3}{R_s + r_1 + r_2 + r_3}. \tag{8.1}$$

The nonlinear nature of the dynamic resistances as a function of the voltage leads to a nonlinear weighting factor of the relative contributions among i_{p1}, i_{p2}, and i_{p3} to I_{photo}. An example of a two-color QWIP is shown in Fig. 8.9. The sample consists of a two-QWIP stack with 25 GaAs/AlGaAs quantum wells in each QWIP. The well width and Al fraction are 4.8 and 0.19, and 5.8 and 0.31 nm, for the two QWIPs, respectively. All barriers are 35-nm thick. The full detail is found in [270]. The two colors of this device are within the LWIR region. Based on the same principle, a two-color device working in

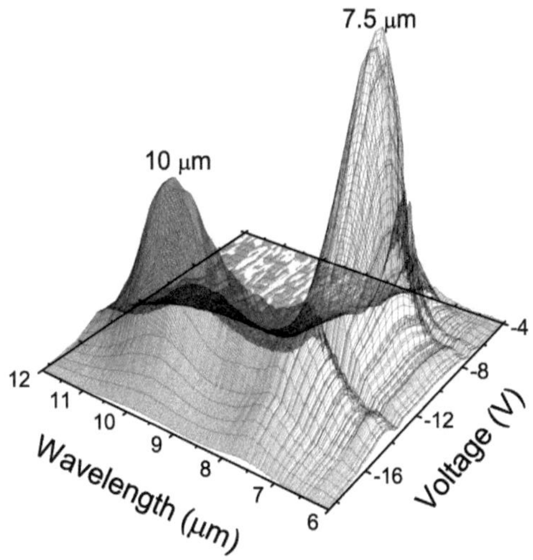

Fig. 8.9. A two-color QWIP response vs. applied voltage and wavelength at 80 K. In the voltage range around −8 V, the dominant response is from the 7.5 µm QWIP; whereas at around −14 V, the 10 µm QWIP becomes strong

LWIR and MWIR was also demonstrated [271]. A similar multicolor QWIP relying on high and low field domains has also been demonstrated [272].

The advantage of this approach is that it is simple in fabrication (as it requires only two terminals) and suited for implementing a QWIP with many colors. The drawback is the difficulty to achieve a negligible electrical crosstalk between colors. Other practical issues need to be resolved before such an approach can be implemented into a full imaging device. The measurements cited above were all conducted under nonblip conditions, whereas the devices are normally used under blip or near blip in practical applications. Moreover, the raw curves showing the wavelength switching (e.g., Fig. 8.9) were collected using a blackbody or globar IR source at a temperature of about 1,000 K. The common applications are, however, in an ambient of about 300 K. Noting the blackbody distribution for different temperatures, one would require a higher responsivity for the MWIR region than that for the LWIR region for a common ambient. The situation is reversed for 1,000 K.

Another approach, which avoids the above nonlinear coupling, relies on different intersubband transitions originating from the same quantum well. An example of this approach is based on a wide quantum well with two occupied subbands [273,274]. Figure 8.10a shows the subband energy diagram of a device containing 9.5 nm thick Si-doped GaAs quantum wells with an electron

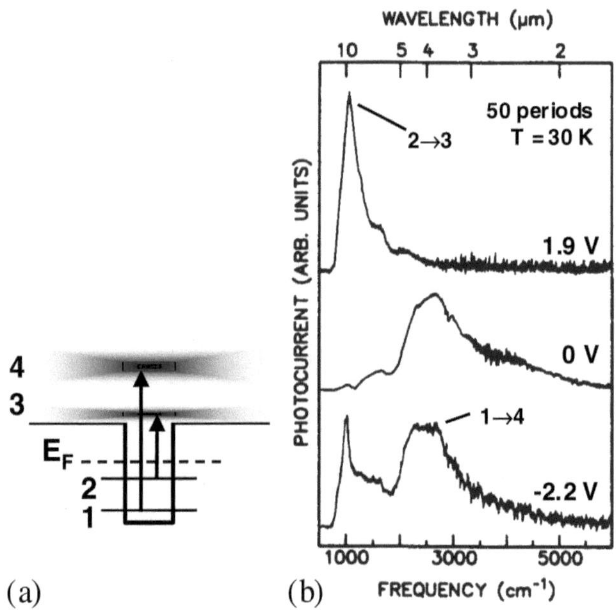

Fig. 8.10. (a) Energy level diagram of a two-color QWIP with the Fermi energy E_F located above the energy of subband 2. *Arrows* indicate the transitions 1→4 and 2→3. **(b)** Low-temperature photocurrent spectra of the device at three different bias voltages

density of $6.5 \times 10^{12}\,\mathrm{cm}^{-2}$ per well, seperated by 50 nm wide $Al_{0.32}Ga_{0.68}As$ barriers. Due to the high carrier density, the lowest two subbands (1 and 2) are occupied. Since two resonances (3 and 4) exist in the continuum, there are two parity-allowed transitions (2→3 and 1→4) which are designed for the 8–12 and 3–5 µm regions, respectively.

Even though both transitions originate from the same well, the associated photocurrents can be accessed independently upon varying the bias. Three representative photocurrent spectra are depicted in Fig. 8.10b. At −2.2 V, both transitions induce photocurrent signals of comparable peak amplitude. Interestingly, only transition 2→3 is visible at 1.9 V while transition 1→4 is dominant at 0 V. In fact, it was possible to change the ratio between these two signal intensities by more than two orders of magnitude simply by adjusting the bias. Phenomenologically, the signal induced by 1→4 shows a photovoltaic behavior, while the component due to 2→3 is photoconductive. This behavior is related to the fact that subband 3 is located close to the conduction band-edge of the AlGaAs barrier, while state 4 is far in the continuum. Apparently, transition 1→4 is associated with a strongly unidirectional emission of photoexcited electrons, which induces a signal at zero bias and blocks it at the wrong polarity. We attribute this asymmetry to an asymmetric distribution of the Si dopants with respect to the well center (see Sect. 7.3). Also, in standard QWIP structures, such dopant distributions are known to cause strong dependencies of the photocurrent and dark current upon the bias polarity.

Unfortunately, this approach has the disadvantage that the noise is the same for both detection wavelengths, i.e., that the noise current is determined by the large dark current associated with the longer wavelength transition. This means that a state-of-the-art detector can be realized for the longer wavelength using this concept – but only a poor detector will be feasible for the shorter wavelength.

A different type of two-color QWIP using asymmetric double quantum wells has been demonstrated by Berger et al. [275]. In this case, a population transfer is induced between the wide and narrow wells by an external voltage, thus giving rise to a corresponding change in the intersubband energy.

8.2.2 Voltage Tuned Multicolor QWIP

The last case (lower part in Fig. 8.6) is a QWIP with its response continuously tuned in a range of wavelengths. The demonstrated examples of approach (3) involve special shapes of quantum wells, e.g., a stepped well [276–278] or an asymmetrically coupled double well [231] so that the response spectrum shifts as a function of applied bias voltage (commonly referred to as the phenomenon of Stark shift). This provides a continuous tuning of the spectrum by moving the intersubband resonance position. A range from 8.5 to 13.5 µm has been achieved using stepped wells [278]. The large continuous tuning capability is the distinct feature of this approach. The difficulty is to ensure a good QWIP performance for all voltages. To achieve a good performance, the transition

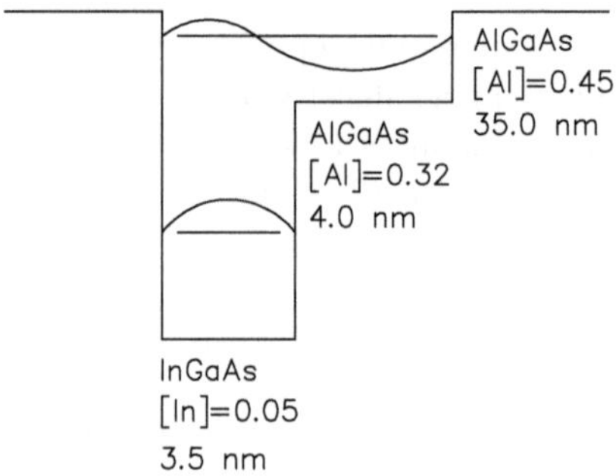

Fig. 8.11. Bandedge profile of stepped quantum well used for voltage tunable QWIP. The device had 32 repeats

final state (usually the first excited state) should be close to the top of the barrier, as discussed in Chap. 4, providing a large intersubband transition strength and, at the same time, an easy escape for the excited carriers. These two conditions are difficult to fulfill for all voltages. Another factor which may degrade the QWIP performance, is the use of relatively wide wells as in the case of a stepped well. This may lead to an enhanced trapping probability and hence a shorter carrier lifetime. Generally, electric field-induced changes of the subband structure usually require relatively high external electric fields, thus giving rise to large dark currents and noise.

As an example, Fig. 8.11 depicts the stepped quantum well QWIP that we fabricated. The QWIP had 32 wells with the centers of the InGaAs layers doped with Si to $5 \times 10^{11}\,\mathrm{cm}^{-2}$. The normalized spectra are given in Fig. 8.12. A large tuning of 17% in the range of 220 to 260 meV was obtained. The apparent "noise" in the -17 V curve was due to water absorption in

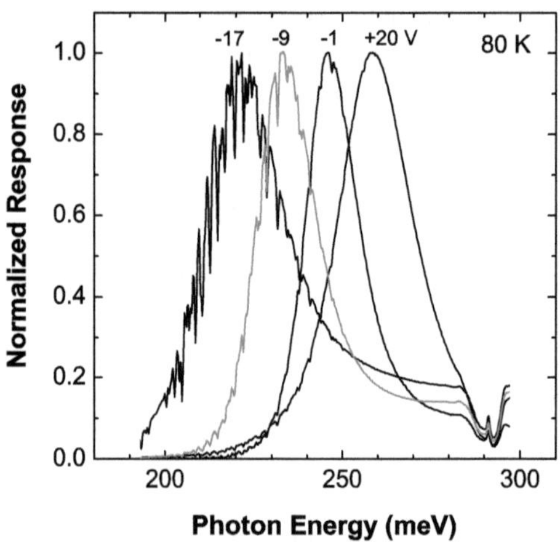

Fig. 8.12. Spectra at different applied voltages for sample in Fig. 8.11 (Data taken by E. Dupont)

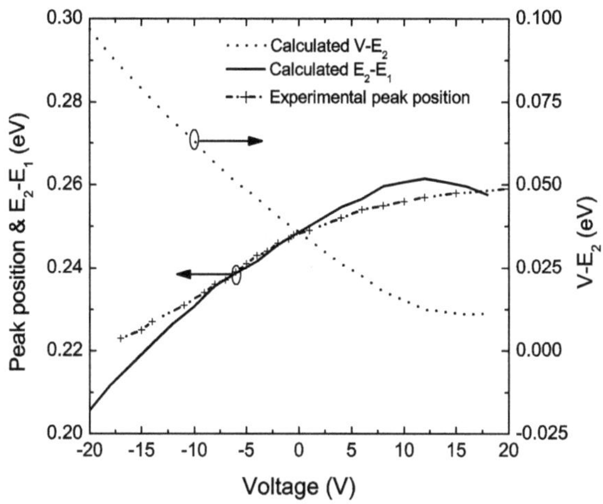

Fig. 8.13. Comparison with calculations. The experimental peak positions are taken from Fig. 8.11. The ground and first excited states are labeled by E_1 and E_2, respectively. Calculated energy difference between the top of the barrier (V) and E_2 is also shown (Calculations made by G.C. Aers)

the measurement optical path. Figure 8.13 shows a comparison between the experimental response peak positions (taken from Fig. 8.12) and calculated eigenenergy difference. A reasonable agreement was obtained. Also shown is the energy difference of the barrier height and the excited level ($V - E_1$). It is clearly seen that the quantity $V - E_1$ varies substantially in the range between 0 and 50 meV. This may be a detrimental effect, as discussed earlier.

As a final comment on multicolor QWIPs, it seems that for all the approaches the device performance has not been fully optimized. Further work is therefore needed both in optimization and in new and better designs. Moreover, multicolor QWIPs require special gratings which have not been studied. Possibilities include the use of quasirandom gratings or V-grooves (as in Chap. 6).

8.3 Interband and Intersubband Dual-band Detectors

For some applications, it is desirable to have a multispectral detector covering not only the middle or far infrared (FIR) but also the visible or near infrared (NIR) spectral regions [279].

8.3.1 Using the Same Quantum Well

So far, we have neglected the valence band. If light with its photon energy larger than the bandgap is present, an interband excitation can occur and a QWIP can operate as an interband detector as well. Together with the

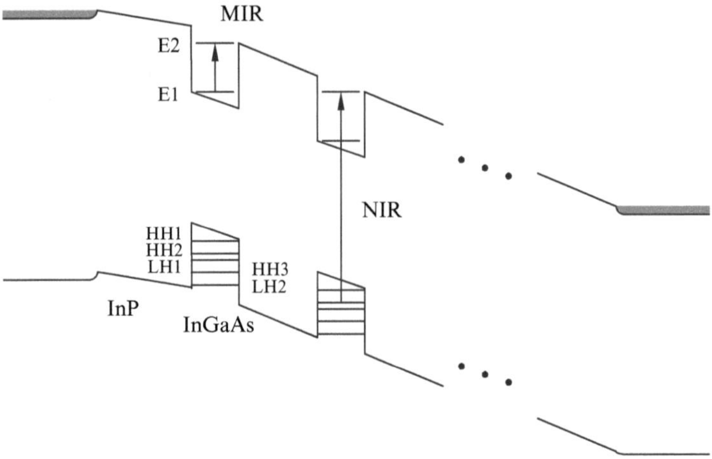

Fig. 8.14. Schematic bandedge profile of the dual-band InGaAs/InP QWIP. The arrow within the conduction band quantum well shows the detection process for the middle infrared; while the arrow from valence band to the conduction band indicates the excitation of electrons by near infrared or visible light

intersubband mechanism, we then have a dual-band detector. The dual-band operation of (1) InGaAs/InP QWIPs for both NIR and middle infrared (MIR) spectra, and (2) GaAs/AlGaAs QWIPs for visible (VIS) and MIR regions have been demonstrated. In the first case [259,280], a usual QWIP structure with InGaAs wells and InP barriers [281] was used for MIR detection. At the same time, NIR can be absorbed in the InGaAs layers resulting in a photocurrent. The contact layers were InP, so that the NIR light could reach the quantum well region without being absorbed. In the second case [282], large bandgap top contacts were used on standard GaAs/AlGaAs QWIPs so that visible light could reach the quantum-well region and be absorbed. Two large bandgap top contacts were investigated, using a high Al fraction AlGaAs and a short period GaAs/AlAs superlattice.

For the NIR and MIR experiments, two 20-well InGaAs/InP QWIPs were used, with 6.6 and 6.0 nm well widths, respectively. The InP barriers were about 30 nm thick. The bandedge profile of the device is schematically shown in Fig. 8.14. The two well widths correspond to bound-to-bound and bound-to-quasibound intersubband transitions.

For VIS and MIR, two samples A and B were tested, both consisting of a standard QWIP and a wide bandgap top contact layer. The QWIP has 40 quantum wells made of GaAs (5.3 and 5.1 nm)/$Al_{0.25}Ga_{0.75}As$(39.5 and 37.8 nm) for samples A and B, respectively. The top contact layers were designed to have a larger bandgap energy than $Al_{0.25}Ga_{0.75}As$. The first design (A) is a 405-nm $Al_{0.7}Ga_{0.3}As$ layer doped with Si to 4×10^{18} cm^{-3}. The second

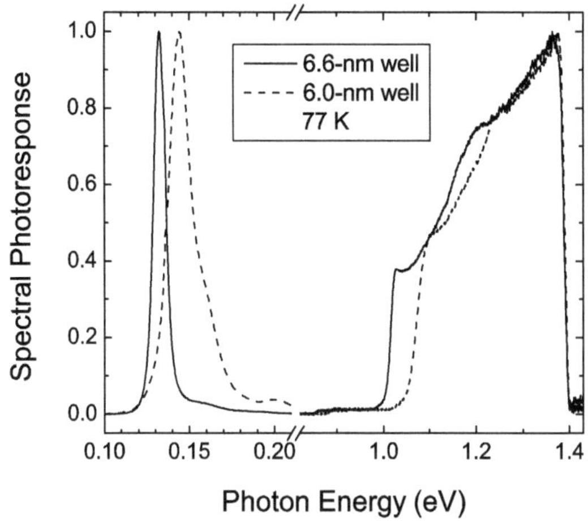

Fig. 8.15. InGaAs/InP device spectral photoresponse curves at a device temperature of about 80 K. The data were taken with two different sets of beamsplitters and light sources. The two parts are separately normalized

design (B) is a 200-period superlattice of GaAs(0.9 nm)/AlAs(1.0 nm) doped with Si to $1.5 \times 10^{18}\,\mathrm{cm}^{-3}$.

Mesa devices were fabricated with alloyed NiGeAu ring or solid contacts. 45° edge facets were polished for light coupling to the intersubband transition. For the NIR and visible spectral response measurements, the illumination was through the facets for the InGaAs/InP devices and through the top windows for the GaAs/AlGaAs devices to avoid the GaAs substrate absorption.

For the InGaAs/InP QWIPs, the spectral photoresponse curves for both MIR and NIR regions are shown in Fig. 8.15. The sharp cutoff at 1.4 eV is due to the InP substrate absorption since the light must pass through the substrate after entering the 45° facet. Figure 8.16 shows the responsivity vs. voltage characteristics at 1.06 and 8.9 μm for the InGaAs/InP QWIP. Both curves show responsivity increasing monotonically with voltage.

For the GaAs/AlGaAs devices, the photoresponse spectra are shown in Fig. 8.17. The MIR response band centered at about 8.3 μm corresponds to the intersubband transition in the QWIP. Two states E_1 and E_2 in the conduction band are confined in the quantum well, with the upper one (E_2) very close to the top of the barrier. The visible response covers a wavelength region from 0.55 to 0.7 μm (Fig. 8.17(top)), or photon energies 1.8–2.3 eV, resulting from interband transitions from both the well and the barrier regions. By considering the confined quantum-well states in both conduction and valence bands as well as the barrier bandgap, all features shown in Fig. 8.17(top) can be accounted for. Since electrons excited from the valence band to E_1 in the quantum wells cannot escape easily, they are likely to recombine with holes before escape, giving rise to a small photocurrent. The photoresponse is

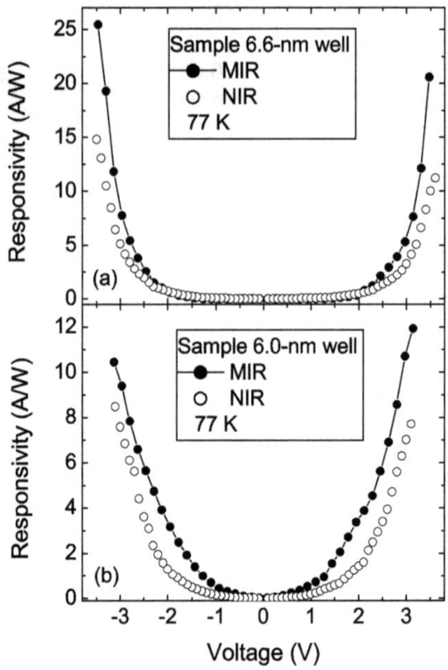

Fig. 8.16. InGaAs/InP device responsivity vs. voltage at a device temperature of about 80 K

stronger for the transition from HH2 to E_2 because E_2 by design is close to the top of the barrier. When the photon energy is higher than the barrier bandgap, i.e., $h\nu > E_g$ for $Al_{0.25}Ga_{0.75}As$, the photocurrent is substantially increased due to the larger barrier thickness in comparison to the well thickness. The sharp peak structure at the onset of the barrier absorption is due to excitons.

It is expected that the top contact in sample A is nearly completely transparent up to about 2.5 eV, corresponding to the direct bandgap of $Al_{0.7}Ga_{0.3}As$; and the indirect minima, although lower than 2.5 eV, should have little absorption for a thickness of 405 nm. For sample B, the effective bandgap is determined by the superlattice minibands in the valence and conduction bands. A simple calculation (including band nonparabolicity) gave an effective bandgap of 2.3 eV. For the conduction band, we used the direct bandgap profile with a barrier height of 1 eV. This probably leads to an overestimate of the conduction miniband, predicting a higher energy as also evidenced in other work [48]. The drop in response at about 2.2 eV for Sample B is attributed to the direct interminiband transition in the superlattice.

For the GaAs/AlGaAs devices, responsivity (Fig. 8.18) in the MIR was measured at 8.3 μm. Results for both devices show a maximum value of about 0.75 A W^{-1}. For the visible region, the responsivity at 0.63 μm is lower than that at 8.3 μm for the same voltage regions ($|V| < 6$ V). A rapid increase in responsivity at 0.63 μm is observed for both devices at high voltages ($|V| > 6$ V), reaching up to about 10 A/W.

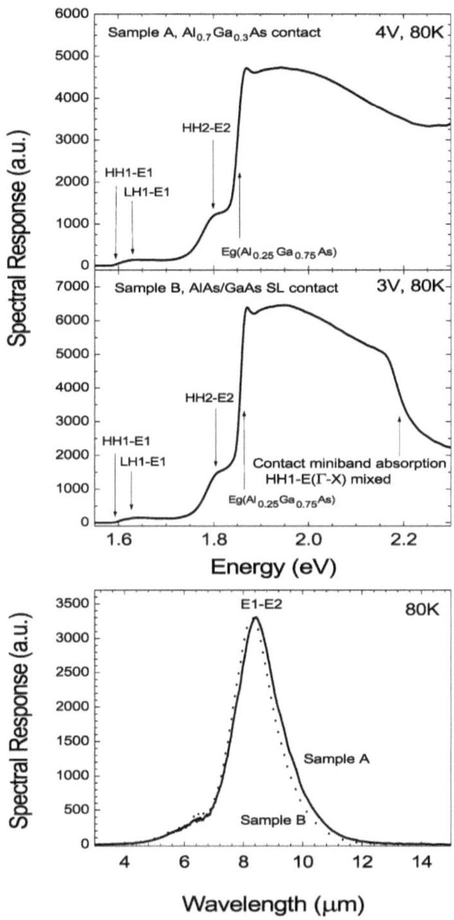

Fig. 8.17. Spectral response in the visible (*top*) and middle infrared (*bottom*) regions for Sample A and B. The visible light is shone through the top window of the mesa at normal incidence. The middle infrared is coupled into the device through the 45° edge facet

To understand the operation qualitatively, we model the detector as a photoconductor for both interband and intersubband excitations, since the device is unipolar (n^+–n–n^+). For the MIR detection, the device works as a standard photoconductive QWIP. For the NIR and visible region, photons excite electrons into the conduction band, leaving holes in the valence band. The holes, if not already in the wells, are trapped into the quantum wells on a subpicosecond time scale. Due to the heavy effective mass, the holes are assumed to be trapped in the well. If electrons are excited into the wells, their escape will contribute to photocurrent. If electrons are in the barriers, their flow under bias leads to photocurrent. The current will flow until the trapped holes recombine with electrons.

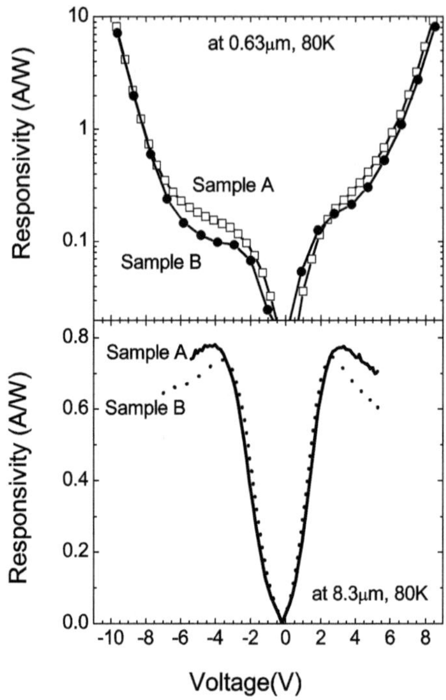

Fig. 8.18. GaAs/AlGaAs device responsivity at 0.63 μm for visible light detection region (*above*) and at the peak (8.3 μm) for middle infrared detection region (*below*) vs. applied bias voltage

For a photoconductor, the current responsivity is given by

$$\mathcal{R}_i = \frac{e}{h\nu}\eta g, \tag{8.2}$$

where ν is the photon frequency, η is the absorption quantum efficiency, and g is the photoconductive gain related to the excited carrier lifetime and transit time by

$$g = \frac{\tau}{\tau_{\text{trans,tot}}} p_{\text{e}}, \tag{8.3}$$

where p_{e} is the escape probability for a photoexcited electron from the well, and $\tau = \tau_{\text{c}}$ the capture time for the intersubband process, or $\tau = \tau_{\text{recomb}}$ the recombination time for the interband case. For a device where the excitation is to a bound state for both the intersubband and interband transitions, the factor p_{e} must be included. For weak absorption, which is valid for these structures, the absorption efficiency can be written as

$$\eta = \alpha L, \tag{8.4}$$

where α is the absorption constant and L is the device length in the growth direction. Substituting (8.3) and (8.4) into (8.2), one obtains:

$$\mathcal{R}_i = \frac{e}{h\nu}\alpha p_\mathrm{e}\tau v, \tag{8.5}$$

where we have used $v = L/\tau_{\mathrm{trans,tot}}$ for the drift velocity.

The electric field dependence of the various quantities in (8.5) determines the characteristics shown in Figs. 8.16 and 8.18. In the InGaAs/InP case, for the MIR response where the 6.6 nm well QWIP uses a bound-to-bound intersubband transition, we observe a substantial delay in the turn-on of the photoresponse as a function of voltage [142]. That is, the escape probability p_e is small at low field and gradually turns on due to field assisted electron tunneling escape. The rise in response thereafter follows the increase in the drift velocity v. One would expect a saturation of the responsivity when v saturates, but this is not observed in these devices up to the maximum voltage (which is limited by device heating). In fact, the value of the QWIP responsivity here is substantially larger than devices of the GaAs/AlGaAs system. The high responsivity is partly due to an intraband impact ionization process as discussed in previous chapter. For the NIR response, the situation is similar since the excitation is also due to the confined states in the well. For the GaAs/AlGaAs devices, the MIR response vs. voltage is typical of a standard QWIP.

As the QWIP IR performance is generally well understood, the key question is whether such a NIR or VIS detector *can* be made highly sensitive. Since the noise current is common to both detection processes, in order to have a higher detectivity for the NIR or VIS region, one needs to have a higher responsivity. For reference, a good NIR/VIS detector is about 100 times better than the background limited 9 μm detector in terms of detectivity [34].

Let us make an estimate using typical values and (8.5). The interband absorption constant is about 10 times larger than that for the intersubband transition. The photon energy is about a factor of 10 larger for the NIR/VIS than the MIR. These two factors approximately cancel when comparing the responsivities. The escape probability p_e and drift velocity v are common for interband and intersubband cases. The recombination time for the interband process is roughly 500 ps, and is therefore a factor of about 100 larger than the capture time in the intersubband process. So, this estimate shows that the responsivity for the interband photoconductor could be 100 times higher than that for the QWIP. This is very encouraging – implying that in the best case one could obtain the desired enhancement value of 100.

For the InGaAs/InP QWIP, looking at the absolute values in Fig. 8.16, the observed NIR responsivity is quite high, comparable to a commercial NIR detector, for example, a Ge one. However, the desired 100-fold projected responsivity enhancement is not seen. The GaAs/AlGaAs devices show a lower interband responsivity (Fig. 8.18) for the low voltage regions ($|V| < 6$ V), most likely due to a short hole recombination time with the abundant electrons in the well. Larger values are only seen at high voltages ($|V| > 6$ V), which could be an indication of the expected long interband recombination time.

To achieve a high interband responsivity, it may help to employ different quantum wells in order to obtain a long interband recombination time. One possible way is to spatially separate the minority holes and the electrons by using an appropriately designed double-well structure. However the first experimental trial to achieve this goal was not successful [283].

Finally, we comment on how the dual-band detector can be implemented as a 2D imaging device or a focal plane array. In the standard configuration (see Chap. 9), the light is incident from the backside. For a QWIP array, the substrate is usually completely removed after the GaAs or InP chip is hybridized with the readout circuit chip. For the fabrication of a dual-band detector into a hybrid array, both the top and bottom contacts should be wide bandgap materials to eliminate absorption.

8.3.2 Stacked QWIP and PIN

Another straightforward way to realize a dual-band device is by stacking a QWIP structure and a photodiode. The proof-of-the-concept device was arranged in a p–i–n–i–n configuration [284]. High-performance IR detection *at both detection wavelengths* was realized because of the highly nonlinear dark current characteristics of the photodiode.

The dual-band detector operation is schematically indicated in Fig. 8.19. The detector was designed such that the photodiode is sensitive at short wavelength and the QWIP detects at long wavelength. Under a sufficiently high forward bias (Fig. 8.19a), the differential resistance of the p-i-n photodiode is negligible as compared to that of the QWIP structure, so that both the pho-

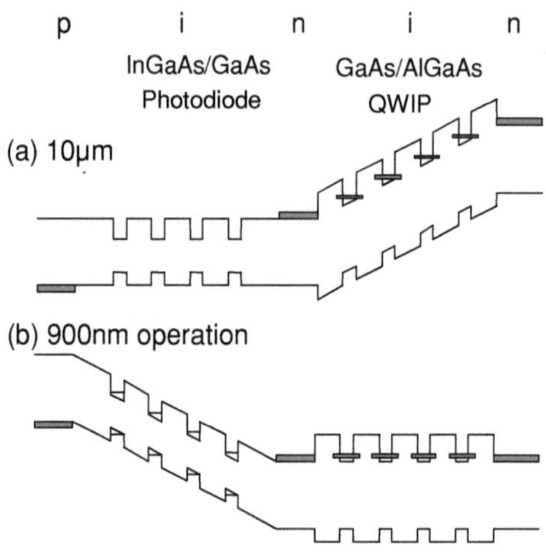

Fig. 8.19. Conduction and valence band distribution of a p–i–n–i–n dual-band detector consisting of a QWIP and a photodiode. **(a)** and **(b)** indicate the detection modes in which the operation is controlled by the QWIP and by the photodiode, respectively

toresponse and the differential resistance are controlled by the QWIP. Under reverse bias (Fig. 8.19b), the photodiode resistance is much larger than the QWIP resistance, and consequently, the detector response is determined by the photodiode. The series resistance of the QWIP only plays a role if the short-wavelength power is high and the reverse bias is simultaneously too small.

To demonstrate the detection scheme of Fig. 8.19 experimentally, a device containing a GaAs/AlGaAs QWIP and a photodiode with InGaAs/GaAs quantum wells in the intrinsic region was fabricated. The InGaAs wells were introduced in order to make the photodiode compatible with an illumination from the substrate side. The detector structure consists (from top to bottom) of a 500-nm p-type GaAs ($1 \times 10^{19}\,\mathrm{cm}^{-3}$ Be) contact layer, a 20-period multiple quantum well with 5-nm $In_{0.08}Ga_{0.92}As$ wells and 20-nm GaAs barriers, 200-nm n-type GaAs ($1 \times 10^{18}\,\mathrm{cm}^{-3}$ Si), a 20 period QWIP structure containing 4-nm GaAs wells ($1 \times 10^{18}\mathrm{cm}^{-3}$ Si) separated by 50-nm $Al_{0.24}Ga_{0.76}As$ barriers, and a 1 µm thick n-type GaAs contact layer ($1 \times 10^{18}\,\mathrm{cm}^{-3}$ Si). The device was grown by MBE on semi-insulating (100)-oriented GaAs substrates and processed into mesa diodes of $0.04\,\mathrm{mm}^2$ area by optical lithography and wet chemical etching.

Figure 8.20 shows the responsivity spectra for the dual-band detector at 77 K. The QWIP responsivity, in the standard 45° facet geometry, has a maximum of about $0.5\,\mathrm{A\,W}^{-1}$ at 2 V bias and at a photon energy of 153 meV

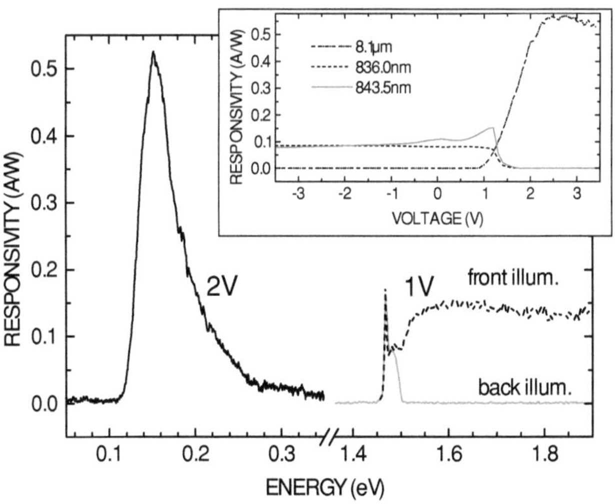

Fig. 8.20. Responsivity under illumination from the GaAs substrate vs. photon energy at different applied voltages as indicated (*full lines*). The *dashed line* indicates the shape of the photocurrent spectrum under front illumination. The *inset* shows the responsivity vs. voltage under monochromatic illumination at 0.153 eV (8.1 µm), 1.470 eV (843.5 nm), and 1.483 eV (836 nm)

(8.1 μm). This responsivity value is essentially the same as observed in conventional QWIP structures [15].

At 1 V forward bias, the responsivity of the photodiode shows a pronounced exciton peak at 1.47 eV, giving rise to a peak responsivity of 0.18 A W^{-1}. This value agrees reasonably well with the responsivities of similar InGaAs/GaAs photodiodes studied in a previous work [285]. Under back illumination, the spectrum shows a cut-off at around 1.5 eV because of the absorption of the 0.5 mm thick GaAs substrate. The spectral shape of the spectrum (obtained by front-illumination of a different mesa with a ring-shaped contact) is also shown for comparison. Here the response at energies above 1.5 eV is drastically improved due to the thinner absorbing layer. The exciton peak disappears under reverse bias, due to field dissociation.

The voltage dependence of the photoresponse is plotted in the inset of Fig. 8.20. The responsivity at 153 meV, representative for the QWIP, sets in at about 1 V forward bias. It shows the usual voltage dependence as in conventional QWIP structures [15], except for the 1 V threshold. The responsivity disappears completely under reverse bias. The threshold behavior is a consequence of the built-in voltage of the photodiode. The responsivity of the photodiode (at 1.483 and 1.470 eV) is almost constant at voltages below 1 V. In this regime, the photocurrent gain is equal to one, indicating complete collection of the photoexcited carriers at the contacts. The residual voltage dependence of the responsivity is caused by electroabsorption effects. The NIR photocurrent disappears completely above 1.5 V forward bias, as that is larger than the built-in voltage of the photodiode.

The observed bias dependence of the photoresponse is consistent with the dark current of the device, which rapidly drops off to less than 1 pA below 1 V, indicating the large resistance of the p–n junction under reverse bias [284]. Above 1.5 V, the dark current shows the typical voltage dependence of a QWIP structure, which is again offset by the built-in voltage of the photodiode. From the responsivity data, we calculate a detectivity of about $4 \times 10^9\,\mathrm{cm}\sqrt{\mathrm{Hz}}\,\mathrm{W}^{-1}$, for the peak wavelength of 8.1 μm, an applied voltage of 2 V, and a temperature of 77 K. The detectivity of the photodiode is already limited by the photocurrent noise at weak illumination levels.

For focal-plane-array cameras hybridized with Si readout electronics, further optimization of this type of detector is straightforward: Diffraction gratings for the QWIP structure can be fabricated in the usual way. Substrate thinning will improve the responsivity of the QWIP [15] and broaden the spectral window of the photodiode (similar to the dashed line in Fig. 8.20). The responsivity of the photodiode can be raised further by introducing a larger number of InGaAs wells and by lowering the effective bandgap.

This dual-band detector represents a detection scheme which is also promising for different material systems. In particular, the combination of a 10-μm QWIP with a photodiode operating in the 3–5 μm window would be interesting for temperature mapping. Such a combination can be realized in the InAs/GaSb/AlSb material system, e.g., by using GaSb/AlGaSb for the

QWIP and bulk InAs or an InAs/GaSb superlattice for the intrinsic region of the photodiode. The only restriction for making the concept work is that the photodiode should be designed for the higher of the two detection energies.

We would like to point out that this approach is also suitable for balanced dual-band detection. Looking at the responsivity data of Fig. 8.20, we see that the dual-band detector is sensitive to both bands at around 1.3 V. Since the photocurrents associated with the two bands have different signs, the sign of the total photocurrent depends on the ratio of the relative illumination intensities. In the case of a dual-band detector operating at 10 and 3–5 μm (or at two different wavelengths within a single atmospheric window), this type of balanced detection can be used for temperature discrimination. As the sign of the total photocurrent is determined by the photon flux ratio associated with the two colors, and this flux ratio is given by the target temperature, irrespective of the particular field of view. It is even possible to select the threshold temperature, by adjusting the bias voltage.

Finally, we note that a similar detector behavior is expected when using a p–i–n–i–p (or n–i–p–i–n) structure containing two photodiodes with different bandgaps. Such a device has been realized in the HgCdTe materials system [286]. Another approach involving integrating a HBT (heterojunction bipolar transistor) with a QWIP is also being explored [287], where the HBT is used with its base floating; i.e., a photoconductor with a large gain or a phototransistor.

8.4 Integrated QWIP–LED

To utilize the flexibility of quantum wells, the integration of QWIPs with light emitting diodes (LEDs) has been investigated. The integrated QWIP–LED device is intended for thermal imaging applications requiring large-area devices in the MIR and FIR region for wavelengths longer than 3 μm.

The basic idea [288] of an integrated QWIP–LED is to epitaxially grow a QWIP with an LED on top. The QWIP can be either n- or p-type. Figure 8.21 shows a GaAs/AlGaAs n-QWIP with an InGaAs/GaAs quantum well LED. Only two contacts are made – to the heavily p-doped LED contact layer and the heavily n-doped QWIP emitter. A forward bias is applied to this serial QWIP – LED, which should be large enough to turn both the QWIP and the LED to their operating bias conditions. For concreteness, we assume that the QWIP detects M&FIR light of wavelength greater than 3 μm and the LED emits in the NIR region at wavelengths 800–1,000 nm. The QWIP is a photoconductor, so that under MIR and FIR light illumination its resistance decreases, which leads to an increase in the voltage drop across the LED and therefore an increase in the amount of NIR emission. This device is therefore an M&FIR to NIR converter. For 77 K operation, a well-designed QWIP can be quite efficient easily with higher than 10% absorption, and the LED technology is well developed. An optimized QWIP–LED therefore can be

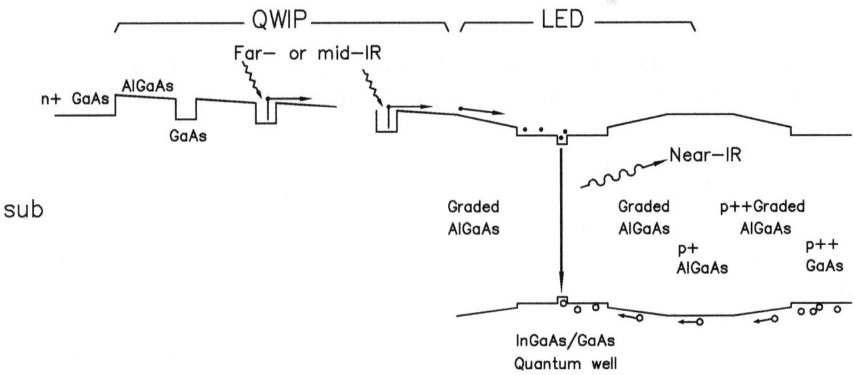

Fig. 8.21. Bandedge profile of the integrated QWIP with LED. For the QWIP part, only the conduction bandedge is shown. A forward bias is applied so that both the QWIP and the LED are in operating conditions. Photocurrent electrons from the QWIP recombine with injected holes in LED, giving rise to an increase in LED emission

very efficient with little or no loss of performance compared with the QWIP used alone as a MIR and FIR detector. The advantage of this integrated QWIP–LED is technologically important. In this scheme, one can make large format 2D imaging devices *without* the need of making any circuits on the device chip, and *without* the need of hybrid bonding with another readout chip (normally a Si IC). The resulting NIR emission can be easily imaged using the well-developed Si-CCD which has a spectral response covering the NIR wavelength.

The first experimental demonstration of a QWIP – LED was made by Liu et al. [288] The concept was independently proposed by Ryzhii et al. [289] The first device used a standard 50-well GaAs/AlGaAs QWIP peaked at 9 µm and a 7.0 nm $In_{0.2}Ga_{0.8}As$ quantum well LED emitting at 927 nm for 77 K operation. InGaAs was used as the well material so that the emitted light was not absorbed in any of the other layers. The design of the LED for this first test of the concept was borrowed from that of a state-of-the-art quantum well laser. The LED emission power increased linearly with bias current, and the measured external efficiency was about 1%, limited entirely by the device geometry. This implied that the LED *internal* efficiency was 100%, within the experimental uncertainty.

An intriguing possibility for the QWIP–LED is to fabricate a continuous image conversion device – a pixelless large-area imaging device [290]. Since the entire active layer of a QWIP–LED in the growth direction is very thin (normally less than 4 µm) the up-conversion process (i.e., photoexcitation in the QWIP, carrier transport to the LED, and radiative recombination in the LED) should have little in-plane spreading. Theoretical analyses [291,292] show that the intrinsic spreading and crosstalk are negligible for a practical QWIP with a large number of wells ($N > 20$). As an order-of-magnitude estimate, the in-plane spreading in the QWIP is limited by the diffusion of excited carriers within their lifetime; and that in the LED is related to the bipolar diffusion and the radiative interband recombination lifetime. Taking some typical numbers (10 and 1 cm^2 s^{-1} for the diffusion constants, and 5 ps and 1 ns for the lifetimes, in the QWIP and LED, respectively) we get spreading lengths (given by the square root of the product of diffusion constant and lifetime) of 0.1 and 0.3 μm, respectively. These values both are negligible since they are much smaller than the wavelength of the IR to be imaged.

The initial demonstration of the pixelless QWIP–LED used a p-type QWIP because of its Simplicity, by avoiding the fabrication of a grating [290,293]. The first pixelless device, although clearly demonstrating the concept, was only able to image a hot object of about 1,000 K. Thereafter, efforts were concentrated on n-type QWIPs and steady improvements were made [191, 294,295].

Apart from improving both the QWIP and the LED efficiencies, a challenge for achieving a pixelless QWIP–LED imaging device – with negligible distortion, smearing, and crosstalk – is related to extrinsic optical effects. One must ensure that incident IR light and the emitted NIR light both have negligible crosstalk and smearing. Removing the entire substrate helps to minimize the optical crosstalk. One way to improve the LED external efficiency is by the use of the photon recycling effect [296,297]. There is however a tradeoff of LED efficiency and smearing if the photon recycling effect is used. Another major issue relates to the material quality: the pixelless device is more demanding on materials. Any leakage point in the QWIP would cause a "hot spot" in the LED emission, and any leakage point through the entire structure would render the whole device useless. Femtosecond laser micromachining [298] has been used to remove these hot spots. Finally, a CCD with sufficient full well capacity is required, since in thermal imaging one always needs to subtract the background. The latest imaging result [191] is shown in Fig. 8.22. Finally, two of the envisioned final camera system layouts are sketched in Fig. 8.23.

There are other potential advantages of this up-conversion approach. These include (a) the ease of implementing multicolor imaging devices in a pixelless geometry [271], (b) the low cooling power requirement since the readout chip (CCD) does not need to be cooled, and (c) the simplicity of fabricating ultralarge-area imaging devices. The concept of up-conversion may be expanded to cover the other technologically important regions, e.g., 3–5 μm [299] and 1.5 μm [211,300–302].

8.5 Quantum Dot Infrared Photodetector

Quantum dot infrared photodetectors (QDIP) have many similarities to QWIPs, and have attracted a lot of interests [303–317]. An *ideal* QDIP is expected to be substantially superior to a QWIP. The area of QDIP research has therefore been very active in recent years. This section discusses key is-

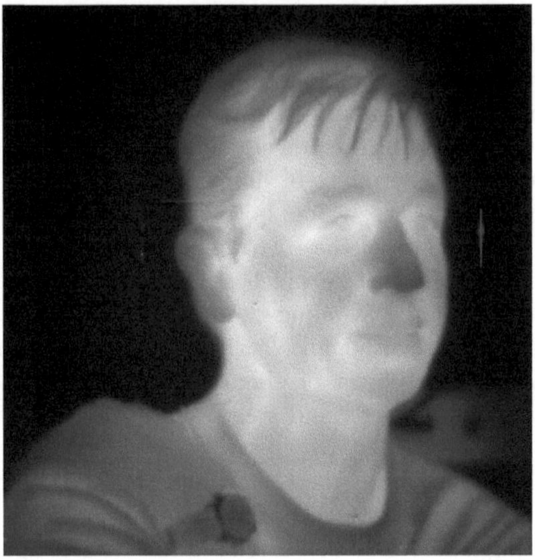

Fig. 8.22. Thermal image of a person (See Dupont et al. [191] for details)

sues related to QDIPs and the potential advantages of the QDIP over the QWIP–the normal incidence response, the dark current, and the responsivity and detectivity. We attempt to address the following questions of what is the

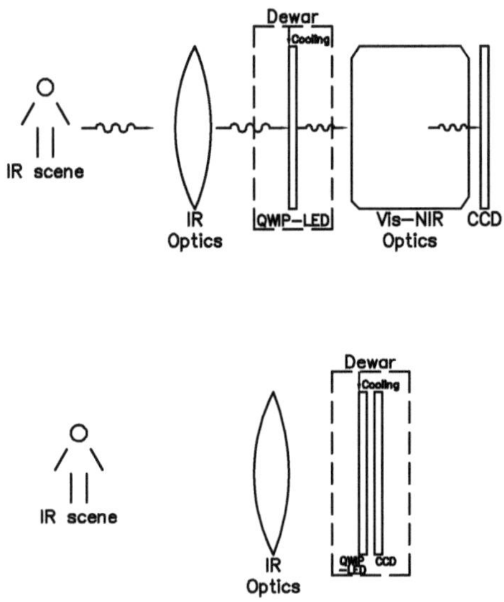

Fig. 8.23. Possible final imaging system configurations. (*Top*): The IR scene is mapped onto the transmissive QWIP–LED device by the IR optics, the device is mounted in the dewar, the output emission image is mapped onto the CCD by visible or near-IR (Vis–NIR) optics. The Vis–NIR optics could be a lens or a fiber bundle. (*Bottom*) The CCD is mounted in the dewar in proximity with QWIP–LED

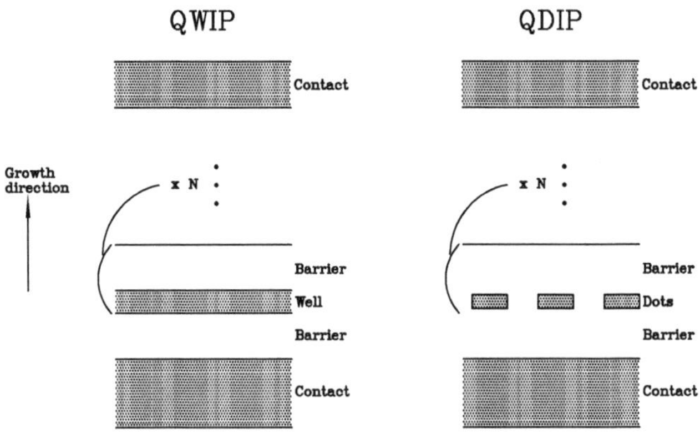

Fig. 8.24. Schematic layers of a QWIP (*left*) and QDIP (*right*). In the QDIP case, the dot cross-section is shown as rectangular, approximating the shape of dots grown under some growth conditions. The wetting layer is neglected

QDIP's potential, what is lacking, and what is needed to make a QDIP viable in practical applications. It is argued that the present QDIPs have not fully demonstrated their potential advantages. Representative experimental results are compared with typical QWIP characteristics, and areas that need improvements are pointed out.

QDIPs are similar to QWIPs, with the quantum wells replaced by quantum dots. The quantum dots discussed have size confinement in all spatial directions. A schematic of the layered structures of a QWIP and an (ideal) QDIP is shown in Fig. 8.24. The most widely studied QDIPs are made of self-assembled InAs dots on GaAs substrates. For these dots there is commonly a thin wetting layer of InAs; however, in the discussion here we neglect any explicit effects of the wetting layer. The detection mechanism in both QWIPs and QDIPs relies on the intraband photoexcitation of electrons from confined states in the conduction band wells or dots into the continuum. If one draws the (bandedge) potential profile along the growth direction, QWIPs and QDIPs would have a similar shape. If the dots were aligned in the growth direction, the potential profile would be exactly the same. However, because the barriers are usually wide to suppress dark current, dots are not correlated between layers. All discussions and estimates here are independent of the position correlation among the dots. We also assume that barriers between any adjacent dots are sufficiently wide so that tunneling can be neglected.

8.5.1 Anticipated Advantages and Current Status

One of the major selling points is that "QDIPs allow normal incidence." That is, the incident light normal to the wafer along the growth direction is expected to cause intraband absorption, unlike the standard n-type QWIPs. The normal incidence property is advantageous because it avoids the need of fabricating a grating coupler as for the standard QWIP imaging arrays. The grating coupler not only adds at least one extra fabrication step but also causes difficulties in realizing a wide and multiple wavelength coverage and in fabricating a short wavelength coupler because of the required small grating features. Indeed, normal incidence response in QDIPs has been reported in several publications [303,305,307,310,313]. Most publications, however, do not show a polarization dependence of the photocurrent spectra, and some [306,308,311] show dominant P-polarized response in the 45° facet geometry. This is very similar to QWIPs measured in the same geometry under flood illumination. In one publication [318] on absorption measurements, clear evidence of absorption features due to in-plane confined quantum dot levels was reported. It seems that a *dominant* normal incident response in present QDIPs has not been achieved.

This is also the conclusion of our experiments on a number of QDIPs where the dominant response comes from light polarized in the growth direction

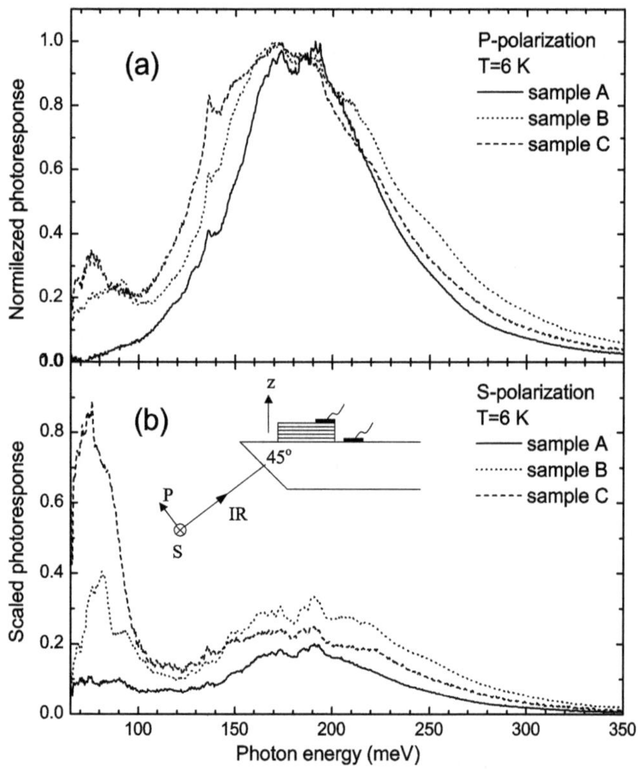

Fig. 8.25. P- and S-polarized spectral response curves in the 45° facet detector geometry. The QDIPs have 50 layers of InAs dots separated by 30-nm GaAs barriers. The dot density is about $5 \times 10^9\ \mathrm{cm}^{-2}$. The number of electrons is estimated to be 2, 5, and 7 per dot for Sample A, B, and C, respectively, due to the delta-modulation Si doping in the barriers

[314,316]. Figure 8.25 shows spectra of three of our samples under both P and S-polarized light in the 45° facet geometry. Clearly the P-polarized response is much stronger than that for S.

The problem is believed to be due to the fact that the self-assembled quantum dots grown so far for QDIPs are wide in the in-plane direction (~20 nm) and narrow in the growth direction (~3 nm). The strong confinement is therefore in the growth direction, while the in-plane confinement is weak, resulting in several levels in the dots. The transitions between in-plane confined levels give rise to the normal incidence response. From the ground state, the transition oscillator strength is reduced for higher final states. In other words, the transitions within the dots (which do not result in a detection photocurrent because the excited electrons cannot escape) exhaust most of the in-plane oscillator strength. In contrast, in the growth direction, the high oscillator strength transition is the one to the continuum, resulting in the dominant photocurrent. This point is illustrated in Fig. 8.26. The strong confinement in the growth direction is represented by a narrow well; whereas the in-plane weak confinement leads to several states. For conceptual simplicity, the confinement potentials are represented by one-dimensional wells separately in z and x–y directions.

Another potential advantage of QDIPs over QWIPs is that "QDIPs have lower dark currents" [319]. Since dark current causes noise, a lower dark current leads to a higher detector sensitivity. The simplest way to estimate dark current is by counting the mobile carrier density in the barrier. The current is then given by multiplying by the carrier velocity (see Sect. 4.1). We use the

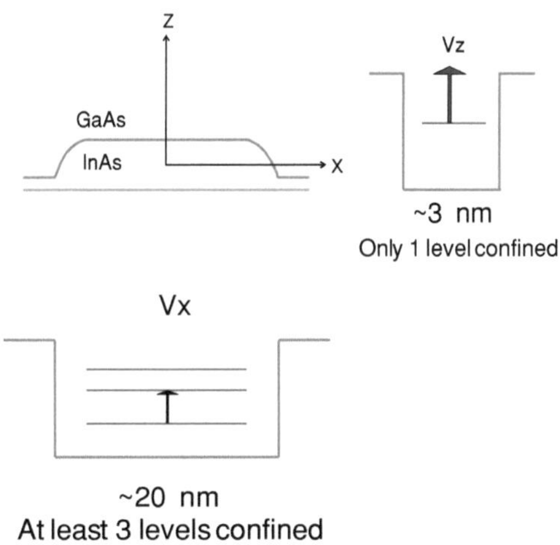

Fig. 8.26. Illustration of transitions under polarized light in the growth direction (z) or in the in-plane directions (x or y). The strong confinement in the growth direction is represented by a narrow well; whereas the in-plane wide potential well leads to several states. The upward arrows indicate the strongest transitions for z and x polarized lights

following expression:

$$j_{\mathrm{dark}} = evn_{\mathrm{3D}}, \tag{8.6}$$

where v is the drift velocity and n_{3D} is the 3D density, both for electrons in the barrier. Equation (8.6) neglects the diffusion contribution. The electron density can be estimated by

$$n_{\mathrm{3D}} = 2(m_{\mathrm{b}}k_{\mathrm{B}}T/2\pi\hbar^2)^{3/2}\exp(-E_{\mathrm{a}}/k_{\mathrm{B}}T), \tag{8.7}$$

where m_{b} is the barrier effective mass and E_{a} is the thermal activation energy which is the energy difference between the top of the barrier and the Fermi level in the well or dot. We have assumed that $E_{\mathrm{a}}/k_{\mathrm{B}}T \gg 1$, appropriate for most practical cases. For similar barriers in a QWIP or a QDIP (i.e., where v and m_{b} are comparable), the difference in E_{a} gives rise to a difference in dark current. If we neglect the field induced barrier lowering effect in E_{a} which makes the estimation valid for low applied fields (but not so low that diffusion must be considered), the activation energy is related to the detection cut-off wavelength (λ_{c}) by

$$E_{\mathrm{a}}^{\mathrm{QWIP}} = hc/\lambda_{\mathrm{c}} - E_{\mathrm{f}}, \tag{8.8}$$

for a QWIP with a bound-to-continuum detection scheme, where E_{f} is the Fermi level in the well, and for a QDIP

$$E_{\mathrm{a}}^{\mathrm{QDIP}} = hc/\lambda_{\mathrm{c}}. \tag{8.9}$$

The term E_{f} in (8.8) is due to the subband nature of QWIP quantum wells: the intersubband transition is from all electrons in a subband at the same energy whereas the thermal activation energy is from the top of the Fermi sea. Given that the optimal design for a QWIP is having $E_{\mathrm{f}} = 2k_{\mathrm{B}}T$ for maximizing detectivity or $E_{\mathrm{f}} = k_{\mathrm{B}}T$ for maximizing operating temperature [19] (as in Sect. 4.3), the reduction in dark current in QDIPs vs. QWIPs for the same cut-off wavelength and barrier material is only in the range of a factor of about 3–7, in the ideal case.

The devices tested so far are far from ideal, and have shown much higher dark currents than the ideal estimate. Figure 8.27 shows a comparison of current–voltage characteristics at 77 K. For a "fair" comparison, the chosen QWIP has the same cut off wavelength of 8.6 μm as one of the QDIPs. Even the shorter wavelength QDIP ($\lambda_c = 8.1$ μm) with lower electron occupation has a substantially higher dark current than that for the QWIP. Effects such as ionized dopant induced potential fluctuations and strain induced distortion of the potential well could be the cause of the excess dark current [320].

The final advantage relates to the potentially long excited electron lifetime or capture time (τ_{c}). It has been anticipated [321] that the relaxation of electrons is substantially slowed when the interlevel spacing is larger than the phonon energy a "phonon bottleneck." This effect has been investigated extensively [322], and the topic is still under debate and controversial. If the phonon

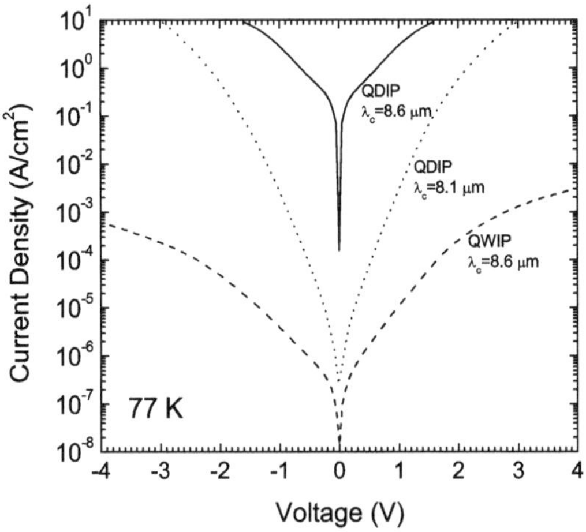

Fig. 8.27. Dark current characteristics for a QWIP and QDIPs with the same cutoff wavelength. The QWIP has 32 GaAs wells center delta-doped with Si to 5×10^{11} cm^{-2} and undoped AlGaAs barriers, with a total device thickness of 1.58 μm. The QDIPs are similar to those in Fig. 8.25

bottleneck can be fully implemented in a QDIP, the long excited electron lifetime directly leads to a higher responsivity, higher operating temperature, and higher dark current limited detectivity. The reason is very simple, since a photoconductor responsivity is given by

$$\mathcal{R} = \frac{e}{h\nu}\eta g, \tag{8.10}$$

where ν is the photon frequency, η is the absorption efficiency, and g is the photoconductive gain:

$$g = \tau_c/\tau_{\text{trans,tot}}, \tag{8.11}$$

with $\tau_{\text{trans,tot}}$ the transit time across the device. A long capture time thus directly translates into a large $\mathcal{R}$. However, high gain does not automatically imply high operating temperature or high detectivity since it also leads usually to a large dark current (see the discussion after (4.6)). These advantages will only come into play if the scattering rate is really reduced due to some effects, for example the phonon bottleneck.

Ryzhii et al. [323] analyzed the detectivity and made comparisons between QWIPs and QDIPs having the same ground state ionization energy (not the same cut-off wavelength). Their main conclusions are similar to those reached here.

Experimentally, the situation is encouraging in view of the high measured responsivity values [310,313,316], often comparable to or even larger than those for QWIPs, despite the small absorption efficiency which is very often not directly measurable; this is an indirect evidence of the long lifetime.

8.5.2 Areas for Improvement

To realize a strong and dominant normal incident response, first and foremost one should make the dots small so that the in-plane confinement leads to only one or two bound states. For two bound states, the second state should be very close to the top of the barrier: this will allow strong and dominant normal incidence absorption. If a broader response spectrum is desired, one could have two or more states in the dots, all occupied with electrons; but no unoccupied states which are deep in the dot potential should exist.

To have a good detector, the absorption efficiency must be high, and this requires a high dot density. To have a comparable absorption as in QWIPs, the electron density per layer of dots should be in the range of $2–10\times10^{11}\,\mathrm{cm}^{-2}$. If there are two electrons occupying every dot, the dot density should be in the range of $1–5\times10^{11}\,\mathrm{cm}^{-2}$, and six electrons per dot, this reduces to $3–17\times10^{10}\,\mathrm{cm}^{-2}$. The dot densities commonly achieved are in the range of $0.01–1\times10^{10}\,\mathrm{cm}^{-2}$, so some improvement is therefore needed. The desired high dot density necessarily requires small dot size. For example, in the extreme case, for a dot density of $5\times10^{11}\,\mathrm{cm}^{-2}$, the dot size must be smaller than 14 nm in diameter.

To populate dots with electrons one needs doping. In a QWIP, this can be simply done by directly doping the wells. Since the doping density is high and degenerate, the effect of random dopant distribution is minor and is expected to lead only to a broadening in the absorption linewidth. In a QDIP, however, if the doping is done in the same layer as the dots, some dopants would reside in the wetting layer and each dot would have a random number of dopants. This random distribution could lead to a significant potential fluctuation. Moreover, if the doping is done in the barriers (modulation doping), the random distribution of the ionized dopants could lead to a leakage current path. Similarly, ionized dopants in the wetting layers could also lead to a leakage path. Detailed modeling and doping control are needed to fully account for the effect of doping and to realize the lower dark current.

There are alternative QDIP designs to further enhance the photoconductive gain such as those in [324–326] involving in-plane transport. These are interesting directions (not available to QWIPs) that explore the uniqueness of quantum dots.

To close this section, we are still far from realizing all the projected advantages, but QDIPs have very attractive potential if the growth/fabrication technology delivers the design requirements. The key areas for improvement are: producing smaller and denser dots, and having better doping controls.

8.6 Single Well and Blocked Miniband QWIPs

A QWIP with only a single quantum well is interesting because it is the simplest structure, and therefore relatively easy to accurately model and simulate.

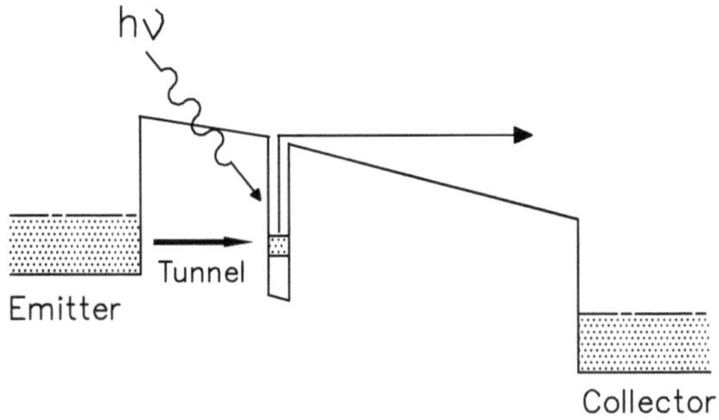

Fig. 8.28. Schematic bandedge profile of a single well QWIP with tunneling refill of the well

In the case of narrow barriers, so that the current in and out of the quantum well is by tunneling, the structure is the well-known resonant tunneling diode – RTD (see, e.g., [327] for a review). Even before the first experimental demonstration of QWIPs, the single well structure was proposed and analyzed for high speed detection [9]. A schematic of a single well QWIP with tunneling injection is shown in Fig. 8.28. Studies of single well QWIPs have been carried out since the early years of research in this field. Intersubband photocurrents were measured and detailed self-consistent calculations were performed [133,134,328,329]. ISBT induced switching was demonstrated in a suitably biased RTD [330]. Single well QWIPs provide a model system to study the noise mechanism [331–333]. Single well QWIPs have been further explored for high speed detection [334–336] – with a rapid refilling of the well by tunneling [9], they are thought to offer the ultimate intrinsic speed.

Obviously, the absorption efficiency in a single well QWIP is limited. As an improvement, the single well is replaced by a superlattice [166,337] schematically shown in Fig. 8.29. This "blocked miniband" QWIP relies on the intersubband transition between superlattice minibands [64]. This approach increases the absorption, however, only by a limited amount since only the region immediately next to the blocking barrier (within a mean free path) is effective. The low absorption is the main problem with these single stage QWIPs. However, an interesting new development in strong coupling to an ISBT in waveguides (see Sect. 6.3) offers the possibility of achieving high absorption

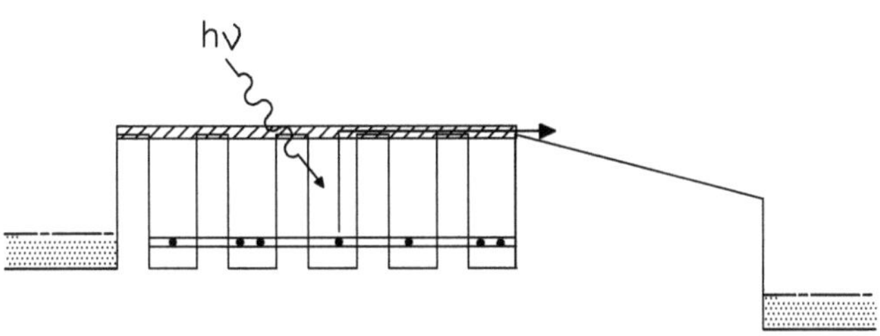

Fig. 8.29. Schematic bandedge profile of a superlattice blocked miniband QWIP

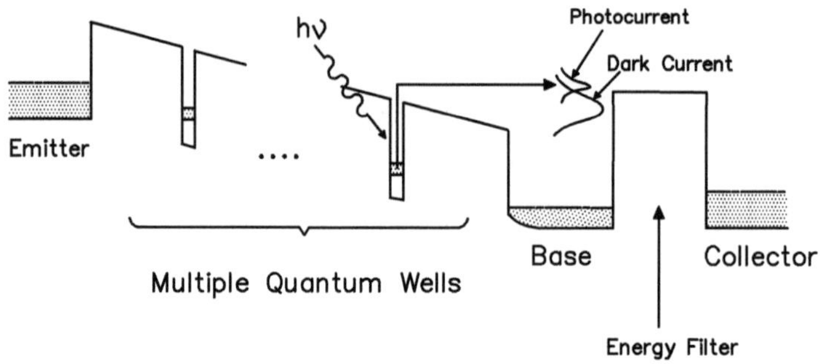

Fig. 8.30. The operation principal of IHET. For an in-depth discussion, see Cho [27]

with one or only a few quantum wells. Another drawback is the low responsivity due to the absence of photoconductive gain. Single well and blocked miniband QWIPs are also interesting for low background and low temperature applications [337,338] since they do not have the problem of slow dielectric relaxation like behavior associated with the recharging of the quantum wells (see discussions in Sect. 10.1).

8.7 Transistors and Monolithic Integration

The IR hot electron transistor (IHET) was invented by Choi and its physics was discussed in detail [27]. IHETs are well suited for the study of various relevant physical processes, such as carrier relaxation and hot electron transport. In addition, they have been shown to improve the QWIP sensitivity by filtering out a fraction of the dark current and yet having most of the photocurrent reach the collector [27,339]. Figure 8.30 shows a schematic of the IHET operation.

Potentially, GaAs based QWIPs can be monolithically integrated with GaAs circuits. Unfortunately, very limited work has been undertaken. A concept based on the integration of a QWIP with a high electron mobility transistor (HEMT) was demonstrated [340]. Various other transistor ideas have been proposed to either improve the QWIP performance (e.g., by providing high gain [341]) or to achieve monolithic integration (e.g., by integrating with an heterojunction bipolar transistor (HBT) [342]). Clearly, a monolithically integrated focal plane array (FPA) is conceptually simple and extremely attractive. However, the GaAs technology is not developed in the area of readout circuits. Substantial development work needs to be done in order to establish the technology for the fabrication of a monolithic FPA based on QWIPs. QWIP FPAs are made so far with the traditional indium bump hybrid technology as discussed in Chap. 9.

Part II

Applications

9

Thermal Imaging

This chapter starts with some general concepts of thermal imaging involving 2D detector arrays, with specific emphasis to QWIPs. Subsequently, particular QWIP arrays, including their technology and performance, are described. They include arrays for both one and two spectral bands. Application fields in thermal imaging which QWIP arrays are particularly suited for are the topic of an additional section. Finally, the potential of QWIP technology for infrared detector arrays with novel functionality is discussed.

9.1 Signal, Noise, and Noise-Equivalent Temperature Difference

9.1.1 Signal Detection

A thermal-imaging camera relies on detecting the thermal background radiation emitted by warm objects. We therefore need to know the spectral power density $\mathrm{d}P_\nu$, incident on a detector element of area A through the aperture of a cold shield with f-number $F_\#$, that is associated with the thermal background at a temperature T_B. An expression for $\mathrm{d}P_\nu$ (2.6) has been given in Chap. 2.

Let us assume that the detector is sensitive between 8 and 9 µm. From (2.6), the signal power P_B within this spectral regime at $T = 300\,\mathrm{K}$ is derived as $P_\mathrm{B} = A/(4f^2+1)\cdot 29\,\mathrm{W\ m}^{-2}$. Values of P_B obtained for some typical pixel areas, f-numbers, and other system parameters are summarized in Table 9.1. For a given responsivity $\mathcal{R}$ according to (2.8), the signal current of the detector is readily calculated from the relation $I_S = \mathcal{R}P_\mathrm{B}$. In this chapter, we omit the subscript of g_photo.

In order to collect the signals from a detector array, each detector element has to be accessed electrically using a readout integrated circuit (ROIC), which is usually based on silicon CMOS technology. On a pixel level, the ROIC includes a circuit like the one shown in Fig. 9.1, with field-effect transistors

Table 9.1. Parameters for infrared detectors with $22 \times 22\,\mu m^2$ and $37 \times 37\,\mu m^2$ area, respectively, at 20 ms integration time, and f/2 optics

Parameter	Symbol	256 × 256	640 × 512
pitch		$40\,\mu m$	$24\,\mu m$
pixel area	A	$37 \times 37\,\mu m^2$	$22 \times 22\,\mu m^2$
filling factor		85%	84%
f-number	$F_{\#}$	2.0	2.0
spectral range		8 to 9 μm	8 to 9 μm
frame rate		50 Hz	50 Hz
integration time	τ_{int}	<20 ms	<20 ms
storage capacity	Q_c	4.5×10^7 e	8×10^6 e
incident power at 300 K	P_B	2.3 nW	0.82 nW
critical responsivity	$\mathcal{R}_c$		
300 K scene		$156\,mA\,W^{-1}$	$78\,mA\,W^{-1}$
318 K scene		$113\,mA\,W^{-1}$	$58\,mA\,W^{-1}$
Thermal resolution	NETD		
Ideal detector (PC, $\eta = 10\%$)		5.5 mK	8.9 mK
Ideal detector (PV, $\eta = 10\%$)		3.9 mK	6.3 mK
Readout-limited (PC, $g = 0.5$)		8.0 mK	19 mK

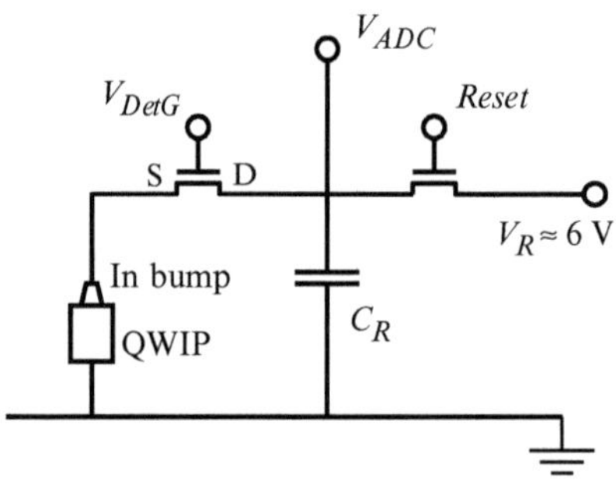

Fig. 9.1. Typical circuit used for operating a QWIP detector element in an ROIC

(FETs) for integration and reset. Here a measurement is initiated through a reset pulse, which charges the readout capacitor C_R to the voltage V_R. During the measurement cycle, C_R is discharged via the photocurrent (and dark current) of the detector element. A second FET is used to keep a constant voltage drop across the detector element during integration. After the integration time τ_{int} has passed, the voltage present at C_R is digitized. Additional electronics, not shown in Fig. 9.1, are used for individually addressing each pixel of the array and for an optional impedance match.

In a 2D array, the area available on the ROIC for each QWIP detector element is limited by the pitch of the array. This lack of area puts restrictions on the readout capacitor, with some tradeoff between its capacitance (on the order of 1 pF) and the breakdown voltage (of a few volts), which can be adjusted by the thickness of the dielectric between the capacitor plates, leading to typical storage capacities of the order of 10^7–10^8 electrons. The limited storage capacity usually imposes restrictions on the operating conditions, in particular on the integration time, in order not to saturate the readout cell. Even though this readout limitation might be circumvented by some electronic impedance transformation, the "direct charging" approach as in Fig. 9.1 is usually preferred since it is a low-noise circuit which is also simple enough to comply with the lack of space on the ROIC chip.

9.1.2 Detector Noise

Generation-Recombination Noise

In general, the most important noise contribution is generation-recombination (g-r) noise, caused by the fluctuation of the number of mobile carriers in the detector and by the statistical variation of their drift lengths before being re-captured. The g-r noise of a QWIP with N periods is given by (5.1)

$$i_{\mathrm{n,gr}}^2 = 4e\left(g - \frac{1}{2N}\right) I\Delta f = \frac{4e}{Np_{\mathrm{c}}}\left(1 - \frac{p_{\mathrm{c}}}{2}\right) I\Delta f. \tag{9.1}$$

Here $I = I_{\mathrm{th}} + I_{\mathrm{ph}}$ is the total current, which is composed of a thermally excited (or dark) component I_{th} and the photocurrent I_{ph}, and Δf the bandwidth of the noise measurement. In a photoconductive QWIP where $p_{\mathrm{c}} \ll 1$, (9.1) is thus reduced to the conventional [15,343] expression $4egI\Delta f$ (as in 2.12).

In the case of a low-noise QWIP, we have $p_{\mathrm{c}} \approx 1$, so the noise induced by statistical variations of the drift length ("recombination noise") is suppressed. Due to this deterministic capture, (9.1) transforms into $i_{\mathrm{n,gr}}^2 = 2egI\Delta f$, and the noise current is a factor of $\sqrt{2}$ smaller than in the previous limit, as already discussed in more detail in Sect. 5.2.3. It is this superior noise performance which is addressed by the term "low-noise QWIP," and which contributes to the excellent thermal resolution achievable with these detectors.

Johnson Noise

Johnson noise (also called Johnson–Nyquist noise) is associated with the differential conductivity $\mathrm{d}I/\mathrm{d}V$ of the detector, induced by random fluctuations of the thermal energy of the carriers. At the bias voltage V, Johnson noise gives rise to the noise current $i_{\mathrm{n,J}}^2 = 4k_{\mathrm{B}}T(\mathrm{d}I/\mathrm{d}V)\Delta f$.

Since g-r noise and Johnson noise are statistically independent, the total noise i_{n} of the detector can be expressed as $i_{\mathrm{n}}^2 = i_{\mathrm{n,gr}}^2 + i_{\mathrm{n,J}}^2$. Since Johnson noise is already present at thermal equilibrium, this noise source is dominant at sufficiently low bias and (simultaneously) weak illumination. Detailed investigations of the noise behavior of QWIP structures indicate that Johnson noise represents only a minor contribution to the total detector noise at bias voltages where the detectivity is maximized [178]. While Johnson noise has been taken into account in order to accurately determine the g-r noise and the photoconductive gain in an experiment, it can be neglected in the discussion of system properties.

9.1.3 System Noise

System noise becomes particularly relevant if the noise generated by the detector is low. Specific noise sources include the analog-to-digital converter (ADC), the input amplifier, the cooling machine (mostly a Stirling cooler), and stray electromagnetic fields. The latter two sources depend strongly on the specific layout of the system and can be suppressed sufficiently by proper construction. Good temperature regulation of the detector is crucial since the dark current of the sensor is strongly temperature dependent.

Optimum system performance also requires an appropriate ADC. Thermal imagers are mostly operated at about 50% of the full storage capacity in order to maintain a reasonable dynamic range, since the thermal background at around 9 µm increases by 100% if T_{B} is raised from 300 to 340 K. In the case of a 14 bit ADC, the least significant bit (LSB) thus corresponds to a relative accuracy of 1/8000, which is equivalent to $\Delta T = 7\,\mathrm{mK}$ at a signal wavelength of 8–9 µm and a 300 K background (see also (227)). Similarly, less sophisticated systems containing a 12 bit ADC are limited to a thermal resolution of 28 mK, which is already poorer than that of a state-of-the-art QWIP focal-plane array (FPA). For the same accuracy, somewhat better performance is achievable at MWIR wavelengths, with $\Delta T = 4\,\mathrm{mK}$ (16 mK) at $\lambda = 5\,\mu\mathrm{m}$ for a 14 bit (12 bit) ADC. Since ADC noise gives rise to an equivalent noise charge Q_{ADC} that does not depend on the charge collected in the readout capacitor, the resulting noise current can be expressed as $i_{\mathrm{n,ADC}} = 2Q_{\mathrm{ADC}}\Delta f$. Usually Q_{ADC} is identical to or slightly larger than the charge corresponding to the LSB.

Amplifier noise describes the degradation of the signal-to-noise ratio induced by amplification. Most conveniently, amplifier noise is expressed in terms of an equivalent noise contribution $i_{\mathrm{n,amp}}$ at the input port of the amplifier, which adds quadratically to the noise of the detector. It is given by [31]

$$i_{\mathrm{n,amp}} = \sqrt{\frac{4k_{\mathrm{B}}T_{\mathrm{A}}\Delta f}{R_{\mathrm{A}}}}, \tag{9.2}$$

which is equivalent to the Johnson noise of a load resistor R_{A} at the noise temperature T_{A}. Due to the strong analogy with the Johnson noise of the detector, both contributions are sometimes combined into the expression [31] $i_{\mathrm{n,J}}^2 + i_{\mathrm{n,amp}}^2 = 4k_{\mathrm{B}}T_{\mathrm{N}}\Delta f/R_{\mathrm{N}}$, with T_{N} being an effective noise temperature and R_N the parallel combination of the detector and load resistances. Since amplifier noise is usually of the same order as Johnson noise, it usually does not represent any significant restriction for a state-of-the-art thermal imaging system.

Additional noise sources arise from mechanical vibrations and from fluctuations of the detector temperature. Since most of the system noise is independent of the detector signal and of the integration time, it affects the signal-to-noise ratio in particular for short integration times and/or small incident photon fluxes.

9.1.4 Thermal Resolution

The noise-equivalent temperature difference (NETD) as defined in Sect. 2.2 can be expressed as $\mathrm{NETD} = i_\mathrm{n}/\mathcal{R}(\mathrm{d}P_\mathrm{B}/\mathrm{d}T)$, where $\mathcal{R}$ is the responsivity and i_n the noise current [31]. For the photon energies considered here, the Bose–Einstein distribution function (see (2.6)) can be approximated by an exponential dependence. In this way, we obtain the relation $\mathrm{d}P_\mathrm{B}/\mathrm{d}T = h\nu P_\mathrm{B}/k_\mathrm{B}T^2$ (see (2.25)), which results in the following expression for the NETD,

$$\mathrm{NETD} = \frac{k_\mathrm{B}T^2}{h\nu}\frac{i_n}{\mathcal{R}P_\mathrm{B}}. \tag{9.3}$$

This equation allows us to derive relatively simple expressions for NETD values NETD_α limited by different noise currents i_α in the case of narrow-band detectors where the detected IR radiation can be assumed to be monochromatic. If several stochastically independent noise sources α are present, the resulting total NETD is readily obtained by quadratic summation, i.e., $\mathrm{NETD} = \sqrt{\sum_\alpha \mathrm{NETD}_\alpha^2}$.

Detector-limited NETD

Starting with the NETD given by the signal and noise of a background-limited detector with quantum efficiency η, we use the standard expression $\mathcal{R} = e\eta g/h\nu$ and obtain from (9.1) and (9.3)

$$\mathrm{NETD}_\mathrm{DET} = 2k_\mathrm{B}T_\mathrm{B}^2\sqrt{\frac{\Delta f}{h\nu\eta P_\mathrm{B}}\left(1-\frac{p_\mathrm{c}}{2}\right)}. \tag{9.4}$$

This expression will be referred to as the *detector-limited* NETD of a QWIP. The expression generalizes the cases $p_\mathrm{c} = 0$ which holds for a standard photoconductor (and is the usual approximation used for a photoconductive QWIP) and $p_\mathrm{c} \approx 1$ which describes the NETD of a low-noise QWIP.

Assuming a rectangular gate function for the readout process, Δf is related to the integration time τ_int via [31] $\Delta f = 1/2\tau_\mathrm{int}$. In actual circuits, the bandwidth Δf for a given τ_int is usually larger than in this relation, which results in a somewhat higher value of the NETD.

Readout-Limited NETD

In thermal imagers, an extremely important system factor is charge storage capacity. Since the photocharge of a 2D FPA is stored in the ROIC, the charge storage capacity Q_c of the readout cell is relatively small since the lateral size of the storage capacitor is limited by the pitch of the detector array. Readout limitation occurs if the integration time has to be reduced below its desired value in order to prevent the readout cell from overload. We therefore define the *critical responsivity* $\mathcal{R}_c$ by the condition that the detected photocharge within the integration time τ_{int} equals the storage capacity Q_c of the readout cell. In addition to the case $T_B = 300\,\text{K}$, Table 9.1 also summarizes the $\mathcal{R}_c$-values at an upper scene temperature of 318 K (45°C), where P_B is about 40% higher than at 300 K. Since the resulting responsivities are easily achieved with most detectors, it is clear that readout limitation is an important issue for 2D FPAs. Additional constrictions appear if there exists a significant dark current, since the system performance of the camera usually degrades more strongly due to the extra charging of the ROIC by the dark current than by the reduced detectivity of the detector.

Different restrictions apply in the case where the responsivity of the detector exceeds $\mathcal{R}_c$, since the integration time has to be reduced according to the storage capacity Q_c of the ROIC. Assuming a 100% background limited detection, we thus obtain in the *readout-limited* case

$$\text{NETD}_{\text{RL}} = \frac{k_B T^2}{h\nu}\sqrt{\frac{2e(g-1/2N)}{Q_c}}. \tag{9.5}$$

The relation $\text{NETD}_{\text{RL}} \sim \sqrt{g/Q_c}$, which holds for $g \gg 1/2N$, implies that a small gain improves the NETD since it enables long integration times and thus increases the effective storage capacity of the sensor. In fact, each detected photon charges the readout capacitor by g electrons, such that a reduced gain implies a reduced number of noise electrons and an improved dynamic range of the sensor. Small responsivities can thus be tolerated for camera applications if the gain is small and if "snapshot" capability with very short integration times is not required. To give a practical example, Table 9.1 lists the NETD-values according to (9.4) and (9.5) for the geometrical parameters of the respective arrays.

NETD Limited by ADC Noise

Although the system noise of a thermal imager usually does not impose any limitation on the NETD, it becomes important if the integration time is chosen to be exceptionally short (such that the collected photocharge is almost zero), and/or if the detector noise is exceptionally low. System noise therefore already becomes relevant for low-noise QWIP FPAs at moderately reduced integration times. The contribution due to the readout noise of the ADC can be expressed as

$$\mathrm{NETD}_{\mathrm{ADC}} = \frac{k_{\mathrm{B}}T^2}{h\nu} \frac{Q_{\mathrm{ADC}}}{\tau_{\mathrm{int}}\mathcal{R}P_{\mathrm{B}}}. \tag{9.6}$$

This expression shows the penalty that occurs at small detector signals, leading to a strong increase of the NETD proportional to $1/\tau_{\mathrm{int}}$.

NETD Limited by Amplifier Noise

For the sake of completeness, we finally write down the NETD limited by amplifier noise (see (9.2)),

$$\mathrm{NETD}_{\mathrm{amp}} = \frac{k_{\mathrm{B}}T^2}{h\nu\mathcal{R}P_{\mathrm{B}}} \sqrt{\frac{4k_{\mathrm{B}}T_{\mathrm{A}}\Delta f}{R_{\mathrm{A}}}}. \tag{9.7}$$

This component bears the practical difficulty that it has the same dependence on the integration time as g-r noise and Johnson noise, and so discrimination between these components cannot be accomplished by varying τ_{int}.

9.1.5 Fixed-Pattern Noise and NETD of an Array

While the previous noise contributions relate to the temporal noise of each individual pixel, fixed-pattern noise is caused by the spatial variation of the pixel photo response across the whole FPA. In standard thermal imaging cameras, nonuniformity correction algorithms are applied to digitized data sets. The variation of the total response thus does not only include the inhomogeneities of the detector array, but also those of the ROIC and the optical system. Inhomogeneities of the ROIC are mostly due to some scatter in the operating points and transconductance of the transistors and in the readout capacitors. Even for perfectly homogeneous arrays, some inhomogeneity is always present since the fields-of-view of the cold shield and the objective vary across the FPA.

The most common algorithm for nonuniformity correction is the so-called two-point correction. Here the output signal $S_i(T_j)$ of each pixel i is determined at two different temperatures T_j ($j = 1, 2$) of a homogeneous blackbody reference scene. From these calibration data, coefficients a_i and b_i are determined such that the corrected signal $S_i^{\mathrm{C}}(T_j) = a_i + (1 + b_i)S_i(T_j)$ equals the mean value $\langle S_j \rangle$ averaged over the FPA. In order to obtain accurate coefficients without too many uncertainties originating from the noise of the calibration measurement, multiple exposures are accumulated at each temperature in order to reduce the temporal noise. By analogy with this algorithm, different correction procedures can be applied, which include three-point corrections with quadratic interpolation and gradual spatial corrections accounting for the geometry of the cold shield [344].

After the calibration parameters are determined, subsequent image data are numerically corrected in real time. The NETD of the corrected FPA then relates temperature-induced changes of the photon flux with temporal and spatial variations of the signal. For a quantitative treatment, we go back to the definitions of the NETD (9.3) and the noise (2.9) and apply these to the corrected output signal S^C. Treating S^C as a stochastic variable, we thus obtain the NETD of an array,

$$\mathrm{NETD_{FPA}} = \frac{k_\mathrm{B}T^2}{h\nu}\frac{\sqrt{\mathrm{var}(S^\mathrm{C})}}{\langle S^\mathrm{C}\rangle}. \tag{9.8}$$

For a single detector, (9.8) is equivalent to the previous definition (9.3), while it also contains the spatial noise when applied to an array.

To obtain a more intuitive approach to spatial inhomogeneities, we assume that temporal noise can be eliminated by averaging over a set of measurements, resulting in the data set $\overline{S^\mathrm{C}}$. Then we can define the relative residual spatial inhomogeneity $u = \sqrt{\mathrm{var}(\overline{S^\mathrm{C}})}/\langle\overline{S^\mathrm{C}}\rangle$. In this way, we obtain the inhomogeneity-equivalent temperature difference (IETD)

$$\mathrm{IETD} = \frac{k_\mathrm{B}T^2}{h\nu}u. \tag{9.9}$$

We thus reach the conclusion that the relation $\mathrm{NETD}^2_\mathrm{FPA} = \langle\mathrm{NETD}\rangle^2 + \mathrm{IETD}^2$ should hold to a good approximation, where $\langle\mathrm{NETD}\rangle$ denotes the average over the temporal NETDs of the individual pixels.

It is thus clear that residual inhomogeneity u is an important figure of merit for a FPA, and that it should be kept smaller than the signal-to-noise ratio of the detector elements. Good homogeneity of the detector array before correction is a necessary condition to achieve a good residual inhomogeneity – of the order of 10^{-4} for high-performance FPAs.

Nonlinearity, Drift, and 1/*f* Noise

If the FPA properties are temporally constant, spatial nonuniformity should be completely eliminated at the particular background temperatures used for the calibration. At other temperatures, always present when looking at an IR scene, a necessary condition to guarantee the absence of fixed-pattern noise is that the calibration parameters should not be too sensitive to the incident signals. Also, for perfectly linear detectors, some residual nonlinearity is usually induced by the characteristics of the transistors in the ROIC. For QWIP arrays, the nonlinearity is usually small enough to achieve excellent homogeneity after two-point correction.

Another important advantage of QWIPs over most interband detectors is the fact that FPA properties are sufficiently constant in time, i.e., the "drift" of the FPA is negligible. Calibration procedures thus need not be repeated during FPA operation, and in many cases, fixed parameters for nonuniformity correction work reliably even for repetitive cooling cycles of the array.

The most important cause of drift is $1/f$ noise, where certain "bad pixels" or, to a lesser extent, the pixels of the whole array change slightly their properties, so that the inhomogeneity correction is no longer accurate after some time. If this is the case, the nonuniformity correction has to be repeated regularly in order to maintain the performance of the array. This effect is relevant for interband semiconductors, particularly at long detection wavelengths. The low bandgap, e.g., of HgCdTe photodiodes in the LWIR, makes the performance of interband photodetectors very sensitive to surface currents induced by lateral space charge fields at the edges. QWIP FPAs which are intersubband detectors have no surface current effects, and therefore $1/f$ noise is usually negligible.

According to (9.4) we have NETD $\sim \sqrt{1/\tau}$, so the NETD goes to zero at large integration times τ, or after averaging over many measurements. This is no longer the case in the presence of $1/f$ noise, as the increasing noise amplitude prevents the variance of the signal from falling below a certain value. Because of their negligible $1/f$ noise, QWIP FPAs are therefore good candidates for all applications where long-term image accumulation is needed.

9.1.6 Modulation Transfer Function

Certainly the most important parameter of a staring detector array (we will not discuss here any scanned arrays) in any imaging application is the spatial part of the signal, i.e., the variation of the signal throughout the array. In general, the geometrical properties of the array do not only affect the noise – leading to fixed-pattern noise) – they also influence the signal. This influence can be expressed quantitatively in terms of the modulation transfer function (MTF). The MTF of an array is defined as the contrast (or modulation) obtained at a particular spatial frequency ν_s. More precisely, if the array is illuminated with a signal $S_{\nu_s}(x) = S_0(1 - \cos(2\pi\nu_s x))$, then the MTF along the x direction is given by

$$\mathrm{MTF}(\nu_s) = \frac{S_{\max} - S_{\min}}{S_{\max} + S_{\min}}, \tag{9.10}$$

where $S_{\max}$ and $S_{\min}$ are the maximum and minimum signals detected by the pixels of the array. The characteristic spatial frequency of the array, the *Nyquist frequency* ν_N is obtained if the maxima and minima of S_v (along the x-direction) match the pixels of the array, i.e., if $1/2\nu_s$ equals the pitch of the array. The y-direction is, of course, analogous.

For an "ideal" array (filling factor of 100%, no crosstalk between adjacent pixels), $S_{\min}$ and $S_{\max}$ can be obtained by integrating S_{ν_s} over one pitch of the array, which leads to the relation $MTF(\nu_s) = (2\nu_N/\pi\nu_s)\sin(\pi\nu_s/2\nu_N)$. At the Nyquist condition of $\nu_s = \nu_N$, this gives an MTF of $2/\pi = 0.64$.

In order to determine the MTF of an array, two different approaches can be employed. One method is to use calibration targets comprising stripes with well-defined spatial frequencies in order to estimate the MTF at this

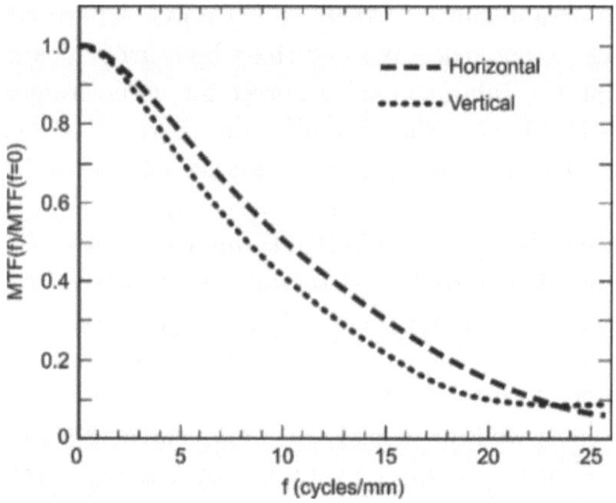

Fig. 9.2. Horizontal and vertical MTF of an MWIR imaging system based on a 1,024 × 1,024 pixel QWIP MWIR camera (courtesy of S.D. Gunapala [345])

particular frequency. An alternative method is to illuminate one point of the FPA with a small light spot then calculate the spatial Fourier transform of the array signal in order to obtain the full MTF. In both cases, it is important to take into account the influence of all system components which can affect the MTF. To a good approximation, the MTF of the system can then be expressed as a product of the MTFs associated with the optics, electronics, FPA, and cables [345].

Using a well-collimated 20 μm diameter spot, the latter approach has recently been exploited by Gunapala et al. [345] in order to obtain the MTFs of QWIP megapixel imagers for MWIR and LWIR wavelengths. An example for a measured system-MTF is shown in Fig. 9.2. The 19.5 μm pitch of this array corresponds to $\nu_{\mathrm{N}} = 25.6$ cycles mm^{-1}. Since the MTF of the spot scanner optics at ν_{N} is 0.2, the as-measured system MTF of about 6% and 9% along the horizontal and vertical axes translates into an MTF of the FPA at the Nyquist condition of 30% and 45%, respectively. In fact, these values are not far from the theoretical limit of 0.64, which emphasizes the excellent spatial resolution obtained by this array.

Minimum Resolvable Temperature Difference

A related, practically important quantitiy is the minimum resolvable temperature difference (MRTD), which is the temperature difference that can still be resolved by an observer equipped with a thermal imager. Clearly, the MRTD depends on the actual size and shape of the region (or target) of slightly different temperature. This influence can be taken into account by using test targets patterned with different spatial frequencies. As the measured temperature difference is degraded by a finite MTF, the MRTD can be expressed as $MRTD_\nu = \mathcal{K}_\nu \times \mathrm{NETD}/MTF_\nu$. The calibration factor $\mathcal{K}_\nu$, which has to be determined in field studies, reflects the visual ability of test persons to distinguish between signal and noise patterns. The concept of the MRTD is widely exploited to model detection ranges of thermal imagers [346].

9.2 QWIP Cameras

In this section we will describe the fabrication, technology, and performance of QWIP-based thermal imagers. Although we will concentrate here almost exclusively on QWIP imagers realized by our own teams, a variety of excellent FPAs have been developed by many other groups. These include the first arrays realized at today's Lucent Technologies (Murray Hill, USA) in the early days, and those of the Jet Propulsion Laboratory (Pasadena CA, USA), Thales Research and Technology (Palaiseau, France), ACREO (Kista, Sweden), and BAE systems (former Lockheed-Martin), among others. Focus varies from new approaches to commercialization. Several university groups, e.g., at Northwestern University (Evanston IL, USA), Jerusalem College of Technology (Jerusalem, Israel), and Middle East Technical University (Ankara, Turkey) have also demonstrated QWIP FPAs.

9.2.1 Fabrication of QWIP FPAs

Mature GaAs-based process technology has made it possible to realize large and homogeneous detector arrays. GaAs substrates are available with diameters up to 8 in., although processing lines for QWIP array processing are usually set for a full-wafer process on 3 or 4 inch GaAs substrates. Epitaxial growth is the first step in array fabrication. In the case of n-type GaAs-based QWIP FPAs, the typical growth sequence includes an etch-stop layer (usually AlGaAs) used for substrate removal, the QWIP active region sandwiched between two n-type GaAs contact layers, and an etch stop followed by a sacrificial layer for the grating. For optical coupling, most groups equip their QWIP FPAs with 2D reflective diffraction gratings (as in Chap. 6). Figure 9.3 shows the cross section of a typical detector element as well as an SEM micrograph of some pixels from a detector array with 40 µm pitch [17,347,348].

Process technology includes selective etching (usually reactive ion beam etching) for patterning the grating coupler into each pixel. The pixels are defined by mesa etching through the QW region into the bottom contact layer. This step can be accomplished using wet chemical (Fig. 9.3c) or dry etching techniques (Fig. 6.6); and Table 9.1 gives further geometrical parameters. Essentially the same processes are used for the fabrication of devices for wavelengths in the MWIR and LWIR, except that the mesas are defined by reactive ion beam etching (RIE) for MWIR QWIPs and by chemically assisted ion beam etching for LWIR QWIPs. This modification is necessary since the MWIR QWIPs are based on InGaAs/AlGaAs multiple quantum

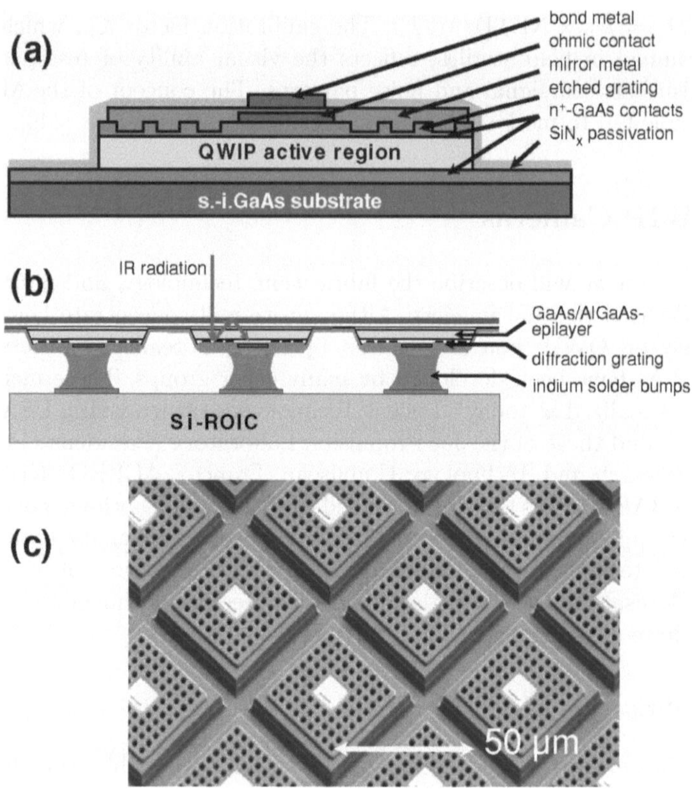

Fig. 9.3. Cross-section of a detector element in a QWIP array **(a)** and of a QWIP array hybridized to a Si ROIC **(b)**; SEM micrograph **(c)** of pixels from a detector array with 40 µm pitch. The 2.95 µm period of the 2D grating is optimized for optical coupling of 8–9.5 µm wavelength

wells, whereas the LWIR devices only contain In-free GaAs/AlGaAs epilayers. Diffraction gratings for the MWIR with 1.65 µm period were successfully fabricated with contact photolithograpy and RIE, similar to the LWIR gratings with 2.95 µm period (as in Fig. 6.6).

Ohmic contacts to the upper and common lower contact layer are evaporated and alloyed by rapid thermal annealing. The grating on each pixel is covered by a metallization. It is advantageous to use a seperate reflector metallization for this purpose rather than the ohmic contact metal in order to increase IR absorption in the QW region. The surface of the array is passivated with silicon nitride. Openings in the nitride are formed to provide electrical contact to each detector element. A separate bond metallization is evaporated in order to facilitate the hybridization. The process layout also has to provide electrical connections to the common lower contact layer, which are usually located adjacent to the actual detector array.

After dicing the wafers into single chips, the detector arrays are hybridized to Si-CMOS ROIC with In bumps. Subsequently, the GaAs substrate is removed in order to reduce mechanical stress arising from the different thermal expansion coefficients of the two chips. Substrate removal is also required to prevent optical crosstalk arising from light propagation between individual pixels. Removal of the GaAs substrate is accomplished using a sequence of mechanical lapping, wet chemical polishing, and a selective wet chemical etching process which stops at a dedicated etch-stop layer deposited during epitaxial growth.

Figure 9.3b depicts the cross section of a typical QWIP FPA, comprising a GaAs-based detector array and a silicon CMOS ROIC. The top contact and active region of the individual detectors (pixels) are separated electrically by a trench, while the back contact is common to the whole array. As indicated in the figure, IR radiation enters from the backside of the chip. After being diffracted by the reflective grating, light propagates in oblique direction to satisfy the polarization selection rules.

9.2.2 System Integration

Setting up a thermal imaging system involves a variety of different components and subsystems in addition to the FPA which comprises the QWIP array and the ROIC. Figure 9.4 summarizes device packages used by AIM Infrarot-Module GmbH (AIM), Heilbronn, Germany. The main subsystem is the integrated detector cooler assembly (IDCA) which comprises FPA, dewar, cooler, and electronic boards. AIM thermography systems [349] are based on a 14 bit ADC which allows digitizing the output of high-performance FPAs without

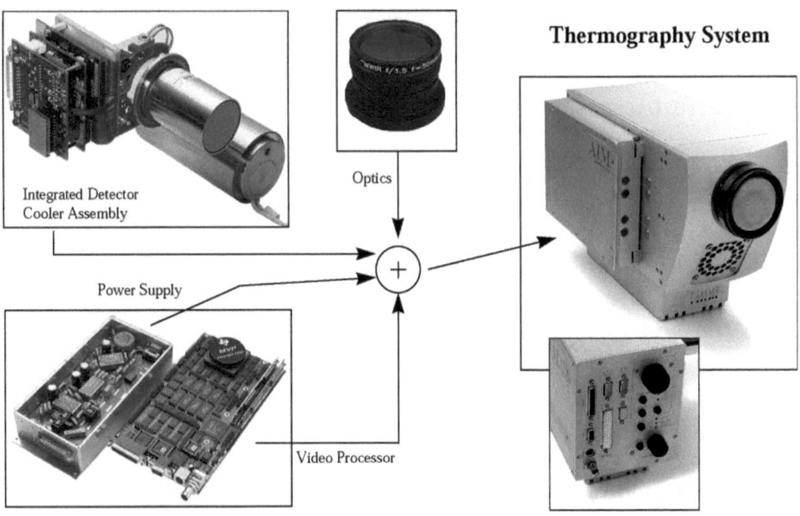

Fig. 9.4. Modular concept of thermography systems [349]

any loss of information. The image sequence is either processed in a video processor board that selects the 8 bit information to be displayed on the built-in liquid crystal display or on an external monitor, or directed to a personal computer through a high-speed digital data interface. The LWIR and MWIR QWIP FPAs developed in a cooperation between AIM and the Fraunhofer-Institute of Applied Solid-State Physics (IAF) are cooled down by Stirling coolers to about 60–65 K and 88 K, respectively. The complete camera platform is equipped with a standard serial interface for easy connection with a personal computer and with a fast parallel port that allows real-time data transfer of the full 14 bit digitized images of the 256×256, 384×28, and 640×512 QWIP FPAs at frame rates up to 200, 120, and 60 Hz, respectively.

9.2.3 Camera Performance

For thermal imaging applications, the extremely low noise levels of low-noise QWIP FPAs enable a higher dynamic range, longer integration time, and improved thermal resolution as compared to conventional photoconductive QWIPs, in particular at a high photon flux. In fact, an NETD as low as 5.2 mK has been observed for a 256×256 low-noise QWIP FPA camera with 40 µm pitch. Photoconductive QWIPs, on the other hand, are best suited if short integration times (5 ms and below) are required. In this case, thermal resolution is not limited by the storage capacity of the readout but by the finite external quantum efficiency η of the detector. Therefore, QWIP FPAs with high quantum efficiency have been developed [18], where a higher carrier concentration of $4 \times 10^{11}\,\mathrm{cm}^{-2}$ electrons per quantum well (about four times higher than for "standard" photoconductive QWIP FPAs) is used in order to increase η. The advantage of high η comes with the penalty that slightly lower (by a few Kelvin) detector temperatures are necessary for background-limited operation. Photoconductive QWIPs with even higher carrier concentration ($2 \times 10^{12}\,\mathrm{cm}^{-2}$ electrons per quantum well) are exploited for MWIR wavelengths where background-limited performance are still obtained at about 90 K.

In order to address the performance of QWIP FPAs, we present representative image data from two FPA types – low-noise and MWIR FPAs, with 640×512 pixels [18]. The experimentally determined NETD histogram of a 640×486 low-noise QWIP camera system with 24 µm pitch is depicted in Fig. 9.5a. At 30 ms integration time, we observe a NETD as low as 9.6 mK. To the best of our knowledge, both NETD values represent the best temperature resolutions ever obtained for thermal imagers operating in the 8–12 µm regime. The main noise contribution in the system stems from the ADC, which also indicates the extremely low noise level and high dynamic range achieved by the low-noise QWIP. The correctability has been characterized by determining the IETD of a 15°C scene after gain calibration at 10°C and 30°C and offset calibration at 20°C scene temperature. Under these conditions, IETD $<$ NETD/2 is found [350] even for this record low NETD,

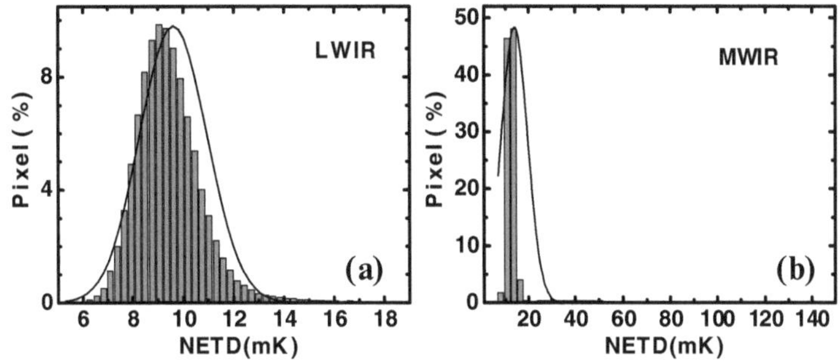

Fig. 9.5. NETD-histogram of a 640 × 512 LWIR low-noise QWIP FPA for f/2 and 30 ms **(a)**, and of a 640 × 512 MWIR QWIP FPA for f/1.5 and 20 ms (b) (after [18])

which demonstrates the excellent linearity and correctability, and thus the low fixed-pattern noise of this array.

As a second example, Fig. 9.5b depicts the NETD histogram of a typical 640 × 512 MWIR QWIP FPA at a detector temperature of 88 K, indicating an excellent NETD value of 14.3 mK. In addition, this FPA shows very good correctability, with an IETD of less than 7 mK after the same calibration procedure as described above [351]. High-performance IR sensors for the MWIR can thus be realized using QWIP technology, which provides an attractive alternative to HgCdTe, InSb, and PtSi.

Figure 9.6 shows sample images taken with both QWIP FPAs. In addition to the excellent thermal resolution and contrast, both thermal images show high detail which indicates a small optical crosstalk between adjacent pixels and a good modulation transfer function. A closer inspection of the imagery reveals a better sharpness for the LWIR image, which is not unexpected as the absorption quantum efficiency of the low-noise QWIP FPA is at least twice as high as that of the MWIR QWIP FPA.

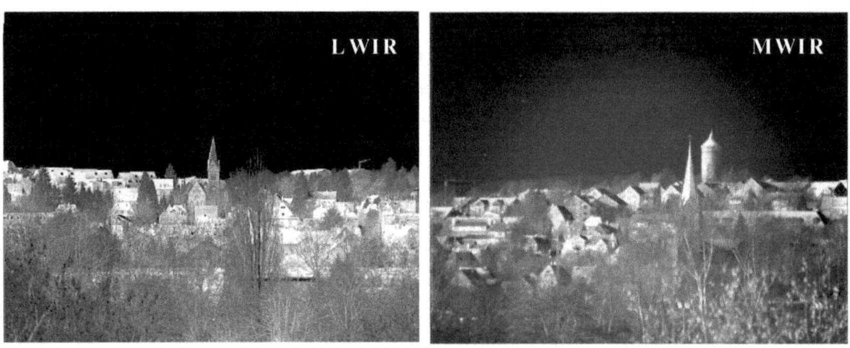

Fig. 9.6. 640 × 486 thermal images taken by a low-noise QWIP camera (*left*) and by a MWIR QWIP camera (*right*) [18]

Table 9.2. Properties of demonstrated QWIP FPAs (PC photoconductive, LN low-noise, HQE high quantum efficiency, τ_{int} integration time)

FPA Type	Array Size	Pitch (µm)	λ (µm)	$F_{\#}$	τ_{int} (ms)	NETD (mK)
256 × 256 PC	256 × 256	40	8–9.5	f/2	16	10
640 × 512 PC	640 × 486 512 × 512	24	8–9.5	f/2	16	20
256 × 256 LN	256 × 256	40	8–9.5	f/2	20 40	7 5
384 × 288 LN	384 × 288	24	8–9.5	f/2	20	10
640 × 512 LN	640 × 486 512 × 512	24	8–9.5	f/2	20	10
384 × 288 PC-HQE	384 × 288	24	8–9.5	f/2	**1.5**	40
640 × 512 PC-HQE	640 × 486 512 × 512	24	8–9.5	f/2	**1.5**	40
640 × 512 PC-MWIR	640 × 486 512 × 512	24	**4.3–5**	f/1.5	20	14

Key properties of our demonstrated QWIP FPAs are summarized in Table 9.2. The best thermal resolution is achieved with low-noise QWIP FPAs, with NETD values representing the best temperature resolutions ever obtained for thermal imagers operating in the LWIR. Specifically, for short integration times, "high-quantum-efficiency" QWIP arrays with higher doping ($4\times10^{11}\,\text{cm}^{-2}$ per QW) and an increased number of periods ($N = 35$) achieve a thermal resolution of 40 mK at a short integration time of only 1.5 ms.

9.3 MWIR/LWIR Dual-Band QWIP FPA

Increasing efforts are being put into camera systems with multiple detection wavelengths, operating either as dual-color FPAs within the LWIR or the MWIR, or as dual-band FPAs at two different (LWIR and MWIR) bands [16,18,268,352–355]. These dual-band or dual-color FPAs provide several advantages, including the ability of remote absolute temperature measurement, operation in a wider range of ambient conditions, and better distinction between targets and background clutter. Because of their enhanced functionality, dual-band FPAs represent a new generation ("third generation") of thermal imagers. QWIP technology is ideally suited for dual- and multispectral FPAs since the intrinsically narrow absorption width $\Delta\lambda/\lambda$ (10–20%) of QWIPs leads to a negligible spectral crosstalk, and the QWIP active region for the longer wavelength band is transparent at shorter wavelengths. In addition, mature GaAs-based technology facilitates the fabrication of the required elaborate FPA topologies with multiple etch depths, metallizations, and passivations. Most military applications involving dual- or multi-spectral IR sensors, however, demand high quantum efficiency. Comparing the different technologies, there is presently a tradeoff between the higher quantum efficiency achievable with HgCdTe-based photodiodes and the better spectral separation achievable with QWIPs.

The first QWIP FPA with simultaneous, pixel-registered operation in the mid-wavelength IR (MWIR, 3–5 μm) and LWIR has been described by Goldberg et al. [353]. With an array size of 256 × 256 pixels, their FPA has achieved an NETD of 30 mK in the MWIR and 34 mK in the LWIR.

9.3.1 Detector Concept

Our approach to dual-band QWIP FPA is summarized in Fig. 9.7, which depicts the schematics of a detector element of the hybridized array. The sensor pixel comprises two QWIP active regions sandwiched between three n-type GaAs contact layers. The detector element is connected with the ROIC by three electrical contacts, with the intermediate n-GaAs layer used as the ground electrode. This configuration allows for simultaneous, pixel-registered detection in both bands.

For optical coupling of the incident thermal radiation according to the polarization rules for intersubband transitions, a 2D reflective grating is located underneath the top contact ("top" refers to the location prior to flipping the GaAs chip). The MWIR QWIP is adjacent to the grating (in the near field

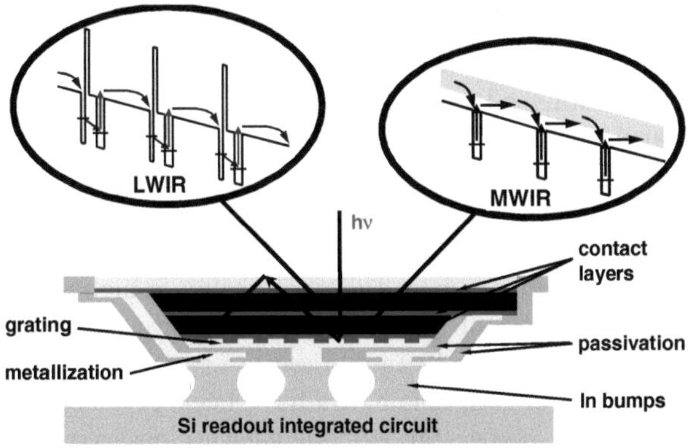

Fig. 9.7. Schematics of MWIR/LWIR dual-band QWIP FPA pixel after flip-chip hybridization to a silicon-based readout integrated circuit and GaAs substrate removal. Insets indicate the potential distributions and transport mechanisms of a photovoltaic low-noise QWIP (*left*) for the LWIR band and of a photoconductive QWIP (*right*) for the MWIR band. Note that the MWIR QWIP is adjacent to the grating (near-field regime)

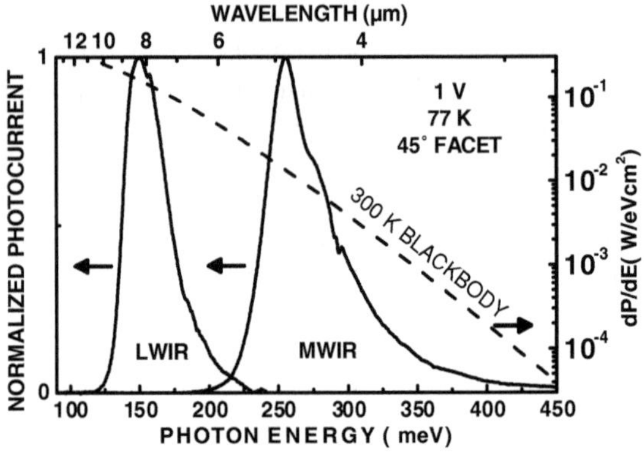

Fig. 9.8. Normalized spectral sensitivity of dual-band QWIP test devices with 45° facet coupling. The dashed line indicates the spectral power density dP/dE of a Planck radiator at 300 K in a log-scale [355]

region, see also Fig. 9.7), giving rise to efficient optical coupling of MWIR radiation.

The whole device relies on two different QWIP types (as in the insets to Fig. 9.7). For the MWIR band, a photoconductive QWIP is used (right inset), comprising 20 periods of 2.6 nm wide $In_{0.3}Ga_{0.7}As$ quantum wells, doped to a carrier density of $2.1 \times 10^{12}\,cm^{-2}$ per well, and 24 nm wide $Al_{0.32}Ga_{0.68}As$ barriers. The barrier width is significantly smaller than in previous MWIR QWIP FPAs [18], so the active region is only 0.56 µm thick. Detection in the LWIR band is accomplished by a photovoltaic QWIP.

The spectral dependence of the dual-band QWIP photocurrent is shown in Fig. 9.8. For comparison, the spectral power density dP/dE for a 300 K thermal background according to Planck's radiation law is also plotted. The sensitivity of the MWIR QWIP has its peak at the long-wavelength end of the 3–5 µm atmospheric window where the photon flux is higher than at shorter wavelengths. The spectrum of the LWIR QWIP, however, is matched to the short-wavelength end of the 8–12 µm regime in order to avoid excessive dark currents. While the spectra in this figure have been obtained using 45° facet coupling, almost identical spectral dependencies are observed for small detector elements with a grating [354]. As can be seen from Fig. 9.8, the photon flux in the LWIR spectral band is about an order of magnitude higher than in the MWIR, this large difference in photoconductive gain (by about a factor of ten) for the two QWIP structures thus compensates for the difference in photon flux. Similar signal levels are therefore obtained for the LWIR and MWIR bands, if the grating has similar coupling efficiencies for both bands.

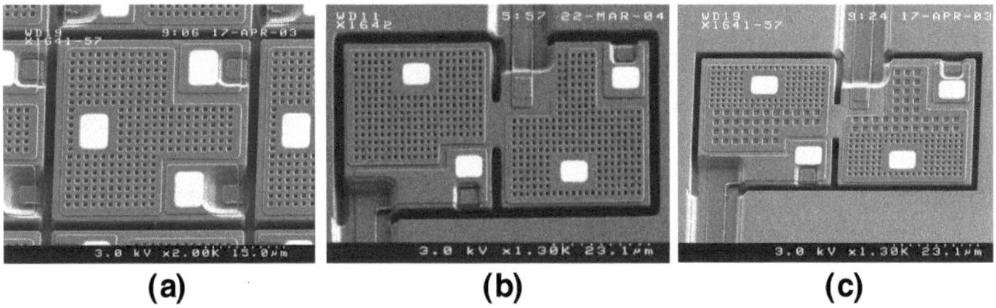

Fig. 9.9. SEM images of a single pixel in a processed dual-band QWIP array (**a**) and test structures, equivalent to two FPA pixels, allowing optoelectronic characterization (**b**), (**c**). The grating period in (**a**) and (**b**) is 1.65 μm, while a segmented grating with periods of 1.65 and 2.8 μm was used in (**c**). Pixel size corresponds to a pitch of 40 μm in all cases [355]

9.3.2 Array Fabrication and FPA Layout

The dual-band QWIP layer structures were grown by molecular beam epitaxy on 3-in. semi-insulating GaAs substrates. The epitaxial layer comprised two QWIP active regions as described above, which were sandwiched between three n-type GaAs contact regions. The growth was completed by an AlGaAs etch stop layer followed by a GaAs sacrificial layer, which defined the depth of the diffraction grating. A full-wafer process based on standard optical lithography was used to fabricate arrays with 384×288 detector elements and 40 μm pitch, with 12 detector arrays per wafer.

An SEM image of a single FPA pixel is shown in Fig. 9.9a. The pixel comprises three contact lands (bright areas) for In-soldering with the ROIC. The etched holes close to the right-hand edge of the pixel in Fig. 9.8a are needed for contacting the two buried n-GaAs contact layers.

In addition to the detector arrays, various test devices were processed simultaneously on the same wafer. They included mesa structures without gratings for standard 45° facet coupling, and detector mesas with gratings. These test devices are available with gratings optimized to the MWIR (1.65 μm grating period, see Fig. 9.9b) and LWIR (2.8 μm grating period, not shown) detection bands as well as in a segmented version (Fig. 9.9c) which contains both gratings.

The pixel layout of Fig. 9.9a has been chosen in order to obtain a hexagonal symmetry for the contacts of the entire detector array. This hexagonal lattice ensures maximum distance between adjacent contact lands for a given FPA pitch. In this way, a nearest-neighbor spacing of approximately 24 μm is obtained for a 40 μm pitch. As can be seen from Fig. 9.10, the hexagonal symmetry appears if the detector array is composed of columns with alternating, vertically flipped symmetry. The rectangular unit cell of such a lattice is indicated in Fig. 9.10a. It includes two pixels.

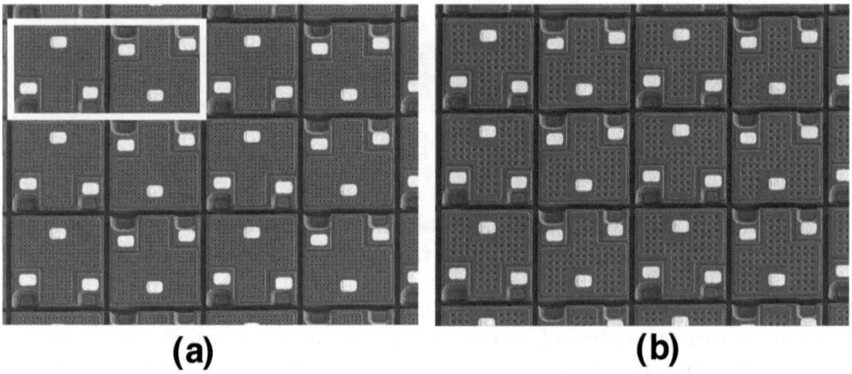

Fig. 9.10. SEM images of processed MWIR/LWIR QWIP arrays with 1.65 µm **(a)** and 2.8 µm **(b)** grating period. The white rectangle in **(a)** indicates the two-pixel unit cell of the detector array. Pixel size corresponds to a pitch of 40 µm [355]

9.3.3 Properties of Dual-Band QWIP Test Devices

Gratings for optical coupling in dual-band QWIP FPAs are subject to several tradeoffs, since optimization for one operating wavelength generally implies reduced performance for the other detection band. This difficulty results from incompatible periods (1.65 vs. 2.8 µm) and depths (0.35 vs. 0.6 µm) that would be needed for the MWIR and LWIR wavelength bands. In addition, only one QWIP stack can be placed within the near field of the grating where the coupling efficiency is substantially enhanced [267]. For these reasons, the coupling efficiency in a dual-band QWIP FPA is necessarily smaller in at least one detection band as compared to independently optimized single wavelength QWIP FPAs. The coupling efficiency also depends on the size of the detector element [201] and on the presence of irregularities such as etched trenches.

The test structure geometries of Figs. 9.9 are used to investigate the dual-band performance of gratings and their coupling efficiency. Figure 9.11 summarizes the results for an MWIR QWIP. We observed the following trends:

1. The photocurrent density increases with the size of individual detector elements. Indeed, a "large" mesa ($120 \times 120\,\mu m^2$ area) with 1.65 µm period grating gives rise to 70% higher coupling efficiency than a "twin pixel" with the same grating. Most of this increase is attributable to a spectrally narrower coupling efficiency achievable with large mesas, which in turn improves the resonant enhancement. This trend is also consistent with the observation [13] that only minor deviations exist between the photocurrent spectra measured for twin-pixels and for 45° facet coupling.
2. A more than threefold increase, as compared to the case without gratings, is observed if the grating period is optimized to the "wrong" (LWIR)

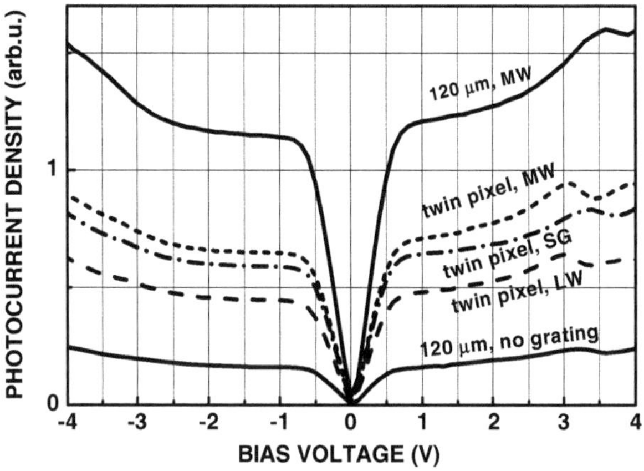

Fig. 9.11. Photocurrent density of MWIR QWIP obtained from different mesa geometries as indicated vs. applied bias [355]. Labels indicate the device geometry and grating type (MW: 1.65 µm, LW: 2.8 µm grating period, SG: segmented grating)

wavelength band. Apparently, the grating enables some residual coupling since its depth is still optimized to MWIR wavelengths.

3. There is still a significant coupling efficiency (13% as compared to an MWIR grating) for mesas without grating. This residual coupling, which is attributed to stray light, is believed to be significantly enhanced by total internal reflections. Even though substrate thinning might have some influence on the coupling efficiency of dual-band QWIP FPAs, we expect similar trends to exist for devices after substrate removal.

Similar results to those in Fig. 9.11, for the LWIR active region, indicate a two-fold increase of the LWIR signal if the "correct" grating with a 2.8 µm period is used, rather than a 1.65 µm period.

9.3.4 System Integration and Dual-Band QWIP FPA Performance

The detector arrays are hybridized to a custom-designed ROIC which exhibits equal storage capacities for the readout cells for the MWIR and LWIR bands. Figure 9.12 shows a section of a dual-band QWIP FPA after GaAs substrate removal. Dark lines between the individual pixels indicate complete removal of the semiconductor material down to the passivation. The procedure thus minimizes thermally induced stress and optical crosstalk. The hybrid is finally mounted in a dewar and cooler assembly.

While results for a first prototype of a MWIR/LWIR dual-band QWIP FPA can be found [354], we show here results for an improved version [355]. The previous FPA showed excellent performance in the MWIR band (17 mK NETD) but had a comparably high NETD of 43 mK in the LWIR. In accordance with the measured behavior of the test devices as discussed in the

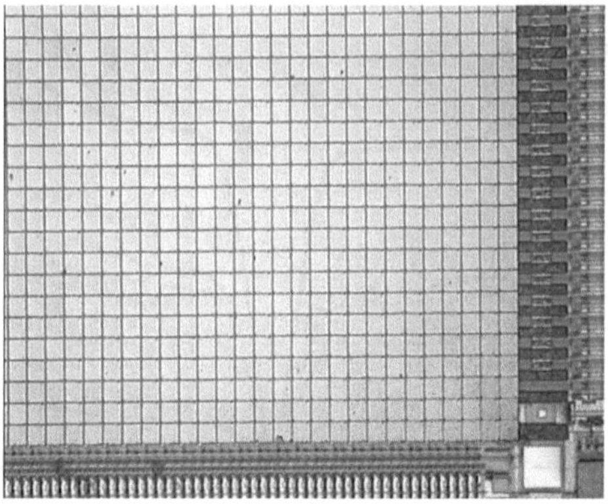

Fig. 9.12. Microscope image of a hybridized dual-band QWIP FPA after substrate removal [355]

previous paragraph, the previous 1.65 μm period grating was replaced by a grating with 2.8 μm period. In this way, the LWIR signal was expected to increase by about a factor two, whereas the MWIR performance should have degraded slightly. Our goal was to achieve similar performance in both bands.

The array histograms in Fig. 9.13 indicate that this goal has been met. In fact, the NETD in the MWIR and LWIR bands, which are plotted together with Gaussian fit functions, indicate an average NETD as low as 20.6 mK in the LWIR and 26.7 mK in the MWIR. Here an integration time of 6.8 ms was chosen, which allows for 100 Hz repetition rate and for a 2×2 microscan with 25 Hz for subpixel resolution (768×576 effective pixels). To our knowledge, this is the first dual-band FPA which achieves NETDs significantly below 30 mK in both bands. The drastic improvement in the LWIR band is not unexpected since the noise associated with the LWIR part is still limited by system noise. This property leads to an essentially linear, rather than square-root, dependence of the NETD on the integrated signal. For the conditions of Fig. 9.13, almost identical signal strengths (with only about 12% difference) were observed in the two detection bands. An excellent pixel yield of >99.5% was obtained.

9.4 Opportunities for QWIP FPAs in Thermal Imaging

QWIP-based thermal-imaging systems compete with those based on other sensor technologies, such as HgCdTe, InSb, PtSi, and uncooled microbolometers. Each technology has advantages and disadvantages, so different detector materials are used according to the specific demands imposed for each application. These requirements determine the required thermal resolution, spatial

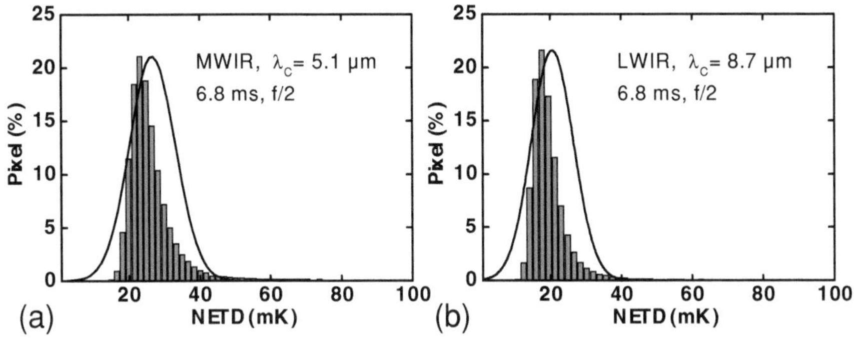

Fig. 9.13. NETD-histograms (*bar plots*) of the MWIR (**a**) and LWIR (**b**) bands of improved dual-band QWIP FPA at 58 K operation temperature and 300 K radiation temperature [355]. Gaussian fit curves (*solid lines*) indicate excellent NETDs of 26.7 mK (MWIR) and 20.6 mK (LWIR)

resolution, wavelength band, integration time, pixel operability, and allowable cost.

At present, QWIP-based thermal imagers are used if high spatial resolution or thermal resolution is needed in the LWIR spectral regime. Further advantages of QWIP arrays include excellent homogeneity, low fixed-pattern noise, low 1/f noise, and low drift. Typical frame rates of QWIP FPAs are in the range of 10 to 100 Hz (as in Table 9.2). For high-performance thermal imaging in the LWIR, the only competing technology is HgCdTe, which is significantly more expensive than QWIPs but allows shorter (sub-ms) integration times and slightly higher operation temperatures.

At very long IR wavelengths above 12 µm and in the terahertz regime, new challenges are waiting for QWIP technology. At these wavelengths, HgCdTe technology becomes extremely difficult, whereas the performance of QWIPs is still comparable to theoretical predictions. In the MWIR regime, high quantum efficiency is very important, and the market for high-performance thermal imaging is presently shared between HgCdTe and InSb. Nevertheless, the MWIR also provides increasing opportunities for QWIPs since quantum efficiencies in excess of 50% have been demonstrated recently [345,356].

Extending their use beyond the military market, QWIP FPAs are particularly promising for civil applications including medical imaging and environmental research. In medicine, the method of dynamic area telethermometry (DAT) has been introduced [357,358] for breast cancer detection and for verifying the efficiency of cancer medication and therapy. The method is based on the detection of subtle temporal changes of the skin temperature, which exhibit characteristic modifications in the vicinity of a tumor. Exploiting both local and global signatures, the method also allows the detection of deeper lying cancerous lesions. To this end, thermal deviations with a modulation frequency from 0.1 to 2 Hz have to be detected. The spatial distribution of temperature modulation on the skin surface is obtained by quantitative numerical analysis of several thousands of subsequent thermal images. Since the amplitude of these temperature fluctuations is only a few mK, low-noise QWIP FPAs are most suitable for this application due to their superior thermal resolution. As a second advantage, the low drift of QWIP arrays is crucial for this application. In contrast to the standard method of x-ray mammography, the new method of DAT enables early detection of cancerous lesions without the involvement of mechanical stress or radiation exposure. The method can therefore be repeated arbitrarily often, thus allowing monitoring of the efficacy of medication. Additional applications in medicine include brain surgery and pre-clinical testing.

An extremely interesting application, for which QWIP arrays are ideally suited due to their narrow spectral width, is the detection of gas leaks. Specifically, the method of "Gas correlation imaging" allows one to locate gas leaks by comparing IR images, which are obtained with and without additional transmission through the targeted gas [359]. This can be achieved by two sequential exposures, by dividing one array into two segments, or by using two separate arrays. After numerical subtraction of the two images, the difference image contains information associated with the absorption lines of the target gas. In this way, the target gas is used to recognize its own specific spectral features, and the gas leak can even be located in real-time by superimposing the difference images with camera signals in the visible. A variety of gases can be detected in this way, the most important ones being methane, ethane, and ammonia. Application areas of this method include the inspection of on- and off-shore oil platforms, gas pipelines, and refineries. The method helps to prevent explosions, fires, production loss, and environmental damage. Due to its ability for real-time detection, the approach largely eliminates motion artefacts, thus allowing for quick inspection of large areas, e.g., by using a helicopter.

The excellent thermal resolution of QWIP cameras also makes them an interesting measurement tool in research. For instance, such a camera was used by Garbe et al. [360] in order to study convection processes on water surfaces and wind–water interactions. This was with the ultimate goal of understanding the exchange of CO_2 between the oceans and the atmosphere. Such investigations also aim towards a better understanding of the greenhouse effect induced by the combustion of fossil fuels.

Further applications of QWIP cameras have been reported recently in semiconductor research. The spatial carrier profile in Si wafers has been measured by shining IR light through the wafers [361]. The excellent thermal resolution of QWIP FPAs has also proven to be very useful in lock-in thermography, where thermal signatures of modulated excitations are detected. The method has recently been applied to capacitance mapping of semiconductor p-n junctions [362].

At present, QWIPs are the only commercially viable technology for the production of large 640×480 and megapixel staring arrays operating in the

LWIR. Further potential also exists in other nonmilitary markets, e.g., production monitoring, nondestructive testing, and fire fighting. Commercialization of dual-band or dual-color thermal imagers will open up further applications, in particular if an absolute determination of the radiation temperature is required.

9.5 Alternative Architecture and New Functionality of QWIP FPAs

Multicolor Arrays

The QWIP approach is nicely suited for multi-band arrays (see Sect. 8.2). Recently, a four-band 640×512 QWIP FPA has been demonstrated [268,363]. The array is divided into four segments with 640×128 pixels, which are sensitive in the 4–5.5, 8.5–10, 10–12, and 13–15.5 µm wavelength bands, respectively. The cross section of the individual pixels is shown schematically in Fig. 9.14. Unwanted bottom detectors are electrically shorted from the outside of the array, which is operated with a common back contact with bias V_c. Varying period numbers and carrier densities are used for the four active layers due to thickness restrictions and due to the common operating voltage. The array, which has been developed for a hyperspectral imaging instrument, was fabricated in the InGaAs/GaAs/AlGaAs material system.

Polarization-Sensitive Arrays

Due to the polarization rule of intersubband absorption, QWIPs are ideally suited to exploit polarization signatures in passive IR thermal imaging. This functionality is very promising for remote sensing and target recognition, in

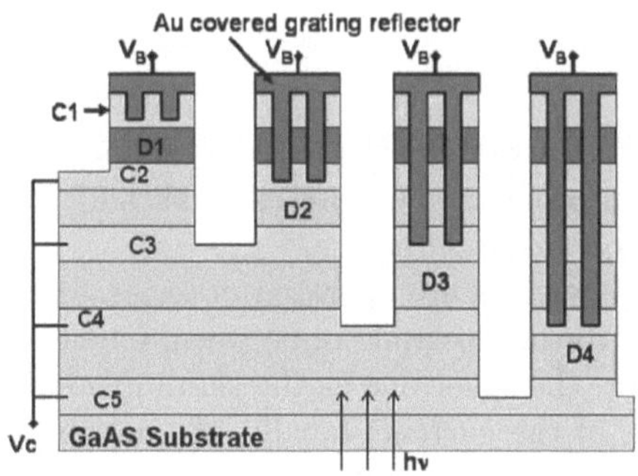

Fig. 9.14. Schematic cross section and pixel isolation scheme of a four-band QWIP array [363]. Gold-coated reflective etched gratings, which also serve as contacts and shorten the unwanted top detectors, are fabricated on each pixel (D1 to D4: QWIP active regions, C1–C5: heavily doped contact layers)

particular of man-made objects which, due to their typically smooth and reflecting surfaces, are mostly associated with stronger polarization signature than the natural scene. The polarization contrast thus provides additional signatures which help to detect specific objects, e.g., land mines [364,365].

The straightforward approach to realize polarization dependent QWIP FPAs is to use pixels equipped with linear gratings which only couple one linear polarization of the radiation. Such an array has been demonstrated by Beekman and Anda [364]. The array comprises pixels with 0°- and 90°-oriented linear gratings which are arranged in a checkerboard pattern. Here the detection of a linear polarization still depends on the relative orientation between the grating and the polarization. A more complete determination of the polarization state, i.e. including the degree of linear polarization and polarization angle, is possible with four kinds of gratings oriented at 0°, 45°, 90°, and 135°. Chen et al. [366] have used corrugated QWIPs oriented in these directions to detect polarized light.

Beyond this lateral architecture for polarization sensitive arrays, it would be advantageous to achieve this functionality on one pixel. In fact, a high degree of co-location for the detection of different polarization is necessary, not only for moving targets but also to obtain polarization information for point objects. From a systems perspective, it will be necessary to detect the degree of polarization with an uncertainty of less than a few percent. If these requirements can be met, superior performance as compared to today's imaging polarimeters, which usually consist of rotating polarization filters in front of the detector array, will be achieved. Serna [365] has conducted computer simulations, and developed a process for integrating a polarimeter on a single pixel. His approach uses four active QWIP regions, seperated by contact layers comprising metallic or dielectric linear gratings with different orientation (with a fourth grating on top of the structure). In this geometry, it is even possible to determine the full Stokes vector of the polarization, i.e., to measure not only linear but also elliptic polarizations.

Skimming QWIP

In order to increase the capacity of the readout circuit, the "skimming QWIP" architecture has been proposed by Costard et al. [267,367]. Figure 9.15 indicates the basic idea of this architecture.

Two QWIP active regions are operated in series, and the readout capacitor C_{int} is connected to the electrode in between. One QWIP structure serves as the active detector, the other one as the skimming element which removes an "unwanted" part of the current. Ideally, the active QWIP is illuminated and contributes a photocurrent and a dark current, whereas the skimming QWIP only produces a dark current. If the two dark currents are equal, then the photocurrent charges the capacitor, while the two dark currents cancel each other. The big advantage of this architecture comes into play in particular at higher operation temperature where the skimming QWIP removes

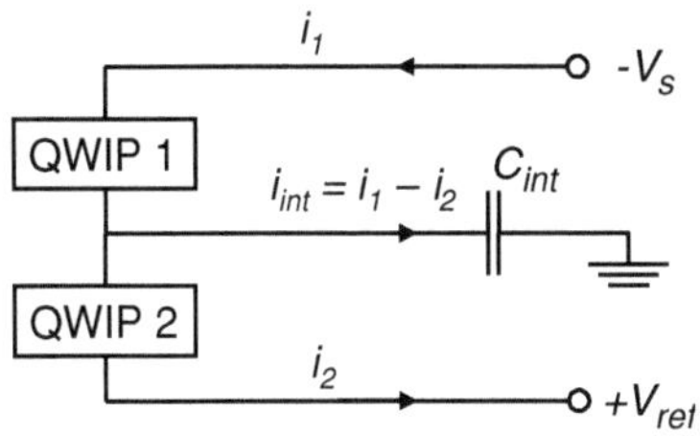

Fig. 9.15. Operation principle of the skimming-QWIP architecture

(or skims) the dark current and thus allows for larger integration times and better signal-to-noise ratios. We should also mention that, in spite of the current cancellation leading to an improved "virtual" readout capacitance, the noise contributions add, such that the absolute noise in a skimmed FPA is somewhat higher than in a normal array.

In practice, it is useful to integrate the two QWIP structures vertically, such that the whole structure resembles a dual-band QWIP (as in Sect. 9.3), albeit with two identical active regions in order to generate identical dark currents. The photocurrent can be selectively generated in the QWIP region adjacent to the grating, since the near-field of the grating is associated with a much higher coupling efficiency than the far-field [201]. In fact, a five times higher photocurrent has been measured in the QWIP region close to the grating as compared to the second active region, even though both active regions were nominally identical [367].

10

Dynamics, Ultrafast, and Heterodyne

Thanks to the intrinsic short carrier lifetime ($\tau \sim 5$ ps), QWIPs are well suited for high-speed and high-frequency applications [368]. The inherent short carrier lifetime was inferred both from heterodyne experiments [215,369], showing a cutoff frequency of about 30 GHz, and from time-resolved photocurrent measurements [148,370,371]. High-speed detectors may also create new applications; e.g., environmental remote sensing of molecules [372] and CO_2 or quantum cascade laser based communication [218], as well as laboratory instrumentation [217,242,373–375] and astronomical studies [376,377]. For these applications, there is commonly a strong signal or a powerful local oscillator, in most cases employing lasers. Under such circumstances, a high dark current can be tolerated to a large degree, and a high absorption and high operating temperature is desirable (discussed in Sect. 8.1). Note also that the polarization sensitivity of the QWIP is no longer a disadvantage for (polarized) laser-based systems.

The intrinsic detector response time is limited by either the photoelectron lifetime or the transit time, whichever is smaller. For a large number of QWs, the total transit time is usually much longer than the lifetime, leading to a lifetime limited intrinsic response time of about 5 ps. If, however, a QWIP has a small number of QWs (e.g., 10 or less), photoexcited carriers will be swept out before capture, resulting in a transit time limited situation. We discuss these dynamical processes, and also review the high speed capability. Presently, QWIPs hold the unique position of having high speed/frequency capability and high absorption for the thermal IR region. There are no competitive alternatives.

10.1 Dynamic Processes in QWIPs

The most direct experiment to investigate the dynamics of the photocurrent in a QWIP is to measure the transient photocurrent after illumination with a short optical pulse [148,370,371]. To get the full picture, we start this section

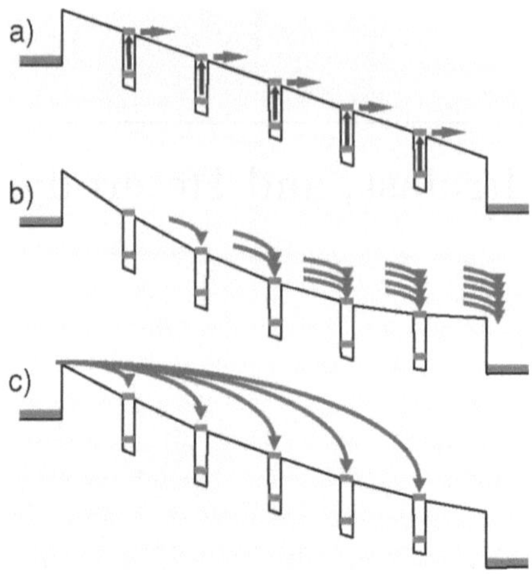

Fig. 10.1. Conduction bandedge distribution of a QWIP structure under an applied voltage, indicating (**a**) photoexcited carrier emission, (**b**) carrier capture, and (**c**) redistribution of space charges. We note that the equilibrium distribution (**a**) corresponds to the assumption of perfectly injecting contact layers; usually space charges are already present without infrared illumination

by investigating the current that is induced by the re-charging of the QWs after photoinduced depletion. We then continue by exploring the signal induced by photoinduced emission and capture processes.

10.1.1 Quantum Well Recharging

The optical response of QWIPs can be separated into a "primary" or "fast" photocurrent j_1 associated with carrier emission and capture, and a "secondary" or "slow" contribution j_2 associated with the recharging of the quantum wells. This two-component behavior influences various effects related to time and frequency dependent photocurrents, frequency dependent noise, photoconductive nonlinearity, and thermal imaging at low background flux.

Figure 10.1 summarizes the different stages of a QWIP structure after pulsed IR illumination. After intersubband excitation, the carriers should leave the quantum well within a very short time, since the intersubband relaxation time is of the order of 1 ps. The emission process is schematically indicated in Fig. 10.1a. An important contribution to this initial transport comes from the fact that the final state involved in the excitation process of a detector structure already carries a finite group velocity, which enables the photoexcited carriers to escape from the well. After emission, carrier capture becomes relevant (see Fig. 10.1b). Due to the short time constant τ_c associated with this process, the duration of the "fast" photocurrent j_1 is of the order of 10 ps.

After carrier capture has occurred, nonequilibrium space charges will remain in the QWIP structure. The redistribution of these space charges gives rise to an additional slow component of the photocurrent, which can be considered as an extra injection current induced by the larger local electric field close to the injection contact (see Fig. 10.1c). The time constant associated with the redistribution of these space charges depends critically on the number of electrons flowing across the QWIP structure and is closely related to the dielectric relaxation time. Depending on the experimental conditions, the associated "slow" photocurrent j_2 can be in the sub-ns regime, or it can last for hours.

Assuming that j_1, j_2, and the dark current of a QWIP with N quantum wells and $N+1$ barriers are associated with the same capture probability p_c, the following relation holds for the time-averaged signals $\langle j_1 \rangle$ and $\langle j_2 \rangle$,

$$\frac{\langle j_1 \rangle}{\langle j_1 \rangle + \langle j_2 \rangle} = 1 - \frac{1}{(N+1)p_c} \left[1 - (1-p_c)^{N+1}\right] . \tag{10.1}$$

This expression properly takes into account the displacement current, i.e., that an electron moving from well j to well k induces a photocharge of $(j - k)/(N+1)$ electron charges in the external circuit. Equation (10.1) is different from a previous expression [154,378], $\langle j_1 \rangle / (\langle j_1 \rangle + \langle j_2 \rangle) = 1 - (1-p_c)^N$, which arises from counting only the carriers that arrive at the collector contact of the QWIP.

To derive (10.1), we assume that s electrons per well are exited into the continuum by a weak optical pulse, giving rise to a photocharge $Q = Q_1 + Q_2 = es/p_c$, which consists of a "fast" photocharge Q_1 and a "slow" photocharge Q_2. Let us now consider the first quantum well, i.e., $j = 1$. The contributions (probabilities weighted by the associated distance relative to the total thickness) are indicated in Fig. 10.2. Summing up the contributions listed in Fig. 10.2, we thus obtain the "fast" photocharge associated with carrier emis-

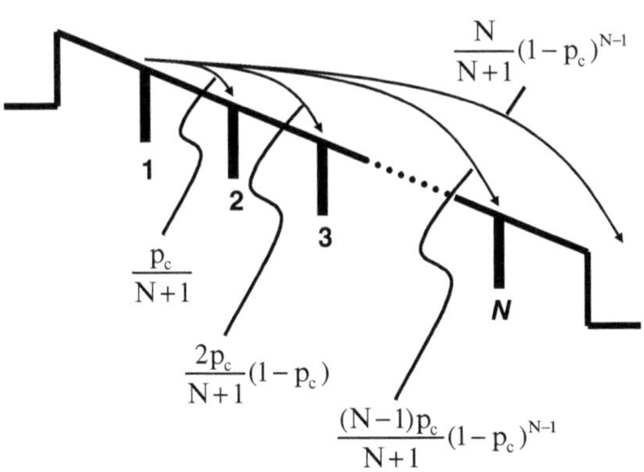

Fig. 10.2. Contributions to the "fast" photocharge Q_1 originating from an electron emitted out of the first well in an N-period QWIP

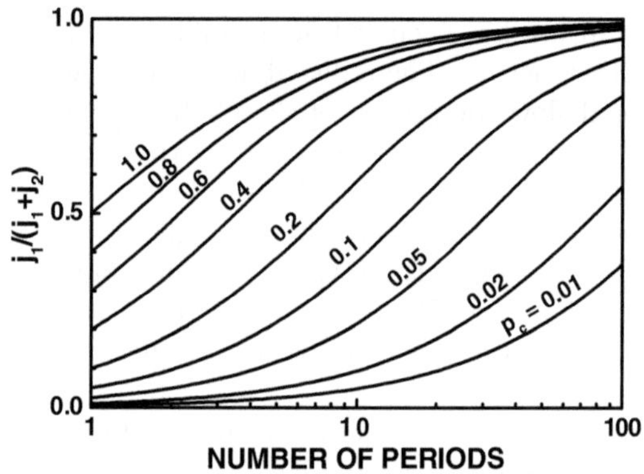

Fig. 10.3. Ratio $\langle j_2\rangle/(\langle j_1\rangle + \langle j_2\rangle)$ vs. QWIP period number at different capture probabilities

sion from the first well,

$$Q_1^{\text{firstQW}} = es\left[\left(\sum_{k=1}^{N-1}\frac{kp_c}{N+1}(1-p_c)^{k-1}\right)+\frac{N}{N+1}(1-p_c)^{N-1}\right]. \quad (10.2)$$

Here the summation represents the electrons captured by the other $N-1$ wells, whereas the last term on the right-hand side stands for the electrons arriving at the collector contact.

The summation over all the N wells now yields[1]

$$\begin{aligned} Q_1 &= \frac{es}{N+1}\left[\left(\sum_{j=1}^{N-1}\sum_{k=1}^{j}kp_c(1-p_c)^{k-1}\right)+\sum_{j=1}^{N}j(1-p_c)^{j-1}\right] \\ &= es\frac{Ng}{N+1}\left[\left(1-\frac{1}{Ng}\right)^N(Ng-1)+N+1-Ng\right], \end{aligned} \quad (10.3)$$

where we have used $g = 1/np_c$. Equation (10.1) readily follows from the relation $\langle j_1\rangle/(\langle j_1\rangle + \langle j_2\rangle) = Q_1/(Q_1+Q_2)$ and (10.3).

To illustrate (10.1), Fig. 10.3 shows the ratio $\langle j_1\rangle/(\langle j_1\rangle+\langle j_2\rangle)$ for different p_c and N. In GaAs/AlGaAs QWIPs, we typically have $p_c \approx 0.1$, such that $\langle j_1\rangle/(\langle j_1\rangle + \langle j_2\rangle) = 80\%$ for a standard 50 period structure. While j_1 usually dominates for $N \geq 50$, j_2 becomes important for QWIP structures with fewer periods. For $N = 1$, we have $\langle j_2\rangle = \langle j_1\rangle$ even for $p_c = 1$, since each electron that is emitted via the collector barrier has to be replaced via the emitter barrier.

Substituting p_c by the gain $g = 1/Np_c$ and taking the limit $N \to \infty$, (10.1) yields the following result, first derived by Ershov et al. [379],

[1] The summations are easily worked out using software for symbolic calculations.

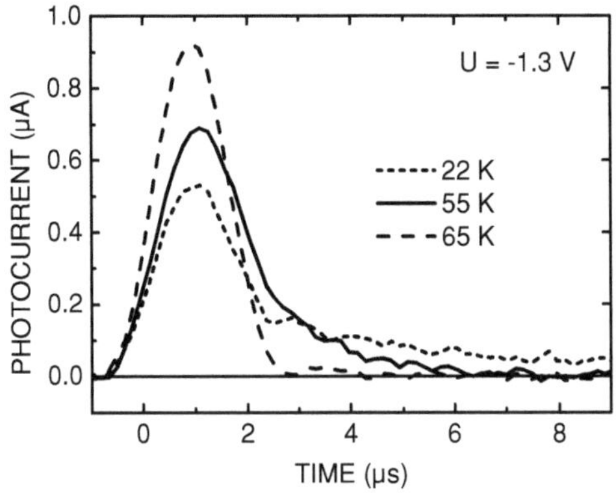

Fig. 10.4. Photocurrent vs. time of an 8-period QWIP structure at $-1.3\,\mathrm{V}$ applied bias and temperatures of 22, 55, and 65 K (after [378])

$$\frac{\langle j_1 \rangle}{\langle j_1 \rangle + \langle j_2 \rangle} = 1 - g\left[1 - \exp\left(\frac{1}{g}\right)\right]. \tag{10.4}$$

The first direct experimental observation of this two-component behavior was reported by Ehret et al. [378] who investigated an 8-period GaAs/AlGaAs QWIP structure. Figure 10.4 shows the time dependence of the photocurrent excited by $1.2\,\mu\mathrm{s}$ IR pulses. The transients consist of fast and slow components, which are directly associated with j_1 and j_2, respectively. While the fast component is limited by the time resolution of the experiment, the duration of the slow component shows a clear temperature dependence. The measurements were performed under extremely weak illumination in order to ensure that the potential distribution was not altered significantly by the illumination. Under this condition, the measured decay times of the slow component were found to be in approximate agreement with the expected dielectric relaxation times.

According to an analytical model by Ershov et al. [379], the time constant τ_2 associated with the slow component j_2 can be obtained from the equation

$$\frac{1}{\tau_2} = \frac{1}{\epsilon_0 \epsilon_r}\left[1 - g\left(1 - \exp\left(-\frac{1}{g}\right)\right)\right]\frac{\partial j_\mathrm{e}}{\partial \mathcal{E}_\mathrm{e}}. \tag{10.5}$$

Here $\partial j_\mathrm{e}/\partial \mathcal{E}_\mathrm{e}$ is the specific differential conductivity at the emitter barrier with the local electric field $\mathcal{E}_\mathrm{e}$. Note that, in general, τ_2 according to (10.5) differs to some extent from the dielectric relaxation time $\tau_D = \epsilon_0\epsilon_r/(\partial j_\mathrm{e}/\partial \mathcal{E})$ of a resistive dielectric. While $\tau_2 \approx \tau_\mathrm{D}$ for $g \ll 1$, we have $\tau_2 \approx 2g\tau_\mathrm{D}$ for $g \gg 1$.

The re-charging of the quantum wells not only influences the photocurrent, but it also influences the frequency dependence of the noise. While noise in QWIPs is generally assumed to be white, there are some frequency dependencies observed in the vicinity of the "recharging frequency" $f_\mathrm{R} = 1/2\pi\tau_2$ associated with the time constant τ_2. This recharging frequency acts as a cut-off frequency for the noise associated with j_2, whereas only j_1 will contribute

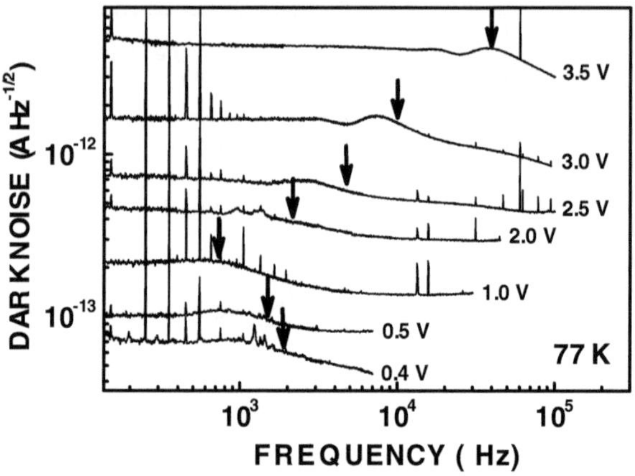

Fig. 10.5. Dark current noise spectra of an InGaAs/GaAs QWIP with 20 periods for different applied bias voltages. The arrows indicate the values of f_R as calculated from experimental dark I–V and gain measurements (after [380])

to the noise at higher frequencies. The frequency dependence of the noise current should thus exhibit a steplike behavior around f_R.

This influence has been studied by Rehm et al. [380] using an InGaAs/GaAs QWIP structure with 20 periods. Figure 10.5 shows the frequency dependent dark current noise at different bias voltages. The spectra clearly reveal a steplike behavior at certain noise frequencies. For comparison, the vertical arrows in Fig. 10.5 indicate the expected values of τ_R according to (10.5). Not only the position, but also the height of the measured steps in the noise spectra agree nicely with the theoretically expected values. Residual deviations between the theoretical and experimental recharging frequencies are attributed to the uncertainty of the precise value of $\mathcal{E}_e$, which was approximated by the average field $\mathcal{E}$ applied to the QWIP structure. While $\tau_2 \approx 0.2$ ms at 1 V in Fig. 10.5, recharging times can easily become as long as hours when working at liquid Helium temperatures.

10.1.2 Picosecond Photocurrent

In order to investigate the fast component of the transient photocurrent in QWIPs, we have used $In_{0.28}Ga_{0.72}As/GaAs$ QWIP structures with $N = 20$ and $N = 100$ periods in the active region. Each period contained a 47.5 nm undoped barrier and a 4.5 nm wide QW with a sheet concentration of 3.8×10^{11} cm^{-2}. The peak wavelength of the photocurrent was 9.2 µm at 8 V.

Subpicosecond IR pulses were generated by difference frequency mixing of the signal and idler beams of an optical parametric oscillator, which was pumped by a mode-locked Ti:Sapphire laser. The optical setup produces MIR pulses with a pulse duration below 200 fs and a wavelength tunability between 6 and 18 µm [179].

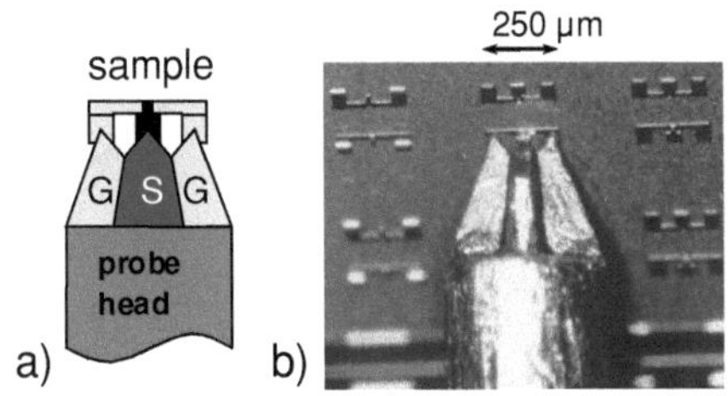

Fig. 10.6. Schematics (**a**) and photograph (**b**) of the ground-signal-ground sample geometry with 100 μm pitch and the microwave probe

A high electrical bandwidth was provided by a ground-signal-ground sample geometry, which is compatible with a coplanar 45 GHz microwave probe with 100 μm pitch (see Fig. 10.6). A semirigid high-frequency cable connected the microwave probe with a bias network (HP 11612B), which separated the high frequency component of the photocurrent signal from the dc dark current. The photocurrent transients were recorded by a sampling oscilloscope (Tektronix CSA 8000) equipped with a 50 GHz sampling head. An external trigger signal was generated by a reflected part of the Ti:Sapphire laser in a fast p–i–n diode.

Figure 10.7 shows the first 9 ns of a typical photocurrent transient of the 20-period InGaAs/GaAs QWIP at 77 K and a bias of 5.2 V. The observed two-component behavior of the photocurrent exhibits a strong "slow" component since the gain under these conditions is very high ($g \approx 7$). The inset also indicates fit functions for j_1 and j_2, which enabled us to separate the two components experimentally.

Figure 10.8a summarizes the transient photoresponse of the device with 100 periods. While the rise time is independent of the bias voltage, the decay

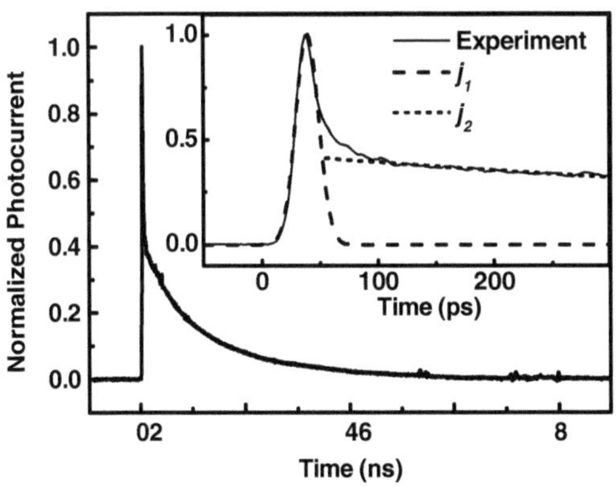

Fig. 10.7. Transient photocurrent of a 20-period InGaAs/GaAs QWIP at 77 K and 5.2 V. The *inset* shows the photocurrent and fit functions for j_1 and j_2 in an expanded time-scale [381]

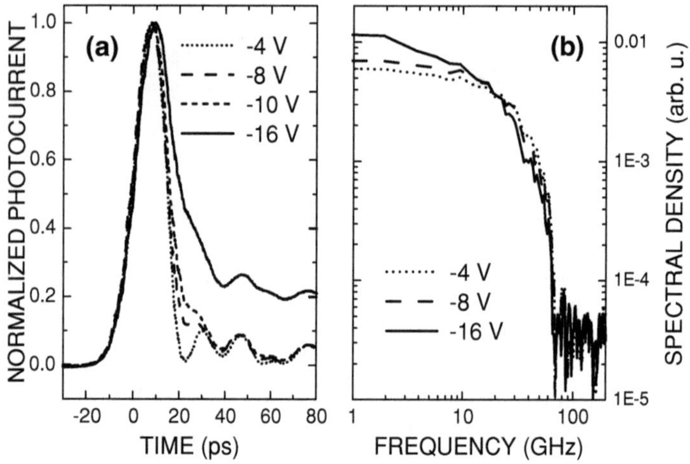

Fig. 10.8. Transient photocurrent (**a**) and corresponding Fourier transforms (**b**) of a 100-period InGaAs/GaAs QWIP device at 77 K and several bias voltages (after [371])

shows some prominent changes. At −16 V bias, contributions of both j_1 and j_2 are clearly observed. Here j_2 yields a pronounced signal since the recharging time is only about 200 ps, short enough to be covered by the temporal measurement window of the experimental setup. To allow a detailed study of the "fast" photocurrent contribution j_1, we thus have to choose conditions, where the slow component can be neglected. In Fig. 10.8a, this requirement is satisfied at bias voltages ≤10 V. This can be seen from the fact that the respective photocurrent traces exhibit the same value at, say, 80 ps, indicating that the slow component takes so long (µs regime) that the associated current becomes negligible.

Figure 10.8b shows the Fourier transform of the photocurrent transients and illustrates the corresponding frequency dependence of the photocurrent. The plot shows two different frequency ranges, with qualitatively different dependencies of the spectral density. At low frequency we observe an increase, and at high frequency a decrease, of the spectral density with increasing bias voltage. This behavior is caused by the fact that the frequency dependence is determined by the total photocurrent $j_1 + j_2$ at low frequency, and by the "fast" photocurrent j_1 at high frequency.

For a quantitative investigation of the dynamics we take into account two characteristic time constants, the capture time τ_c, which describes the recapture of optically excited electrons, and the transit time $\tau_{\mathrm{trans,tot}}$, which is the drift time of a carrier across the whole active region from the emitter to the collector contact. The decay of the photocurrent associated with the captured electrons is exponential and the decay of the photocurrent due to the arrival of optically excited electrons at the collector contact can be described linearly. Therefore the total decay of the "fast" photocurrent $I_1(t)$ can be expressed as [379]

$$I_1(t) = \begin{cases} I_0 \left(1 - \frac{t}{\tau_{\text{trans,tot}}}\right) \exp(-t/\tau_{\text{c}}) & \text{for } t < \tau_{\text{trans,tot}}, \\ 0 & \text{for } t \geq \tau_{\text{trans,tot}}. \end{cases} \quad (10.6)$$

According to (10.6), it is crucial to use a QWIP with a large $\tau_{\text{trans,tot}}$, i.e., large period number N, in order to measure τ_{c} in a reliable way.

To extract the characteristic time constants from the experimental data, we have created a fit-function based on (10.6), which additionally considers the finite electrical bandwidth of the measurement setup. Assuming that the system response, i.e. the response of the electric circuit to a δ-like excitation, can be described by a Gaussian pulse with a finite temporal width. Then, the fit-function for the time dependence is given by the convolution of this Gaussian pulse and (10.6). The Fourier transform of this convolution results in a product of the Fourier transforms of each contribution.

Since the voltage dependent feature of the ps photocurrent transient is small and disturbed by parasitic oscillations (see Fig. 10.8a), the Fourier transform of the transient allows a better analysis. In Fig. 10.9a, the Fourier transform of a transient at 77 K and 7 V and the corresponding fit function are plotted. The fitting procedure was performed between the 2 GHz lower frequency, which is related to the measurement window of 500 ps, and the upper limit of 73 GHz, where any photocurrent contribution from the QWIP to the signal vanishes. To restrict the number of fit parameters we eliminated $\tau_{\text{trans,tot}}$ by the gain g via $g = \tau_{\text{c}}/\tau_{\text{trans,tot}}$. The latter was determined from noise measurements. The corresponding noise gain at 77 K is plotted in Fig. 10.8b.

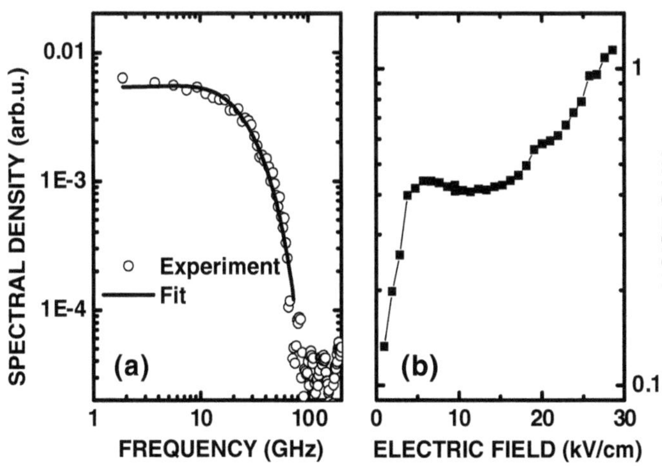

Fig. 10.9. (**a**) Fourier transform of a transient at 77 K and 13.3 kV cm^{-1} and the corresponding fit function. (**b**) Noise gain of the 100 period sample at 77 K as a function of the electric field (after [371])

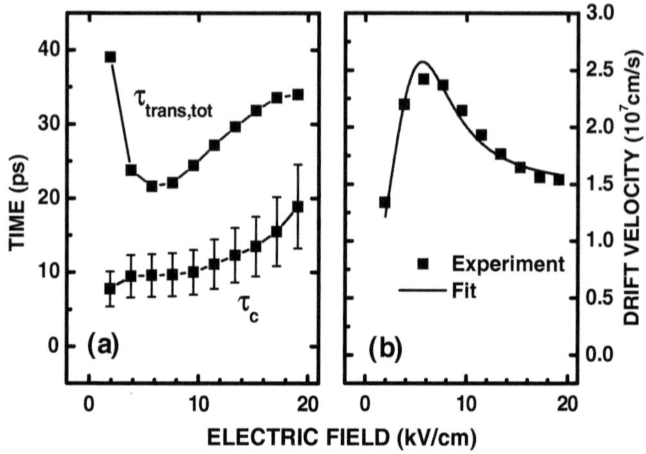

Fig. 10.10. (**a**) Capture time τ_c and transit time $\tau_{trans,tot}$ at 77 K as a function of the electric field. (**b**) Drift velocity extracted from our measurements as a function of the electric field and fit function describing the negative differential behavior (after [371])

The capture time τ_c established by this method is plotted in Fig. 10.10. The error bars indicated in Fig. 10.10a take into account a relative uncertainty of 30%. This uncertainty of the capture time is related to the assumed system response time, i.e. the full-width at half-maximum (FWHM) of the Gaussian pulse. For the lower limit, we estimated an FWHM of 8.5 ps, which corresponds to the specified 3-dB bandwidth of the bias-tee (45 GHz). As an upper limit for the system response time, we took the FWHM of the fastest ever measured transient of 12.7 ps. For our fitting procedure we used an FWHM of 10 ps, which is in between these limits.

The calculated values of the transit times using the measured capture times enable us to determine the bias dependence of the drift velocity v_d via the relation $v_d = D/\tau_{trans,tot}$, where D is the length of the active region. The result is plotted in Fig. 10.10b and shows the typical negative differential behavior of the barrier material GaAs [226], caused by intervalley scattering of electrons (see Sect. 7.2.2). To check whether the drift velocity determined in this way is physically justifiable, we fitted (7.6) to the measured data. Figure 10.10b shows that this fit accurately reproduces the experimental data. In addition, this procedure results in the saturation drift velocity $v_{sat} = 1.49 \times 10^7\,\mathrm{cm\,s^{-1}}$, electron mobility $\mu_e = 6,140\,\mathrm{cm^2\,Vs^{-1}}$, and critical field $F_c = 5.9\,\mathrm{kV\,cm^{-1}}$. This comparison shows that μ_e and v_{sat} are between the values observed in conventional GaAs/AlGaAs-QWIPs and those of bulk GaAs. The observed higher values of μ_e and v_{sat} in the present InGaAs/GaAs-QWIP as compared to standard GaAs/AlGaAs QWIPs are not unexpected since similar deviations exist between the respective barrier materials, i.e., GaAs and AlGaAs [256].

10.2 High Frequency and Heterodyne QWIPs

The measurement of high-speed and high-frequency characteristics can be done in either the frequency [215,219,369] or the time [148,370,371] domain. In this section, we discuss the physics and show results in the frequency domain.

10.2.1 Microwave Rectification

We first discuss the microwave rectification technique [219]. We apply a microwave signal to the QWIP and measure the change in its dc biasing current. This is complementary to the optical heterodyne technique, which involves generating a microwave signal within a QWIP at the difference frequency of two optical beams [369]. The rectification in the QWIP relies on its inherent nonlinear I–V characteristic, and therefore probes its transport properties.

The small-signal rectified dc current is given by

$$I_{\mathrm{rect}} = \frac{1}{4} I'' V_\mu^2, \tag{10.7}$$

where I'' is the second derivative of the I–V curve and V_μ is the amplitude of the microwave voltage applied to the device. Both I'' and V_μ depend on the microwave frequency (ω). The dependence of I'' on ω reflects the frequency roll-off behavior of the intrinsic transport mechanism, and therefore is expected to behave as $1/[1 + (\omega\tau)^2]$, where τ is some characteristic time. This τ is expected to be approximately the excited electron lifetime or the photoconductive lifetime. Given a constant output power from a microwave source, V_μ varies as a function of frequency because of the circuit limited by the device capacitance and differential resistance, and other parasitics. We then rewrite (10.7), separating out the frequency dependences:

$$I_{\mathrm{rect}} = \frac{1}{4} I_0'' V_{\mu 0}^2 \, \alpha(\omega)\beta(\omega), \tag{10.8}$$

where

$$\alpha(\omega) = \frac{1}{1 + (\omega\tau)^2}, \tag{10.9}$$

$\beta(\omega)$ is the circuit dependence, and I_0'' and $V_{\mu 0}$ are the low frequency limiting values of I'' and V_μ, respectively.

The experiment is schematically shown in Fig. 10.11. QWIPs were connected to the end of a 50-Ω coplanar transmission line by a short wirebond. The dc bias was applied through a bias-tee. The microwave power was supplied by a microwave source capable of frequencies up to 40 GHz. The QWIP was modeled by a parallel resistance–capacitance (RC) equivalent circuit and the parasitic inductance (L) was caused by the wirebond. Given the equivalent circuit of the device and its parasitic inductance (Fig. 10.11), the microwave voltage across the device can be straightforwardly found:

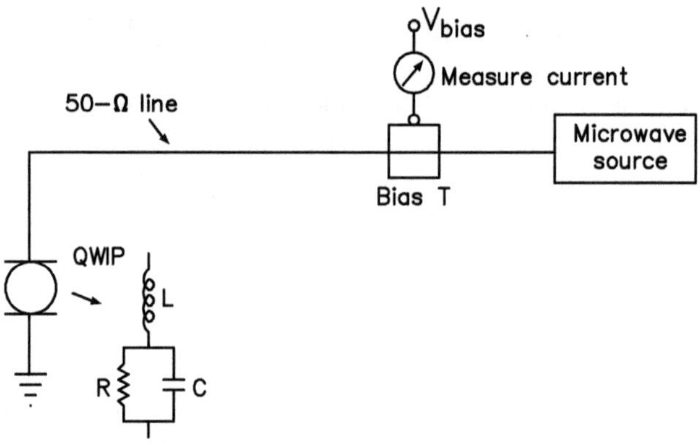

Fig. 10.11. Schematic of the microwave rectification experiment. The QWIP is mounted at the end of a 50-Ω transmission line. The QWIP is modeled by a parallel resistance-capacitance equivalent circuit and the parasitic inductance is caused by the short wirebond

$$V_\mu^2 = \frac{8R_\mathrm{L}}{(1-\omega^2 LC)^2 + \omega^2(R_\mathrm{L}C + L/R)^2} P_\mathrm{out}, \tag{10.10}$$

where $R_\mathrm{L} = 50\,\Omega$ is the line impedance, $R = 1/I'$ is the device differential resistance, I' is the derivative of the I–V curve, and P_out is the output power from the microwave source. We have made the approximation $R + R_\mathrm{L} \approx R$ because $R_\mathrm{L} \ll R$ for a typical QWIP. For $\omega \to 0$ we get

$$V_{\mu 0}^2 = 8R_\mathrm{L}P_\mathrm{out}, \tag{10.11}$$

and by comparison with (10.8)

$$\beta(\omega) = \frac{1}{(1-\omega^2 LC)^2 + \omega^2(R_\mathrm{L}C + L/R)^2}. \tag{10.12}$$

The $L \to 0$ limit of the above gives the usual RC roll-off: $1/[1 + (\omega R_\mathrm{L}C)^2]$.

We show experimental results on the three samples listed in Table 10.1. The samples differ mainly in the number of quantum wells. All experiments were carried out with the sample at 77 K. The relevant device parameters

Table 10.1. Sample structural parameters

Sample	x	L_w (nm)	L_b (nm)	Repeats
16W	0.265	5.7	23.7	16
8W	0.260	5.9	24.6	8
4W	0.276	6.0	24.6	4

(The Si doping density in the two contact layers was $1.5 \times 10^{18}\,\mathrm{cm}^{-3}$ and the center Si δ-doping density in the wells was $9 \times 10^{11}\,\mathrm{cm}^{-2}$ for all samples)

Table 10.2. Sample device parameters

Sample	L_{tot} (μm)	C (fF)	f_{RC} (GHz)	f_{max} (GHz)	τ (ps)
16W	0.49	21.5	148	33	4.8
8W	0.27	39.4	81	≈ 33	≈ 4.8
4W	0.15	72.2	44	≈ 33	≈ 4.8

(The device size is $10 \times 10\,\mu\text{m}^2$. L_{tot} is the total active device thickness, C is the device capacitance, $f_{\text{RC}} = 1/(2\pi R_{\text{L}} C)$, $R_{\text{L}} = 50\,\Omega$, f_{max} is the lifetime limited cutoff frequency, and $\tau = 1/(2\pi f_{\text{max}})$)

are listed in Table 10.2. The expected RC characteristic frequency is $f_{\text{RC}} = 1/(2\pi R_{\text{L}} C)$ for $R_{\text{L}} = 50\,\Omega$ load resistance. The estimated carrier lifetime (τ) and the intrinsic cutoff frequency (f_{max} defined by $\tau = 1/(2\pi f_{\text{max}})$) will be discussed later.

Figure 10.12 shows the measured (dots) and calculated (lines) rectified current versus frequency for the three samples. The 3 dB point from the maximum f_{max} for the 16 well sample (sample 16W) is about 33 GHz. For this sample, the RC-limited frequency is much larger than 33 GHz, and hence the roll-off here directly relates to τ resulting in a value of 4.8 ps. However, for the

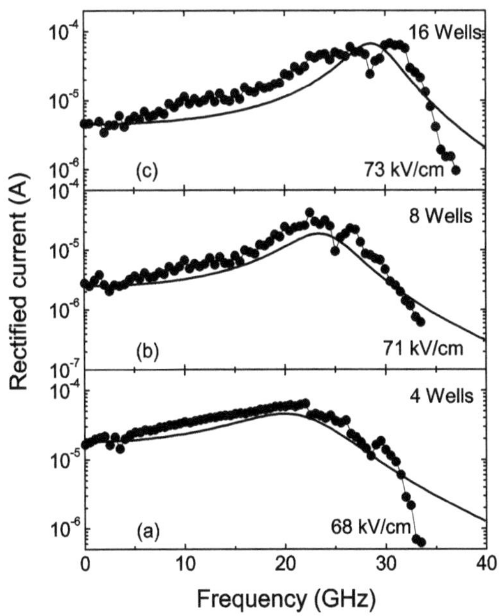

Fig. 10.12. Rectified current vs. frequency for (**a**) the 4 well, (**b**) the 8 well, and (**c**) the 16 well samples. The *solid lines* are calculated. The bias voltages were chosen to give approximately the same electric field of $70\,\text{kV cm}^{-1}$. All devices areas were $10 \times 10\,\mu\text{m}^2$. The device temperature was 77 K. The microwave power was 1.0 mW

4 and 8 well samples, the RC effect is clearly seen. The rectified current starts to roll off at a lower frequency and with a much slower rate. The calculated curves in Fig. 10.12 used the expressions given above with the inductance L determined by the length of the wire-bond. Further details are given in [219].

10.2.2 Heterodyne Detection

Heterodyne detection involves a local oscillator (LO), commonly a laser, at a slightly different wavelength from that of the signal. The difference frequency signal (often referred to as the IF – the intermediate frequency signal) is measured. This method is also used to characterize the high-frequency behavior by beating two lasers and measuring the IF as a function of the frequency.

Although QWIPs are very well suited to heterodyne detection, this has received limited attention [215,369,373,382,383]. The advantage of heterodyne detection is well known and described [31]. The key point is that with a sufficiently high LO power the ideal detection limit can be reached, and the sensitivity (e.g., the noise-equivalent power (NEP)) depends only on the absorption quantum efficiency for a given wavelength. The argument can be easily constructed for a photoconductive QWIP as follows. As established, the detector current responsivity is written as:

$$\mathcal{R}_i = \frac{e}{h\nu}\eta g, \tag{10.13}$$

where η is the absorption quantum efficiency and g is the photoconductive gain. Under a strong LO power (P_{LO}), i.e., the LO induced current (I_{LO}) dominates the dark current, the g-r noise current spectral density is

$$S_i = 4egI_{\mathrm{LO}} = 4eg\mathcal{R}_i P_{\mathrm{LO}}. \tag{10.14}$$

Using (10.13), this becomes

$$S_i = (2eg)^2 \frac{\eta}{h\nu} P_{\mathrm{LO}}, \tag{10.15}$$

and hence the noise current is

$$i_{\mathrm{noise}} = \sqrt{S_i B} = 2eg\sqrt{\frac{\eta}{h\nu} P_{\mathrm{LO}} B}, \tag{10.16}$$

where B is the measurement bandwidth.

The heterodyne current for a signal power of P_{sig} is

$$i_{\mathrm{het}} = 2\mathcal{R}_i \sqrt{P_{\mathrm{LO}} P_{\mathrm{sig}}} = 2eg\frac{\eta}{h\nu}\sqrt{P_{\mathrm{LO}} P_{\mathrm{sig}}}, \tag{10.17}$$

using (10.13).

The minimum detectable signal is when $i_{\rm noise} = i_{\rm het}$. From (10.16) and (10.17), we then have

$$\frac{P_{\rm sig,min}}{B} = \frac{h\nu}{\eta}, \tag{10.18}$$

or equivalently,

$$({\rm NEP})_{\rm het} = \frac{{\rm h}\nu}{\eta}{\rm B}. \tag{10.19}$$

The heterodyne NEP is proportional to the measurement bandwidth or inversely proportional to the measurement time (different from the square root dependence for the usual detection scheme). It is shown from (10.19) that the heterodyne NEP depends only on the absorption quantum efficiency η and the photon quantum $h\nu$, and is independent of gain g. This of course is in the regime where LO induced current is dominant, referred to as photon noise limited. If a similar analysis is carried out for an ideal photodiode, an NEP of a factor of two better is obtained.

For an experimental implementation of heterodyne detection, the LO and signal beams should have the same polarization, and their phase fronts should be parallel. Due to this spatial coherence between the two beams, heterodyne detection is often referred to as "coherent detection."

The direct measurement of the IF signal is limited to frequencies up to the capability of the available spectrum analyzer. To achieve higher frequencies, we can employ mixing in the QWIP to down-convert the IF signal generated by the optical heterodyne [215]. This again (as in the rectification case) relies on the nonlinear I–V characteristic as in rectification. Specifically, we apply to the QWIP not only the two IR beams, as in a conventional heterodyne experiment, but an additional microwave excitation. Furthermore, let $f_{\rm IR1}$ and $f_{\rm IR2}$ denote the two IR frequencies, and $f_{\rm \mu wave}$ the microwave frequency. The IR heterodyne frequency is then $f_{\rm het} = |f_{\rm IR1} - f_{\rm IR2}|$, while the down-converted signal frequency is $|f_{\rm het} - f_{\rm \mu wave}|$. In this scheme, one can reach very high $f_{\rm het}$ frequencies using a standard spectrum analyzer at the output of the QWIP. Of course, when $f_{\rm het}$ and $f_{\rm \mu wave}$ are both sufficiently small we also observe the up-converted signal at $f_{\rm het} + f_{\rm \mu wave}$ on the spectrum analyzer.

The device used in this experiment was a 100-well QWIP with a GaAs well width of 4.5 nm and an $Al_{0.21}Ga_{0.79}As$ barrier width of 40 nm. The center 2.5 nm of each well was doped with Si to 2.5×10^{18} cm^{-3}. The IR sources were either two CO_2 lasers [384], or a CO_2 laser and a lead-salt temperature tunable diode laser (TDL) [385]. The microwave or millimeter-wave radiation was either generated by a microwave source tunable up to 40 GHz or by a Gunn oscillator mechanically tunable in the range of 91–94 GHz. The CO_2 lasers were operated in the neighborhood of 10.3 μm on individual lines selected by gratings with a separation between adjacent lines of about 41 GHz. The TDL was used only for $f_{\rm het} < 26.5$ GHz, which was the limit of the spectrum analyzer. A schematic of the experimental techniques is shown in Fig. 10.13.

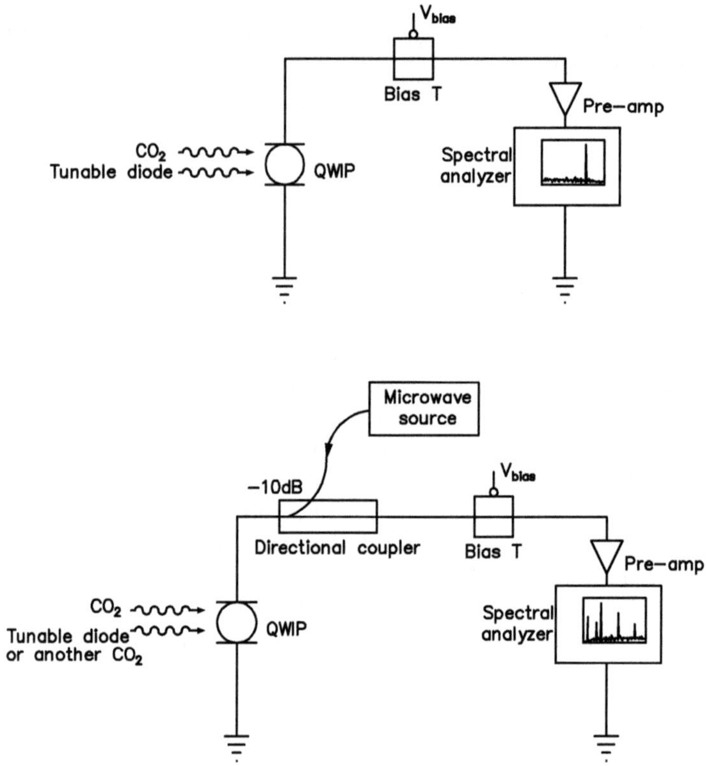

Fig. 10.13. A schematic of heterodyne detection experiment: (*above*) direct measurement of the IF and (*below*) mixing on QWIP for extending the measurement frequencies

The experimental results covering the frequency range of about 1–100 GHz are shown in Fig. 10.14. The curves show the expected roll-off behavior due to the device RC time constant and the photocarrier lifetime, again using $\tau \sim 5$ ps. The optical heterodyne data were taken with a CO_2 laser and the TDL laser [369]. We have normalized the signal for a constant incident power of about 0.2 mW from each of the IR lasers and 0.3 mW from the microwave source. The measured signal in the direct heterodyne case agrees with (10.17) in the low frequency limit. The five data points [215] shown in Fig. 10.14 for the mixing experiment were taken using different sources and different microwave coupling schemes. For the first three points at $f_{\text{het}} = 1.83$, 5.37, and 15.5 GHz, we used one CO_2 laser, the TDL, and the microwave source at frequencies of $f_{\mu\text{wave}} = 9$, 15, and 10 GHz, respectively, and measured the sum signals of f_{het} and $f_{\mu\text{wave}}$. The point at 41.42 GHz was obtained using two CO_2 lasers separated by $f_{\text{het}} = 41.42$ GHz and the microwave source at $f_{\mu\text{wave}} = 20$ GHz, measuring the difference signal at $41.42 - 20 = 21.42$ GHz. The point at 82.16 GHz was obtained using two CO_2 lasers separated by

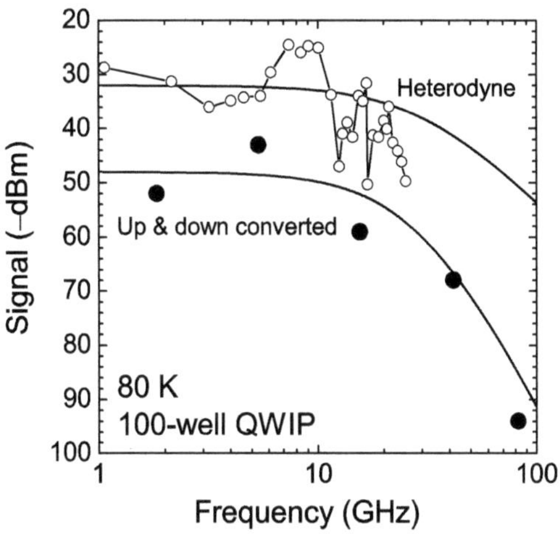

Fig. 10.14. Direct infrared heterodyne and mixed heterodyne frequency with microwave frequency signal vs. heterodyne frequency for a bias voltage of 2 V. The *curves* show the expected roll-off behaviors. The heterodyne frequency is defined as the difference between the two IR frequencies. The incident powers from the two IR sources and the microwave source are normalized to about 0.2 and 0.3 mW, respectively. The device temperature was about 80 K

this frequency and a Gunn oscillator mechanically tuned in the range of 91–94 GHz. The difference signal in the range of 9–12 GHz was measured. For this data point, millimeter-wave radiation was coupled into the device through free space.

To cross-check the consistency between results measured in frequency and time domains, both microwave rectification and time resolve photocurrent measurements were made on one of our QWIPs. The Fourier transform of the time resolved photocurrent is compared with the microwave rectification curve in Fig. 10.15. For this experiment, we used a QWIP very similar to sample QWIP10 in Sect. 8.1, but with a doping concentration of $10^{12}\,\mathrm{cm}^{-2}$. The details of this sample were described in [220]. The plot in Fig. 10.15 exhibits an excellent quantitative agreement between the two experimental methods. There was a defect in the packaging of this device, resulted in a drop in the frequency response at about 8 GHz.

Recently, substantial advances both in fabrication and packaging of high frequency devices and in measurement techniques have been made [386]. Air-bridge and coplanar waveguide have been monolithically integrated with the QWIP, eliminating the wirebond. Direct measurements of the heterodyne signal over coaxial cables have been performed up to 110 GHz and from cryogenic to room temperatures. Data measured on a 100-well QWIP with $10^{12}\,\mathrm{cm}^{-2}$ well doping are shown in Fig. 10.16, representing a substantial improvement over the results in Fig. 10.14.

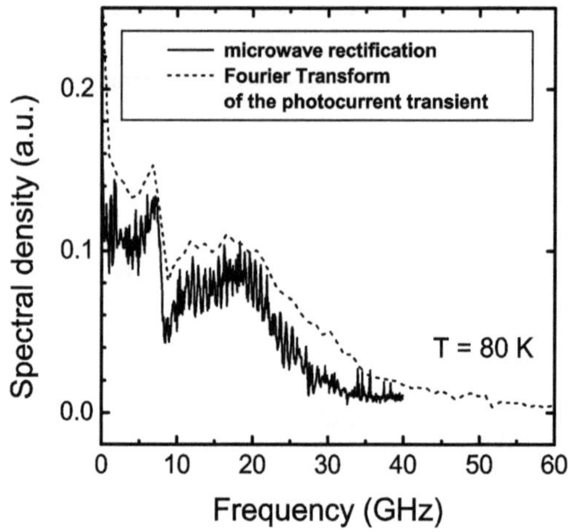

Fig. 10.15. Comparison of the Fourier transform of the photocurrent transient and the frequency dependence obtained from the microwave rectification method

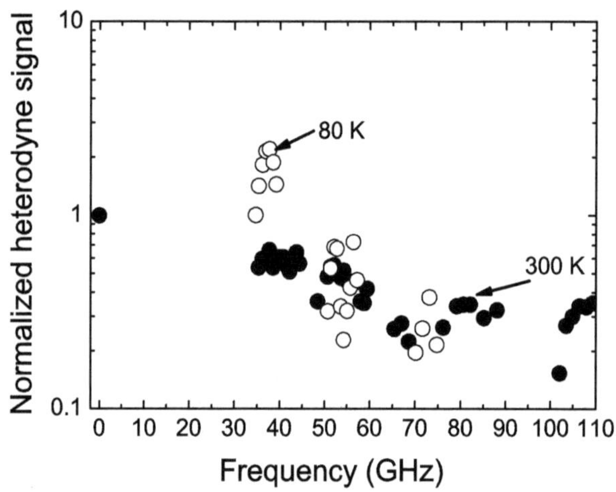

Fig. 10.16. Normalized heterodyne signal taken at 80 K (*circles*) and 300 K (*dots*)

10.3 Two-Photon QWIP

10.3.1 Equidistant Three-Level System for Quadratic Detection

While the detection methods so far discussed in this chapter need to take into account the residual capacitance and inductance of the QWIP device and its packaging, much faster signals can be studied by exploiting the intrinsic nonlinearity discussed in Sect. 7.2.3. Since such nonlinear spectroscopy usually suffers from a lack of detection sensitivity, an artificial three-level system has been realized by energetically equidistant subbands in a semiconductor

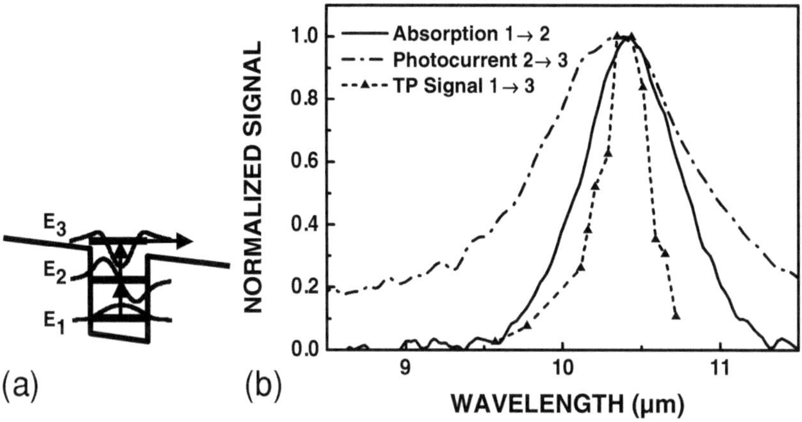

Fig. 10.17. (**a**) Schematics of the two-photon QWIP. (**b**) Normalized absorption at 77 K in Brewster-angle geometry, photocurrent at 130 K due to excitation from thermally populated subband 2 into subband 3, and two-photon photocurrent at 77 K under excitation by a continuous-wave CO_2 laser, vs. excitation wavelength [387]

quantum well [387,388], as indicated in Fig. 10.17a. Subbands 1 and 2 are bound in the QW, while the third state is a continuum resonance located close to the barrier energy. In an external electric field, the carriers excited into the continuum are swept out of the QW and give rise to a photocurrent. According to numerical simulations, a 7.6 nm thick GaAs QW, sandwiched between $Al_{0.33}Ga_{0.67}As$ barriers, is optimized for a transition wavelength of 10.2 μm, whereas operation at 7.9 μm is achieved using a 6.8 nm $In_{0.10}Ga_{0.90}As$ QW and $Al_{0.38}Ga_{0.62}As$ barriers.

Figure 10.17b summarizes the spectral characteristics of a two-photon QWIP comprising 20 GaAs QWs of 7.6 nm width, Si-doped to an electron concentration of $n_{2D} = 4 \times 10^{11}\,\mathrm{cm}^{-2}$, and $Al_{0.33}Ga_{0.67}As$ barriers (for the first device). The spectral dependence of the optical transition from the first to the second subband ($1 \rightarrow 2$) was obtained through intersubband absorption measurements in a Brewster-angle geometry at 77 K. The $2 \rightarrow 3$ transition was studied through photocurrent measurements at 130 K, where thermal population of the second subband causes a signal. The spectral dependence of the two-photon photocurrent at 77 K has been measured with a wavelength-tunable CO_2 laser. Most remarkably, all three experimental curves give identical peak wavelengths (10.4 μm), indicating perfectly equidistant subband energies ($E_2 - E_1 = E_3 - E_2 = 119\,\mathrm{meV}$). The spectrum associated with CO_2 laser illumination is narrower than the absorption spectrum, which reflects the quadratic nature of the two-photon transition [212].

Figure 10.18 shows the photocurrent versus the power density P upon CO_2 laser illumination at 10.2 μm. The photocurrent exhibits a linear increase with P up to about $0.1\,\mathrm{W\,cm^{-2}}$ at 70 K (up to $1\,\mathrm{W\,cm^{-2}}$ at 90 K) followed by a

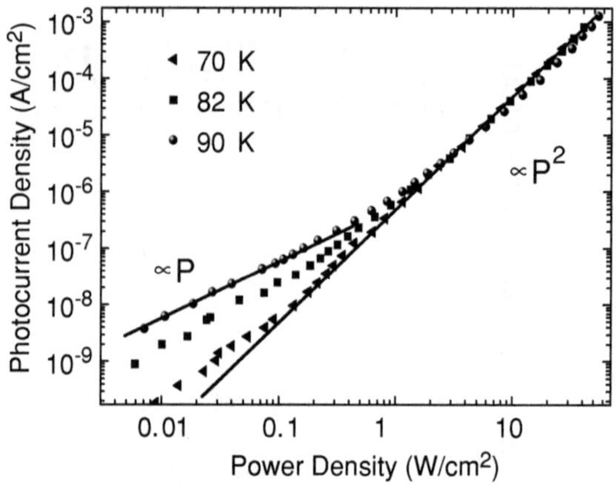

Fig. 10.18. Photocurrent density under continuous wave illumination at a wavelength of 10.3 μm and 1.5 V detector bias versus power density at different temperatures as indicated [387]

quadratic increase. Closer investigation at different temperatures up to 160 K reveals thermally activated behavior of the photocurrent in the linear regime, with an activation energy of 104 meV. This result allows us to conclude that the linear regime is caused by thermal population of E_2. Quadratic behavior can be achieved even below 0.1 W cm^{-2} when operating the device at lower temperature.

We now determine the two-photon absorption coefficient β. In a two-photon detector with an absorbing region of thickness L, the associated quantum efficiency η_{2P} is given by $\eta_{2P} = \beta P L f_\theta$. Here the factor f_θ relates to the selection rule for intersubband transitions. If θ is the angle between light propagation and the sample normal, we have $f_\theta = \sin^4\theta/\cos\theta$ for two-photon absorption, whereas the corresponding factor in the linear case is $\sin^2\theta/\cos\theta$. β is defined via its effect on the power density $P(z)$ along the detector normal z, which obeys $\mathrm{d}P/\mathrm{d}z = -\beta f_\theta P(z)$ in the case of pure two-photon absorption. The approximation $\beta P L \ll 1$, justified up to power densities approaching 10^6 W cm^{-2}, then yields $P(z) \approx P(0) - \beta f_\theta P z$. Using the definition of the responsivity (2.8), the two-photon photocurrent density j_{2P} is thus expressed as

$$j_{2P} = \frac{eg}{h\nu}\beta L f_\theta P^2. \tag{10.20}$$

For a double-pass at $\theta = 45°$, the light traverses $2N = 40$ absorbing QWs with a total thickness of $L = NL_\mathrm{w} \approx 0.3$ μm. The gain as determined from noise measurements amounts to $g = 0.3$ at 1.5 V. Equation (10.20) thus yields a two-photon absorption coefficient of 1.3×10^7 cm GW^{-1}. This nonlinearity is six orders of magnitude higher than in typical bulk semiconductors (Si, ZnSe, GaAs), where values of $\beta \approx 10$ cm GW^{-1} are observed. Our present results also

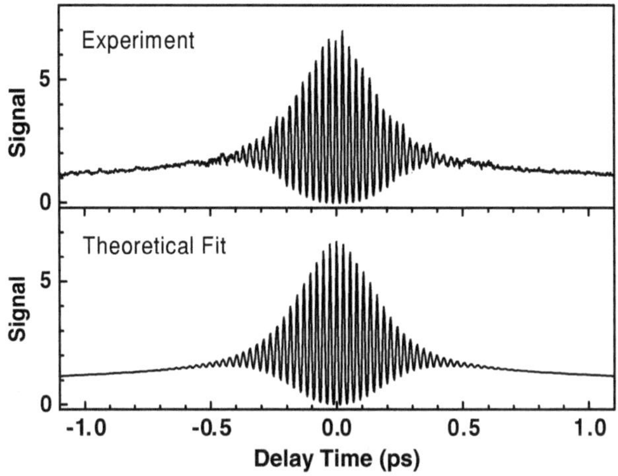

Fig. 10.19. Two-photon photocurrent autocorrelation of fs optical pulses at 8.0 µm (*upper part*) and numerical fit (*lower part*) vs. delay time. The signal is normalized to its value at a large delay time above 10 ps [387]

indicate a huge increase of β as compared to the detuned two-photon QWIP discussed in Sect. 7.2.3, where β takes a value of about 5×10^3 cm GW^{-1} [216].

10.3.2 Autocorrelation of Subpicosecond Optical Pulses

We now exploit this detection scheme for quadratic autocorrelation measurements of ultrashort optical pulses. Using a Michelson-interferometer, the pulses are split into two parts separated by a variable time delay τ. While quadratic *intensity* autocorrelation measurements have been studied in [388], we concentrate here on *interferometric* autocorrelation. Here the interference between the pulses contains additional information associated with the relative phase between the pulses. The optical pulse source used for the experiment is based on difference frequency generation between the signal and idler pulses of an optical parametric oscillator pumped by a fs Ti:Sapphire laser [179]. The pulse source is wavelength tunable from 6 to 18 µm, with 10 pJ pulse intensity, 160 fs pulsewidth, and 76 MHz repetition rate. Due to the high sensitivity of the present two-photon QWIP, interferometric autocorrelation measurements can be performed in spite of the low optical power.

The upper part of Fig. 10.19 shows an autocorrelation trace measured at 77 K on a device with 6.8 nm $In_{0.10}Ga_{0.90}As$ quantum wells and $Al_{0.38}Ga_{0.62}As$ barriers. From these data, important information both on the temporal width of the optical pulses and on the dynamic properties of the two-photon QWIP can be obtained. While the interference fringes around zero time delay correspond to those of a nearly ideal autocorrelation of the optical field, two additional signatures are associated with intrinsic time constants of the detector. First, the amplitude of the interference fringes towards increasing positive and negative delay times shows an exponential rather than Gaussian-like decay. This behavior is attributable to the decay time T_2 (dephasing time) of the coherent intersubband polarization between the first and second subbands. Second, after the disappearance of the fringes, the signal still decreases exponentially with delay time towards its asymptotic value. The associated time constant T_1 is due to intersubband relaxation of electrons from the second back to the first subband.

While the ideal interferometric autocorrelation for Gaussian pulses of width σ is proportional to

$$j(\tau) \sim 1+2\mathrm{e}^{-4\ln 2(\frac{\tau}{\sigma})^2}+4\mathrm{e}^{-3\ln 2(\frac{\tau}{\sigma})^2}\cos(\omega\tau)+\mathrm{e}^{-4\ln 2(\frac{\tau}{\sigma})^2}\cos(2\omega\tau), \qquad (10.21)$$

a proper theoretical description of the experiment is based on the density matrix formalism and the optical Bloch equations [389]. The complexity of this three level system, however, prevents an analytic solution. We therefore use a phenomenologic approach, introduced by Nessler et al. [390] for electron photoemission in $Bi_2Sr_2CaCu_2O_{8+\delta}$, which shows satisfactory agreement with the theory of [389]. Whereas the first term of the right-hand side of (10.21) describes the signal for large time delay τ, the second term is affected by the population decay time T_1, which is taken into account by a convolution with a symmetric exponential decay function, $f(t) = \exp(-|t/T_1|)$. Similarly, the third term is broadened by the phase relaxation time T_2, whereas the broadening of the fourth term can be neglected due to the very short coherence time associated with the level E_3. These convolutions result in an analytical model function which allows for numerical fitting.

The lower part of Fig. 10.19 shows a numerical least squares fit of the measured autocorrelation. The fit yields $T_1 = 0.46$ ps and $T_2 = 0.10$ ps, which agrees with typical values obtained from degenerate four-wave mixing [391]. The value of T_2 corresponds to a broadening with a full-width at half maximum of $\Gamma_{12} = h/\pi T_2 = 13$ meV. Since the absorption linewidth of this sample has a similar value (11 meV), we conclude that the $1\rightarrow 2$ transition is homogeneously broadened. We note that similar time constants, $T_1 = 0.53$ ps and $T_2 = 0.13$ ps, have been obtained for the sample of Fig. 10.17b [388].

Theoretically, two different mechanisms give rise to photoemission involving two absorbed photons. One mechanism is a coherent two-photon transition which, like degenerate four-wave mixing, is associated with the third-order nonlinear susceptibility χ^3. Within second-order perturbation theory, it gives rise to the two-photon absorption coefficient

$$\beta_{\chi^3} = \left(\frac{e^2}{4\epsilon_0 n_\mathrm{r} mc}\right)^2 \frac{n_\mathrm{2D}}{L_\mathrm{w} h\nu} f_{12} f_{23} 2T_2^2 T_\mathrm{e} \qquad (10.22)$$

with the well width L_w, the refractive index n_r, and the dimensionless oscillator strengths f_{12} and f_{23} associated with the $1 \rightarrow 2$ and $2 \rightarrow 3$ transitions, respectively. T_e is the time constant associated with the broadening of state 3, which is attributed to the escape time. Equation (10.22) (together with

(10.20)) can be seen as a generalization of the two-photon quantum efficiency $\eta^{(2)}$ of (7.8) to the case of N quantum wells.

The other mechanism is a *sequential* two-step absorption

$$\beta_{\text{2step}} = \left(\frac{e^2}{4\epsilon_0 n_{\text{r}} mc} \right)^2 \frac{n_{\text{2D}}}{L_{\text{w}} h\nu} f_{12} f_{23} 4T_1 T_2 T_{\text{e}}, \tag{10.23}$$

where the population of subband 2 is either excited into 3 or decays with the time constant T_1.

While the two processes cannot be distinguished in a continuous-wave experiment, they have different signatures in the autocorrelation traces in Fig. 10.19. In fact, the χ_3 process manifests itself by interference fringes, limited with increasing time delay by the decay time T_2 of the coherent intersubband polarization. Sequential absorption produces the "slow" decrease in the autocorrelation trace, prominent at larger time delays when the coherence has been lost.

Interestingly, the expressions in (10.22) and (10.23) take the same value if the decoherence of the $1 \rightarrow 2$ transition is entirely due to relaxation (i.e., absence of pure dephasing [391]) and the relation $T_2 = 2T_1$ holds. In the present experiment, the coherence loss is mainly caused by pure dephasing, such that β_{2step} is almost an order of magnitude larger than β_χ^3. Using an estimated value of $T_{\text{e}} = 50\,\text{fs}$ (consistent with the linewidth of the measured $2 \rightarrow 3$ absorption), (10.23) in fact yields a value of $1 \times 10^7\,\text{cm}\,\text{GW}^{-1}$, which is in excellent agreement with the experimentally observed two-photon absorption coefficient. Even though $\beta_{\chi_3} \ll \beta_{\text{2step}}$, the χ_3 process is still dominant at short delay times, as (10.22) and (10.23) refer to the time-integrated response.

10.3.3 Externally Switchable Quadratic and Linear Response

We have recently reported a QWIP structure which can be switched between perfectly quadratic and linear behavior by reversing the polarity of the bias voltage [392]. The same detector can thus be used for measurements of the linear autocorrelation (or interferogram), second-order intensity autocorrelation, and second-order interferometric autocorrelation. According to Diels et al. [393], these measurements are sufficient to determine the electric field of IR pulses in amplitude and phase. For characterizing IR ps-laser pulses such as those generated by mode-locked quantum cascade lasers [217], the detector presented here combines the possibility of measuring the average output power in the linear regime with the determination of the pulse width in the case of quadratic power dependence.

Figure 10.20 depicts the conduction band edge of the device for both negative and positive bias voltage. The asymmetry with respect to the bias voltage is introduced into the structure by modulation-doping the quantum-wells on one side. In Fig. 10.20a, we have a three-level system where only electrons that absorb two photons reach the continuum resonance E_3 and contribute to

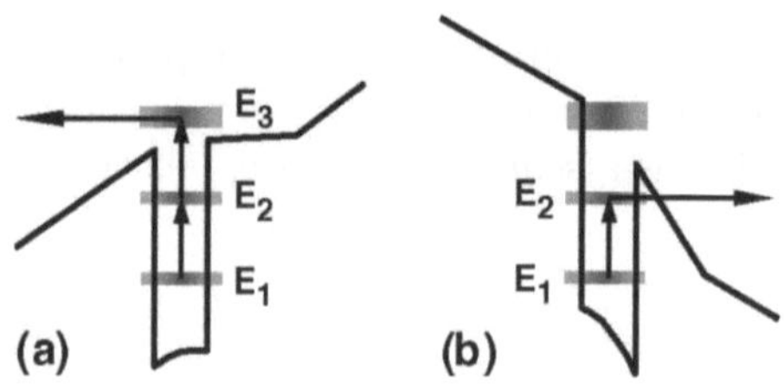

Fig. 10.20. Conduction band diagram under (**a**) negative and (**b**) positive bias voltage. The subbands are indicated as shaded areas [392]

a photocurrent. Radiation resonant with the $E_1 \rightarrow E_2$ transition leads to a strongly enhanced two-photon photoemission, such that the photocurrent is proportional to the square of the power density. Upon switching the bias polarity (Fig. 10.20b), the space charge in the barriers induced by the one-sided modulation-doping lowers the effective barrier height and results in intermediate state electron tunnelling, which becomes significantly more efficient than a two-step excitation. Therefore, the detector scheme under positive bias reduces to a two-level system and exhibits a behavior comparable to a standard (bound-to-quasibound) QWIP.

According to (10.22) and (10.23), the two-photon photocurrent arises from an incoherent two-step photoexcitation and two-photon absorption, which are proportional to T_1T_2 and T_2^2, respectively. Therefore, the highest two-photon sensitivity is expected in modulation-doped QWIPs, where reduced impurity scattering in the quantum-wells leads to larger intra- and inter-subband scattering times as compared to well-doped structures.

The InGaAs/AlGaAs QWIP-device under investigation contains 20 $In_{0.1}Ga_{0.90}$ QWs of 7.3 nm width separated by 46 nm wide $Al_{0.31}Ga_{0.69}As$ barriers. Each quantum-well is modulation-doped on one side, with a 2 nm thick doped region at a 12 nm distance from the quantum well, to a sheet doping concentration of $2 \times 10^{11}\,cm^{-2}$. Measurements of the linear photocurrent in a 45° facet geometry using a Fourier-transform IR spectrometer indicated a peak wavelength of 8.2 µm. Highly efficient tunnelling of intermediate state electrons is observed at positive bias, with more than $100\,mA\,W^{-1}$ responsivity at 9 V, whereas this process is negligible, e.g., at -2 V [392]. At moderate negative bias voltages, the photocurrent is thus quadratic with the incident power.

Figure 10.21a demonstrates the use of the detector as a quadratic autocorrelator using the 165 fs pulse source described in the previous section. The excitation energy was set to match the $E_1 \rightarrow E_2$ transition. Despite the low pulse energy of only a few pJ, the detector shows ideal quadratic behavior as can be seen from the peak-to-background ratio close to the theoretical value of 8:1. As before, the finite values of T_1 and T_2 lead to deviations from an ideal autocorrelation trace, and alow us to determine these time constants through numerical fitting. The thin line in Fig. 10.21a indicates the envelope of such a

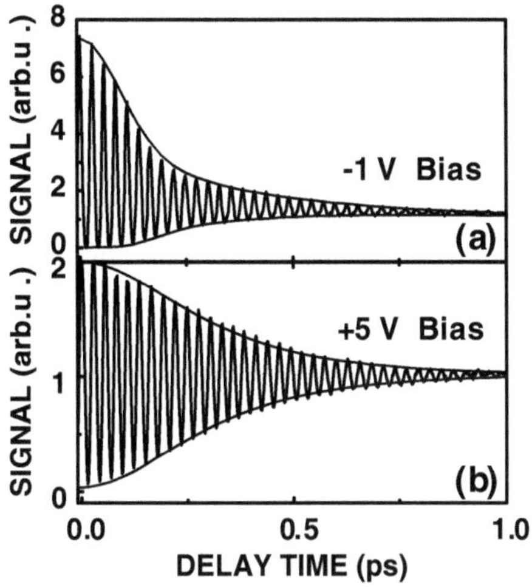

Fig. 10.21. Photocurrent for (a) negative and (b) positive bias voltage as function of the time between two pulses with 165 fs duration and a peak wavelength matched to the $E_1 \rightarrow E_2$ transition. The thin lines show the envelope of a numerical fit function. (after [392])

numerical fit. The obtained relaxation times are $T_1 = 640\,\text{fs}$ and $T_2 = 260\,\text{fs}$, which is consistent with theoretical calculations [91] and independent measurements using degenerate four wave mixing [394].

In contrast, a linear interference pattern with a 2:1 peak-to-background-ratio is found at positive bias (see Fig. 10.21b). Here the signal height is proportional to the intermediate state population, whereas the shape of the interferometric signal is independent of T_1. The numerical fit, which assumes dephasing in a two-level system, yields the same value for T_2 as in the quadratic case.

Comparing this detector with the one of Fig. 10.19, we notice the influence of impurity scattering on observed values of T_1 and T_2. Recent experiments performed on GaAs/AlGaAs two-photon QWIPs resulted in a similar two-fold increase of T_2 and 40%-increase of T_1 in modulation-doped QWIPs with respect to well doped reference structures [395]. These results indicate that impurity scattering is of similar importance as interface scattering (which dominates the dephasing in modulation-doped devices [80]), and that impurity scattering represents a non-negligible channel for intersubband relaxation (even though LO-phonon relaxation is still dominant).

It also becomes obvious from these considerations that two-photon QWIPs are not only useful for infrared pulse diagnostics, but that these devices also provide a convenient and precise experimental access to the dynamics of intersubband relaxation and dephasing.

Concluding this Section, two-photon QWIPs based on resonant optical transitions between three energetically equidistant subbands allow us to study the autocorrelation of weak infrared sources with unprecedented sensitivity and femtosecond temporal resolution. The observed nonlinear absorption ($\beta = 1.3 \times 10^7\,\mathrm{cm\,GW^{-1}}$) is more than three orders of magnitude higher than in previous, non-resonant devices. In this three-level system, both coherent and sequential two-photon absorptions are necessarily present, and manifest themselves through different experimental signatures, which also allow us to determine the dephasing time T_2 and the inter-subband relaxation time T_1. This approach enables quadratic detection at much lower radiation density. In addition, both linear and quadratic detection modes are possible through voltage-switchable devices. Two-photon QWIPs are thus expected to make substantial impact on the characterization and development of novel infrared and terahertz radiation sources.

11

Conclusions and outlook

To conclude, there is no question that the new quantum well approach to IR detection has been one of the success stories in modern semiconductor research. QWIPs presently represent a mature technology for high-performance thermal imaging, in fact the only technology providing large (i.e., 640×480 pixels and up) staring focal plane arrays for LWIR wavelengths.

Of course, each detector material has its pros and cons as compared to the others. This implies that the commercial market will be shared among different detector technologies, and no individual technology is likely to totally dominate. QWIPs will most likely not reach the high-temperature performance of bolometric and pyroelectric detectors, or the high quantum efficiencies of HgCdTe and InSb (although quantum efficiencies >50% at MWIR wavelengths have been reported for QWIPs implemented into focal plane arrays). In comparison with other technologies, however, the QWIP approach is favorable and attractive in many areas such as producibility, thermal resolution, spatial resolution, long-term stability, multicolor and multiband, high speed, and integration. Although there are many issues that need to be studied further, this new technology has found its way into the commercial market. As should be the case for any new successful technology, in addition to competing with existing technology, QWIPs have already created their own applications.

QWIPs are also extremely promising for the field of high-speed/heterodyne detection in the thermal IR, which is still at the beginning stage. In addition, QWIPs may provide practical and convenient devices for the field of terahertz science and technology. In both cases, the QWIP approach is likely to provide new and unique directions and solutions, and we therefore expect that QWIPs will contribute to many emerging applications.

As emphasized in this book, the QWIP approach is very flexible, and has created a wealth of different device structures that can be used for a variety of functionalities and applications. Many of these new directions are hitherto almost unexplored, such that it is likely that future research will give rise to exciting advances and developments. We hope that the present book will contribute to understanding and solving some of the known problems, and to excite new thoughts and ideas leading to novel devices and applications in the coming years.

References

1. H. Kroemer: All that glitters isn't silicon or steel and aluminum revisited. In: *Future Trends in Microelectronics: Reflections on the Road to Nanotechnology. Applied Sciences*, vol E323, ch. 1, ed by S. Luryi, J. Xu, A. Zaslavsky (Kluwer, Dordrecht 1996) pp. 1–12
2. T. Ando, A.B. Fowler, F. Stern, Rev. Mod. Phys. **54**, 437 (1982)
3. L.L. Chang, L. Esaki, G.A. Sai-Halaz, IBM Tech. Disc. Bull. **20**, 2019 (1977)
4. L. Esaki, H. Sakaki, IBM Tech. Disc. Bull. **20**, 2456 (1977)
5. L.C. Chui, J.S. Smith, S. Margalit, A. Yariv, A.Y. Cho, Infrared Phys. **23**, 93 (1983)
6. J.S. Smith, L.C. Chui, S. Margalit, A. Yariv, A.Y. Cho, J. Vac. Sci. Technol. B **1**, 376 (1983)
7. D.D. Coon, R.P.G. Karunasiri, Appl. Phys. Lett. **45**, 649 (1984)
8. D.D. Coon, R.P.G. Karunasiri, L.Z. Liu, Appl. Phys. Lett. **47**, 289 (1985)
9. D.D. Coon, R.P.G. Karunasiri, H.C. Liu, J. Appl. Phys. **60**, 2636 (1986)
10. K.W. Goossen, S.A. Lyon, Appl. Phys. Lett. **47**, 1257 (1985)
11. K.W. Goossen, S.A. Lyon, J. Appl. Phys. **63**, 5149 (1988)
12. L.C. West, S.J. Eglash, Appl. Phys. Lett. **46**, 1156 (1985)
13. A. Harwit, J.S. Harris, Appl. Phys. Lett. **50**, 685 (1987)
14. B.F. Levine, K.K. Choi, C.G. Bethea, J. Walker, R.J. Malik, Appl. Phys. Lett. **50**, 1092 (1987)
15. B.F. Levine, J. Appl. Phys. **78**, R1 (1993)
16. S.D. Gunapala, S.V. Bandara: Quantum well infrared photodetector focal plane arrays. In: *Intersubband Transition in Quantum Wells: Physics and Device Applications I. Semiconductors and Semimetals*, vol. 62, ch. 4, ed by H.C. Liu, F. Capasso (Academic, San Diego 2000) pp. 197–282
17. H. Schneider, M. Walther, C. Schönbein, R. Rehm, J. Fleissner, W. Pletschen, J. Braunstein, P. Koidl, G. Weimann, J. Ziegler, W. Cabanski, Physica E **7**, 101 (2000)
18. H. Schneider, J. Fleissner, R. Rehm, M. Walther, W. Pletschen, P. Koidl, G. Weimann, J. Ziegler, R. Breiter, W. Cabanski, SPIE **4820**, 297 (2003)
19. H.C. Liu: Quantum well infrared photodetector physics and novel devices. In: *Intersubband Transition in Quantum Wells: Physics and Device Applications I. Semiconductors and Semimetals*, vol. 62, ch. 3, ed by H.C. Liu, F. Capasso (Academic, San Diego 2000) pp. 126–196

20. A. Rogalski, J. Appl. Phys. **93**, 4355 (2003)
21. E. Rosencher, B. Vinter, B. Levine (ed.), *Intersubband Transitions in Quantum Wells* (Plenum, New York, 1992)
22. H.C. Liu, B.F. Levine, J.Y. Andersson (ed.), *Quantum Well Intersubband Transition Physics and Devices* (Kluwer, Dordrecht, 1994)
23. S.S. Li, Y.-K. Su (ed.), *Intersubband Transitions in Quantum Wells: Physics and Devices* (Kluwer, Boston, 1998)
24. S.D. Gunapala, H.C. Liu, H. Schneider (ed.), Infrared Phys. Technol. **42**, 113 (2001)
25. G. Sarusi, A. Carbone, S. Gunapala, H.C. Liu (ed.), Infrared Phys. Technol. **44**, 305 (2003)
26. H.C. Liu, S.D. Gunapala, H. Schneider (ed.), Infrared Phys. Technol. **47**, 1 (2005)
27. K.K. Choi, *The Physics of Quantum Well Infrared Photodetectors* (World Scientific, Singapore, 1997)
28. J. Faist, F. Capasso, D.L. Sivco, C. Sirtori, A.L. Hutchinson, A.Y. Cho, Science **264**, 553 (1994)
29. J. Faist, F. Capasso, C. Sirtori, D.L. Sivco, A.Y. Cho: Quantum cascade lasers. In: *Intersubband Transition in Quantum Wells: Physics and Device Applications II. Semiconductors and Semimetals*, vol. 66, ch. 1, ed by H.C. Liu, F. Capasso (Academic, San Diego 2000) pp. 1–83
30. F. Capasso, R. Paiella, R. Martini, R. Colombelli, C. Gmachl, T.L. Myers, M.S. Taubman, R.M. Williams, C.G. Bethea, K. Unterrainer, H.Y. Hwang, D.L. Sivco, A.Y. Cho, M.A. Sergent, H.C. Liu, E.A. Whittaker, IEEE J. Quant. Electron. **38**, 511 (2002)
31. R.H. Kingston, *Detection of Optical and Infrared Radiation* (Springer, Berlin Heidelberg New York, 1978)
32. J.-P. Greffet, R. Carminati, K. Joulain, J.-P. Mulet, S. Mainguy, Y. Chen, Nature **4165**, 61 (2002)
33. A. Rose, *Concepts in Photoconductivity and Allied Problems* (Wiley Interscience, New York, 1963)
34. S.M. Sze, *Physics of Semiconductor Devices* (Wiley, New York, 1981)
35. L.L. Chang, K. Ploog (ed.), *Molecular Beam Epitaxy and Heterostructures*, ch. 1 (Martinus Nijhoff, Dordrecht, 1985)
36. A.Y. Cho (ed.), *Molecular Beam Epitaxy* (AIP, New York, 1994)
37. F. Capasso, Science **235**, 172 (1987)
38. G. Bastard, *Wave Mechanics Applied to Semiconductor Heterostructures* (les éditions de physique, Les Ulis, France, 1988)
39. C. Weisbuch, B. Vinter, *Quantum Semiconductor Structures* (Academic, Boston, 1991)
40. A. Shik, *Quantum Wells* (World Scientific, Singapore, 1997)
41. P. Harrison, *Quantum Wells, Wires and Dots: Theoretical and Computational Physics* (Wiley, New York, 1999)
42. M. Helm: The basic physics of intersubband transitions. In: *Intersubband Transition in Quantum Wells: Physics and Device Applications I. Semiconductors and Semimetals*, vol. 62, ch. 1, ed. by H.C. Liu, F. Capasso (Academic, San Diego 2000) pp. 1–99
43. H.C. Liu: The basic physics of photoconductive quantum well infrared detectors. In: *Long Wavelength Infrared Detectors*, ch. 1, ed. by M. Razeghi (Gordon and Breach, Amsterdam 1996) pp. 1–59

44. H.C. Liu, M. Buchanan, Z.R. Wasilewski, Appl. Phys. Lett. **72**, 1682 (1998)
45. A.G. Steele, H.C. Liu, M. Buchanan, Z.R. Wasilewski, Appl. Phys. Lett. **59**, 3625 (1991)
46. H.C. Liu, J. Appl. Phys. **73**, 3062 (1993)
47. H. Schneider, F. Fuchs, B. Dischler, J.D. Ralston, P. Koidl, Appl. Phys. Lett. **58**, 2234 (1991)
48. H.C. Liu, M. Buchanan, Z.R. Wasilewski, J. Appl. Phys. **83**, 6178 (1998)
49. E. Luna, M. Hopkinson, J.M. Ulloa, A. Guzmán, E. Muñoz, Appl. Phys. Lett. **83**, 3111 (2003)
50. H.C. Liu, T. Oogarah, E. Dupont, Z.R. Wasilewski, M. Byloos, M. Buchanan, F. Szmulowicz, J. Ehret, G.J. Brown, Electron. Lett. **38**, 909 (2002)
51. S.D. Gunapala, K.M.S.V. Bandara, B.F. Levine, G. Sarusi, D.L. Sivco, A.Y. Cho, Appl. Phys. Lett. **64**, 2288 (1994)
52. A.G.U. Perera, W.Z. Shen, S.G. Matsik, H.C. Liu, M. Buchanan, W.J. Schaff, Appl. Phys. Lett. **72**, 1596 (1998)
53. A.G.U. Perera, S.G. Matsik, H.C. Liu, M. Gao, M. Buchanan, W.J. Schaff, W. Yeo, Appl. Phys. Lett. **77**, 741 (2000)
54. H.C. Liu, C.Y. Song, A.J. SpringThorpe, J.C. Cao, Appl. Phys. Lett. **84**, 4068 (2004)
55. H. Luo, H.C. Liu, C.Y. Song, Z.R. Wasilewski, Appl. Phys. Lett. **86**, 231103 (2005)
56. H.C. Liu, D.D. Coon, Superlatt. Microstruct. **3**, 357 (1987)
57. H.C. Liu, M. Buchanan, Z.R. Wasilewski, J. Appl. Phys. **68**, 3780 (1990)
58. H.C. Liu, M. Buchanan, Z.R. Wasilewski, H. Chu, Appl. Phys. Lett. **58**, 1059 (1991)
59. J.S. Blakemore, J. Appl. Phys. **53**, R123 (1982)
60. H.C. Liu, D.D. Coon, O. Byungsung, Y.F. Lin, M.H. Francombe, Superlatt. Microstruct. **4**, 343 (1988)
61. A.G. Steele, H.C. Liu, M. Buchanan, Z.R. Wasilewski, J. Appl. Phys. **73**, 1062 (1992)
62. H. Asai, Y. Kawamura, Appl. Phys. Lett. **56**, 1149 (1990)
63. W. Chen, T.G. Andersson, Phys. Rev. B **44**, 9068 (1991)
64. M. Helm, Semicond. Sci. Technol. **10**, 557 (1995)
65. G.J. Brown, F. Szmulowicz: Normal incidence detection of infrared radiation in P-type GaAs/AlGaAs quantum well structures. In: *Long Wavelength Infrared Detectors*, ch. 5, ed. by M. Razeghi (Gordon and Breach, Amsterdam 1996) pp. 271–333
66. H.C. Liu, F. Szmulowicz, Z.R. Wasilewski, M. Buchanan, G.J. Brown, J. Appl. Phys. **85**, 2972 (1999)
67. A. Shen, H.C. Liu, M. Gao, M. Buchanan, E. Dupont, J. Ehret, G.J. Brown, F. Szmulowicz, Appl. Phys. Lett. **77**, 2400 (2000)
68. N. Iizuka, K. Kaneko, N. Suzuki, T. Asano, S. Noda, O. Wada, Appl. Phys. Lett. **77**, 648 (2000)
69. C. Gmachl, H.M. Ng, A.Y. Cho, Appl. Phys. Lett. **79**, 1590 (2001)
70. K. Kishino, A. Kikuchi, H. Kanazawa, T. Tachibana, Appl. Phys. Lett. **81**, 1234 (2002)
71. D. Hofstetter, S.-S. Schad, H. Wu, W.J. Schaff, L.F. Eastman, Appl. Phys. Lett. **83**, 572 (2003)
72. K.M.S.V. Bandara, D.D. Coon, O. Byungsung, Y.F. Lin, M.H. Francombe, Appl. Phys. Phys. **53**, 1931 (1988); erratum **55**, 206 (1989)

73. L. Hedin, B.I. Lundqvist, J. Phys. C **4**, 2064 (1971)
74. J.P. Perdew, Y. Wang, Phys. Rev. B **45**, 13244 (1992)
75. W.P. Chen, Y.J. Chen, E. Burstein, Surf. Sci. **58**, 263 (1976)
76. H.C. Liu, C.Y. Song, Z.R. Wasilewski, A.J. SpringThorpe, J.C. Cao, C. Dharma-wardana, G.C. Aers, D.J. Lockwood, J.A. Gupta, Phys. Rev. Lett. **90**, 77402 (2003)
77. D.E. Nikonov, A. Imamoglu, L.V. Butov, H. Schmidt, Phys. Rev. Lett. **79**, 4633 (1997)
78. R.J. Warburton, K. Weilhammer, J.P. Kotthaus, M. Thomas, H. Kroemer, Phys. Rev. Lett. **80**, 2185 (1998)
79. I. Waldmüller, J. Förstner, S.-C. Lee, A. Knorr, M. Woerner, K. Reimann, R.A. Kaindl, T. Elsaesser, R. Hey, K.H. Ploog, Phys. Rev. B **69**, 205307 (2004)
80. T. Unuma, M. Yoshita, T. Noda, H. Sakaki, H. Akiyama, J. Appl. Phys. **93**, 1586 (2003)
81. M. Carras, V. Berger, X. Marcadet, B. Vinter, Phys. Rev. B **70**, 233310 (2004)
82. D. Stehr, C. Metzner, M. Helm, T. Roch, G. Strasser, Phys. Rev. Lett. **95**, 257401 (2005)
83. E.O. Kane: The $k \cdot p$ method. In: *Semiconductors and Semimetals*, vol. 1, ch. 3, ed by R.K. Willardson, A.C. Beer (Academic, New York 1966) pp. 75–100
84. U. Rössler, Solid State Commun. **49**, 943 (1984)
85. G. Bastard, Phys. Rev. B **25**, 7584 (1982)
86. H. Schneider, K. von Klitzing, K. Ploog, Superlatt. Microstruct. **5**, 383 (1989)
87. R.P. Leavitt, Phys. Rev. B **44**, 11270 (1990)
88. D.E. Aspnes, S.M. Kelso, R.A. Logan, R. Bhat, J. Appl. Phys. **60**, 754 (1986)
89. D.J. Newson, A. Kurobe, Semicond. Sci. Technol. **3**, 786 (1988)
90. B. Deveaud, A. Chomette, D. Morris, A. Regreny, Solid State Commun. **85**, 367 (1993)
91. R. Ferreira, G. Bastard, Phys. Rev. B **40**, 1074 (1989)
92. P.J. Price, Ann. Phys. **133**, 217 (1981)
93. H. Schneider, E.C. Larkins, Semicond. Sci. Technol. **10**, 1329 (1995)
94. J.Y. Andersson, J. Appl. Phys. **78**, 6298 (1995)
95. B.N. Murdin, W. Heiss, C.J.G.M. Langerak, S.-C. Lee, I. Galbraith, G. Strasser, E. Gornik, M. Helm, C.R. Pidgeon, Phys. Rev. B **55**, 5171 (1997)
96. T. Ando, J. Phys. Soc. Jpn. **54**, 2671 (1985)
97. S.-C. Lee, I. Galbraith, Phys. Rev. B **59**, 15796 (1999)
98. H.C. Liu, L. Li, M. Buchanan, Z.R. Wasilewski, J. Appl. Phys. **82**, 889 (1997)
99. M. Ershov, V. Ryzhii, C. Hamaguchi, Appl. Phys. Lett. **67**, 3147 (1995)
100. M.J. Kane, S. Millidge, M.T. Emeny, D. Lee, D.R.P. Guy, C.R. Whitehouse, in *Intersubband Transitions in Quantum Wells*, ed. by E. Rosencher, B. Vinter, B. Levine (Plenum, New York, 1992), pp. 31–42
101. P. Man, D.S. Pan, Appl. Phys. Lett. **66**, 192 (1995)
102. C.H. Chu, C.I. Hung, Y.H. Wang, M.P. Houng, IEEE Photonics Technol. Lett. **9**, 1262 (1987)
103. H.C. Liu, L. Li, M. Buchanan, Z.R. Wasilewski, G.J. Brown, F. Szmulowicz, S.M. Hegde, J. Appl. Phys. **83**, 585 (1998)
104. F. Szmulowicz, G.J. Brown, Phys. Rev. B **51**, 13203 (1995)

105. H.C. Liu, A.G. Steele, M. Buchanan, Z.R. Wasilewski, J. Appl. Phys. **73**, 2029 (1993)
106. B.F. Levine, C.G. Bethea, G. Hasnain, V.O. Shen, E. Pelve, R.R. Abbott, S.J. Hsieh, Appl. Phys. Lett. **56**, 851 (1990)
107. S.R. Andrews, B.A. Miller, J. Appl. Phys. **70**, 993 (1991)
108. Y. Xu, A. Shakouri, A. Yariv, T. Krabach, S. Dejewski, Electron. Lett. **31**, 320 (1995)
109. D.D. Coon, H.C. Liu, Superlatt. Microstruct. **3**, 95 (1987)
110. M.A. Kinch, A. Yariv, Appl. Phys. Lett. **55**, 2093 (1989); see also a comment by B.F. Levine and the reply in **56**, 2354–2356 (1990)
111. A.G. Petrov, A. Shik, Semicond. Sci. Technol. **6**, 1163 (1991)
112. H.C. Liu, Appl. Phys. Lett. **60**, 1507 (1992)
113. S.V. Meshkov, Sov. Phys. JETP **64**, 1337 (1986)
114. H.C. Liu, Z.R. Wasilewski, M. Buchanan, H. Chu, Appl. Phys. Lett. **63**, 761 (1993)
115. Z.R. Wasilewski, H.C. Liu, M. Buchanan, J. Vac. Sci. Technol. B **12**, 1273 (1994)
116. M. Ershov, C. Hamaguchi, V. Ryzhii, Jpn. J. Appl. Phys. **35**, 1395 (1996)
117. M. Ershov, H.C. Liu, L. Li, M. Buchanan, Z.R. Wasilewski, V. Ryzhii, Appl. Phys. Lett. **70**, 1828 (1997)
118. M. Ershov, H.C. Liu, M. Buchanan, Z.R. Wasilewski, V. Ryzhii, Appl. Phys. Lett. **70**, 414 (1997)
119. M. Ershov, Appl. Phys. Lett. **69**, 3480 (1996)
120. M. Ershov, Appl. Phys. Lett. **73**, 3432 (1998)
121. L. Thibaudeau, P. Bois, J.Y. Duboz, J. Appl. Phys. **79**, 446 (1996)
122. V. Ryzhii, J. Appl. Phys. **81**, 6442 (1997)
123. V.D. Jovanović, P. Harrison, Z. Ikonić, D. Indjin, J. Appl. Phys. **96**, 269 (2004)
124. M. Ryzhii, V. Ryzhii, Appl. Phys. Lett. **72**, 842 (1998)
125. M. Ryzhii, V. Ryzhii, M. Willander, J. Appl. Phys. **84**, 3403 (1998)
126. O.O. Cellek, S. Memis, U. Bostanci, S. Ozer, C. Besikci, Physica E **24**, 318 (2004)
127. O.O. Cellek, C. Besikci, Semicond. Sci. Technol. **19**, 183 (2004)
128. A.G. Petrov, A. Shik, J. Appl. Phys. **83**, 3203 (1998)
129. D. Long: Photovoltaic and photoconductive infrared detectors. In: *Optical and Infrared Detectors*, ch. 4, ed. by R.J. Keyes (Springer, Berlin Heidelberg New York 1980)
130. E.L. Dereniak, D.G. Crowe, *Optical Radiation Detectors* (Wiley, New York, 1984)
131. N. Sclar, Prog. Quant. Elect. **9**, 149 (1984)
132. K.K. Choi, B.F. Levine, R.J. Malik, J. Walker, C.G. Bethea, Phys. Rev. B **35**, 4172 (1987)
133. H.C. Liu, M. Buchanan, Z.R. Wasilewski, Phys. Rev. B **44**, 1411 (1991)
134. H.C. Liu, G.C. Aers, M. Buchanan, Z.R. Wasilewski, D. Landheer, J. Appl. Phys. **70**, 935 (1991)
135. H.C. Liu, J. Li, M. Buchanan, Z.R. Wasilewski, J.G. Simmons, Phys. Rev. B **48**, 1951 (1993)
136. J. Li, H.C. Liu, M. Buchanan, Z.R. Wasilewski, J.G. Simmons, J. Appl. Phys. **75**, 1748 (1994)

137. G. Hasnain, B.F. Levine, S. Gunapala, N. Chand, Appl. Phys. Lett. **57**, 608 (1990)
138. E. Rosencher, F. Luc, P. Bois, S. Delaitre, Appl. Phys. Lett. **61**, 468 (1992)
139. J.Y. Andersson, L. Lundqvist, Appl. Phys. Lett. **59**, 857 (1991)
140. W. Beck, Appl. Phys. Lett. **63**, 3589 (1993)
141. F.L. Serzhenko, V.D. Shadrin, Sov. Phys. Semicond. **25**, 953 (1991)
142. B. Xing, H.C. Liu, P.H. Wilson, M. Buchanan, Z.R. Wasilewski, J.G. Simmons, J. Appl. Phys. **76**, 1889 (1994)
143. C. Schönbein, H. Schneider, R. Rehm, M. Walther, Appl. Phys. Lett. **73**, 1251 (1998); see a comment by B.F. Levine in **74**, 892 (1999)
144. S.D. Gunapala, B.F. Levine, L. Pfeiffer, W. West, J. Appl. Phys. **69**, 6517 (1991)
145. M.C. Tatham, J.F. Ryan, C.T. Foxon, Phys. Rev. Lett. **63**, 1637 (1989)
146. J. Baier, I.M. Bayanov, U. Plödereder, A. Seilmeier, Superlatt. Microstruct. **19**, 9 (1996)
147. J. Faist, F. Capasso, C. Sirtori, D.L. Sivco, A.L. Hutchinson, S.N.G. Chu, A.Y. Cho, Appl. Phys. Lett. **63**, 1354 (1993)
148. S. Ehret, H. Schneider, J. Fleissner, P. Koidl, G. Böhm, Appl. Phys. Lett. **71**, 641 (1997)
149. S.M. Goodnick, P. Lugli, Phys. Rev. B **37**, 2578 (1988)
150. P. Sotirelis, P. von Allmen, K. Hess, Phys. Rev. B **47**, 12744 (1993)
151. P.W.M. Blom, J.E.M. Haverkort, P.J. van Hall, J.H. Wolter, Appl. Phys. Lett. **62**, 1490 (1993)
152. P.W.M. Blom, C. Smit, J.E.M. Haverkort, J.H. Wolter, Phys. Rev. B **47**, 2072 (1993)
153. K. Muraki, A. Fujiwara, S. Fukatsu, Y. Shiraki, Y. Takahashi, Phys. Rev. B **53**, 15477 (1996)
154. H.C. Liu, Appl. Phys. Lett. **61**, 2703 (1992)
155. B.F. Levine, A. Zussman, J.M. Kuo, J. de Jong, J. Appl. Phys. **71**, 5130 (1992)
156. V.D. Shadrin, V.V. Mitin, V.A. Kochelap, K.K. Choi, J. Appl. Phys. **77**, 1771 (1995)
157. V.D. Shadrin, V.V. Mitin, K.K. Choi, V.A. Kochelap, J. Appl. Phys. **78**, 5765 (1995)
158. B.F. Levine, A. Zussman, S.D. Gunapala, M.T. Asom, J.M. Kuo, W.S. Hobson, J. Appl. Phys. **72**, 4429 (1992)
159. L.C. Lenchyshyn, H.C. Liu, M. Buchanan, Z.R. Wasilewski, Semicond. Sci. Technol. **10**, 45 (1995)
160. S.K.H. Sim, H.C. Liu, A. Shen, M. Gao, K.F. Lee, M. Buchanan, Y. Ohno, H. Ohno, E.H. Li, Infrared Phys. Technol. **42**, 115 (2001)
161. B. Ferguson, X.-C. Zhang, Nat. Mater. **1**, 26 (2002)
162. P.H. Siegel, IEEE Trans. Microwave Theory Techn. **50**, 910 (2002)
163. M. Nagel, P. Haring Bolivar, M. Brucherseifer, H. Kurz, A. Bosserhoff, R. Büttner, Appl. Phys. Lett. **80**, 154 (2002)
164. R. Köhler, A. Tredicucci, F. Beltram, H.E. Beere, E.H. Linfield, A.G. Davies, D.A. Richie, R.C. Iotti, F. Rossi, Nature **417**, 156 (2002)
165. B.S. Williams, S. Kumar, Q. Hu, J.S. Reno, Opt. Exp. **13**, 3331 (2005)
166. A. Kastalsky, T. Duffield, S.J. Allen, J. Harbison, Appl. Phys. Lett. **52**, 1320 (1988)

167. H. Schneider, E.C. Larkins, J.D. Ralston, K. Schwartz, F. Fuchs, P. Koidl, Appl. Phys. Lett. **63**, 782 (1993)
168. C. Schönbein, H. Schneider, G. Bihlmann, K. Schwarz, P. Koidl, Appl. Phys. Lett. **68**, 973 (1996)
169. H. Schneider, J. Appl. Phys. **74**, 4789 (1993)
170. H. Schneider, P. Koidl, C. Schönbein, S. Ehret, E.C. Larkins, P. Koidl, Superlatt. Microstruct. **19**, 347 (1996)
171. L. Gendron, M. Carras, A. Huynh, V. Ortiz, C. Koeniguer, V. Berger, Appl. Phys. Lett. **85**, 2824 (2004)
172. L. Gendron, C. Koeniguer, V. Berger, X. Marcadet, Appl. Phys. Lett. **86**, 121116 (2005)
173. H. Schneider, C. Schönbein, M. Walther, K. Schwarz, J. Fleissner, P. Koidl, Appl. Phys. Lett. **71**, 246 (1997)
174. H. Schneider, C. Schönbein, S. Ehret, R. Rehm, M. Walther, K. Schwarz, J. Fleissner, P. Koidl, ECS Proc. **97-33**, 180 (1997)
175. H. Akiyama, H. Sugawara, Y. Kadoya, A. Lorke, C. Tsujino, H. Sakaki, Appl. Phys. Lett. **65**, 424 (1994)
176. C. Schönbein, H. Schneider, R. Rehm, M. Walther, ECS Proc. **98-21**, 158 (1999)
177. H. Schneider, Appl. Phys. Lett. **82**, 4376 (2003)
178. R. Rehm, H. Schneider, C. Schönbein, M. Walther, Physica E **7**, 124 (2000)
179. S. Ehret, H. Schneider, Appl. Phys. B **66**, 27 (1998)
180. H. Schneider, C. Schönbein, M. Walther, P. Koidl, G. Weimann, Appl. Phys. Lett. **74**, 16 (1999)
181. R.W. Ditchburn, *Light* (Academic, London, 1976)
182. M. Ershov, Appl. Phys. Lett. **72**, 2865 (1998)
183. D. Heitmann, J.P. Kotthaus, E.G. Mohr, Solid State Commun. **44**, 715 (1982)
184. J.Y. Andersson, L. Lundqvist, Z.F. Paska, Appl. Phys. Lett. **58**, 2264 (1991)
185. J.Y. Andersson, L. Lundqvist, J. Appl. Phys. **71**, 3600 (1992)
186. G. Sarusi, B.F. Levine, S.J. Pearton, K.M.S. Bandara, R.E. Leibenguth, J.Y. Andersson, J. Appl. Phys. **76**, 4989 (1994)
187. J.Y. Andersson, L. Lundqvist: Grating coupled quantum well infrared detectors. In: *Long Wavelength Infrared Detectors*, ch. 4, ed. by M. Razeghi (Gordon and Breach, Amsterdam 1996) pp. 207–270
188. S.S. Li: Metal grating coupled bound-to-miniband transition III–V quantum well infrared photodetectors. In: *Long Wavelength Infrared Detectors*, ch. 3, ed. by M. Razeghi (Gordon and Breach, Amsterdam 1996) pp. 145–206
189. C.P. Lee, K.H. Chang, K.L. Tsai, Appl. Phys. Lett. **61**, 2437 (1992)
190. K.L. Tsai, C.P. Lee, J.S. Tsang, H.R. Chen, K.H. Chang, IEEE Electron. Dev. Lett. **16**, 49 (1995)
191. E. Dupont, M. Byloos, T. Oogarah, M. Buchanan, H.C. Liu, Infrared Phys. Technol. **47**, 132 (2005)
192. P. Bois, E. Costard, X. Marcadet, E. Herniou, Infrared Phys. Technol. **42**, 291 (2001)
193. T.R. Schmiert, S.L. Barnes, A.J. Brouns, F.C. Case, P. Mitra, L.T. Claiborne, Appl. Phys. Lett. **68**, 2846 (1996)
194. E. Dupont, J. Appl. Phys. **85**, 2687 (2000)
195. G. Sarusi, B.F. Levine, S.J. Pearton, K.M.S. Bandara, R.E. Leibenguth, Appl. Phys. Lett. **64**, 960 (1994)

196. B. Xing, H.C. Liu, J. Appl. Phys. **80**, 1214 (1996)
197. S.I. Borenstain, U. Arad, I. Lyubina, A. Segal, Y. Warschawer, Appl. Phys. Lett. **75**, 2659 (1999)
198. T. Fujii, P. Masalkar, H. Nishino, Y. Miyamoto, Infrared Phys. Technol. **42**, 199 (2001)
199. H. Nishino, P. Masalkar, T. Miyatake, Y. Matsukura, Y. Miyamoto, T. Fujii, Infrared Phys. Technol. **45**, 1 (2004)
200. V. Jandhyala, D. Sengupta, B. Shanker, E. Michielssen, M. Feng, G. Stillman, J. Appl. Phys. **83**, 3360 (1998)
201. A. De Rossi, E. Costard, N. Guerineau, S. Rommeluere, Infrared Phys. Technol. **44**, 325 (2003)
202. Y. Li, D. Abookasis, J. Rosen, Appl. Opt. **40**, 2864 (2001)
203. K. Sakoda, *Optical Properties of Photonic Crystals* (Springer, Berlin Heidelberg New York, 2001)
204. D. Dini, R. Köhler, A. Tredicucci, G. Biasiol, L. Sorba, Phys. Rev. Lett. **90**, 116401 (2003)
205. E. Dupont, H.C. Liu, A.J. SpringThorpe, W. Lai, M. Extavour, Phys. Rev. B **68**, 245320 (2003)
206. F.H. Julien, J.M. Lourtioz, N. Herschkorn, D. Delacourt, J.P. Pocholle, M. Papuchon, R. Planel, G. Le-Roux, Appl. Phys. Lett. **53**, 116 (1988)
207. D.J. Newson, A. Kurobe, Appl. Phys. Lett. **53**, 2516 (1988)
208. K.L. Vodopyanov, C. Chazapis, C.C. Phillips, B. Sung, J.S. Harris, Semicond. Sci. Technol. **12**, 708 (1997)
209. M. Załużny, Phys. Rev. B **47**, 3995 (1993)
210. K. Craig, B. Galdrikian, J.N. Heyman, A.G. Markelz, J.B. Williams, M.S. Sherwin, K. Campman, P.F. Hopkins, A.C. Gossard, Phys. Rev. Lett. **76**, 2382 (1996)
211. H. Luo, D. Ban, H.C. Liu, P.J. Poole, M. Buchanan, IEEE Electron Dev. Lett. **25**, 129 (2004)
212. H.C. Liu, E. Dupont, M. Ershov, J. Nonlin. Opt. Phys. Mater. **11**, 433 (2002)
213. J.Y. Duboz, E. Costard, E. Rosencher, P. Bois, J. Nagle, J.M. Berset, D. Jaroszynski, J.M. Ortega, J. Appl. Phys. **77**, 6492 (1995)
214. J.Y. Duboz, E. Costard, J. Nagle, J.M. Berset, J.M. Ortega, J.M. Gérard, J. Appl. Phys. **78**, 1224 (1995)
215. H.C. Liu, J. Li, E.R. Brown, K.A. McIntosh, K.B. Nichols, M.J. Manfra, Appl. Phys. Lett. **67**, 1594 (1995)
216. A. Zavriyev, E. Dupont, P.B. Corkum, H.C. Liu, Z. Biglov, Opt. Lett. **20**, 1886 (1995)
217. R. Paiella, F. Capasso, C. Gmachl, D.L. Sivco, J.N. Baillargeon, A.L. Hutchinson, A.Y. Cho, H.C. Liu, Science **290**, 1739 (2000)
218. R. Martini, R. Paiella, C. Gmachl, F. Capasso, E.A. Whittaker, H.C. Liu, H.Y. Hwang, D.L. Sivco, J.N. Baillargeon, A.Y. Cho, Electron. Lett. **37**, 1290 (2001)
219. H.C. Liu, J. Li, M. Buchanan, Z.R. Wasilewski, IEEE J. Quant. Electron. **32**, 1024 (1996)
220. H.C. Liu, R. Dudek, A. Shen, E. Dupont, C.-Y. Song, Z.R. Wasilewski, M. Buchanan, Appl. Phys. Lett. **79**, 4273 (2001)
221. T. Oogarah, H.C. Liu, E. Dupont, Z.R. Wasilewski, M. Byloos, M. Buchanan, Semicond. Sci. Technol. **17**, L41 (2002)

222. C. Mermelstein, H. Schneider, A. Sa'ar, C. Schönbein, M. Walther, G. Bihlmann, Appl. Phys. Lett. **71**, 2011 (1997)
223. J.E. Hubbs, D.C. Arrington, M.E. Gramer, G.A. Dole, Opt. Eng. **39**, 2660 (2000)
224. M. Ershov, H.C. Liu, A.G.U. Perera, S.G. Matsik, Physica E **7**, 115 (2000)
225. S.R. Schmidt, A. Seilmeier, H.C. Liu, J. Appl. Phys. **91**, 5545 (1999)
226. B.G. Bosch, R.W.H. Eagelman, *Gunn Effect Electronics* (Pitman, London, 1975)
227. J. Zongfu, M. Benkun, Phys. Rev. B **44**, 11072 (1991)
228. H. Schneider, C. Mermelstein, R. Rehm, C. Schönbein, A. Sa'ar, M. Walther, Phys. Rev. B **57**, R15096 (1998)
229. H. Schneider, C. Schönbein, C. Mermelstein, R. Rehm, A. Sa'ar, M. Walther, ECS Proc. **98-21**, 206 (1999)
230. P.M. Solomon, S.L. Wright, C. Lanza, Superlatt. Microstruct. **2**, 521 (1986)
231. K.K. Choi, B.F. Levine, C.G. Bethea, J. Walker, R.J. Malik, Phys. Rev. B **39**, 8029 (1989)
232. H.T. Grahn, H. Schneider, K. von Klitzing, Appl. Phys. Lett. **54**, 1757 (1989)
233. L.L. Bonilla, H.T. Grahn, Rep. Prog. Phys. **68**, 577 (2005)
234. Z.Y. Han, S.F. Yoon, K. Radhakrishnan, D.H. Zhang, Appl. Phys. Lett. **66**, 1120 (19950)
235. A. Yamashita, R. Nii, Jpn. J. Appl. Phys. **5**, 263 (1966)
236. H. Schneider, C. Schönbein, R. Rehm, M. Walther, P. Koidl, Appl. Phys. Lett. **88**, 051114 (2006)
237. S. Schmidt, J. Kaiser, A. Seilmeier, Physica B **272**, 384 (1999)
238. H.C. Liu, J. Li, E. Costard, E. Rosencher, J. Nagle, Solid State Electron. **40**, 567 (1996)
239. C. Sirtori, F. Capasso, D.L. Sivco, A.Y. Cho: Nonlinear optics in coupled-quantum-well quasi-molecules. In: *Intersubband Transition in Quantum Wells: Physics and Device Applications II. Semiconductors and Semimetals*, vol. 66, ch. 2, ed. by H.C. Liu, F. Capasso (Academic, San Diego 2000) pp. 85–125
240. N. Owschimikow, C. Gmachl, A. Belyanin, V. Kocharovsky, D.L. Sivco, R. Colombelli, F. Capasso, A.Y. Cho, Phys. Rev. Lett. **90**, 043902 (2003)
241. E. Dupont, P.B. Corkum, H.C. Liu, P.H. Wilson, M. Buchanan, Z.R. Wasilewski, Appl. Phys. Lett. **65**, 1560 (1994)
242. R. Paiella, F. Capasso, C. Gmachl, H.Y. Hwang, D.L. Sivco, A.L. Hutchinson, A.Y. Cho, H.C. Liu, Appl. Phys. Lett. **77**, 169 (2000)
243. E.C. Larkins, H. Schneider, S. Ehret, J. Fleissner, B. Dischler, P. Koidl, J.D. Ralston, IEEE Trans. Electron Dev. **41**, 511 (1994)
244. H. Schneider, P. Koidl, F. Fuchs, B. Dischler, K. Schwartz, J.D. Ralston, Semicond. Sci. Technol. **6**, C120 (1991)
245. E. Dupont, P.B. Corkum, H.C. Liu, M. Buchanan, Z.R. Wasilewski, Phys. Rev. Lett. **74**, 3569 (1995)
246. C. Schönbein, H. Schneider, R. Rehm, M. Walther, Phys. Rev. B **60**, R13993 (1999)
247. H. Schneider, C. Schönbein, K. Schwarz, M. Walther, Physica E **2**, 28 (1998)
248. M.I. Sumetskii, H.U. Baranger, Appl. Phys. Lett. **67**, 3560 (1995)
249. F. Capasso: Physics of avalanche photodiodes. In: *Lightwave Communications Technology: Part A, Material Growth Technologies. Semiconductors and Semimetals*, vol. 22, ch. 1, ed. by W.T. Tsang (Academic, New York 1985) pp. 2–173

250. S. Tomic, M. Tadic, V. Milanovic, Z. Ikonic, J. Appl. Phys. **59**, 2885 (1986)
251. F. Capasso, S. Sen, A.C. Gossard, A.L. Hutchinson, J.H. English, IEEE Electron Dev. Lett. **7**, 573 (1986)
252. M. Toivonen, M. Jalonen, A. Salokatve, M. Pessa, Appl. Phys. Lett. **62**, 891 (1993)
253. B.F. Levine, K.K. Choi, C.G. Bethea, J. Walker, R.J. Malik, Appl. Phys. Lett. **51**, 934 (1987)
254. R. Rehm, H. Schneider, M. Walther, P. Koidl, Appl. Phys. Lett. **82**, 2097 (2003)
255. M. Ershov, H.C. Liu, J. Appl. Phys. **86**, 6580 (1999)
256. R. Rehm, H. Schneider, K. Schwarz, M. Walther, P. Koidl, G. Weimann, SPIE **4288**, 379 (2001)
257. J.R. Hayes, A.F.J. Levi, W. Wiegmann, Phys. Rev. Lett. **54**, 1570 (1985)
258. J.R. Hayes, A.F.J. Levi, IEEE J. Quant. Electron. **22**, 1744 (1986)
259. B. Aslan, H.C. Liu, A. Bezinger, P.J. Poole, M. Buchanan, R. Rehm, H. Schneider, Semicond. Sci. Technol. **19**, 442 (2004)
260. D.T. Wilhelm, R.K. Purvis, A. Singh, D.Z. Richardson, R.A. Hahn, J.R. Duffey, W.T. Roberts, J.A. Ruffner, SPIE **2217**, 307 (1994)
261. H.C. Liu, P.H. Wilson, M. Buchanan, Z.R. Wasilewski, S.M. Khanna, SPIE **2746**, 134 (1996)
262. H.C. Liu, L. Li, S.M. Khanna, M. Buchanan, SPIE **3061**, 749 (1997)
263. V.D. Akhmetov, E.B. Gorokhov, N.T. Moshegov, A.I. Toropov, Infrared Phys. Technol. **36**, 837 (1995)
264. H.C. Liu, R. Dudek, A. Shen, E. Dupont, C.-Y. Song, Z.R. Wasilewski, M. Buchanan, J. High Speed Electron. Syst. **12**, 803 (2003)
265. A. Köck, E. Gornik, G. Abstreiter, G. Böhm, M. Walther, G. Weimann, Appl. Phys. Lett. **60**, 2011 (1992)
266. Ph. Bois, E. Costard, J.Y. Duboz, J. Nagle, SPIE **3061**, 764 (1997)
267. E.M. Costard, P.F. Bois, F. Audier, E. Herniou, SPIE **3436**, 228 (1998)
268. S.D. Gunapala, S.V. Bandara, J.K. Liu, S.B. Rafol, J.M. Mumolo, C.A. Shott, R. Jones, J. Woolaway II, J.M. Fastenau, A.K. Liu, M. Jhabvala, K.K. Choi, Infrared Phys. Technol. **44**, 411 (2005)
269. H.C. Liu, J. Li, J.R. Thompson, Z.R. Wasilewski, M. Buchanan, J.G. Simmons, IEEE Electron Dev. Lett. **14**, 566 (1993)
270. L.C. Lenchyshyn, H.C. Liu, M. Buchanan, Z.R. Wasilewski, J. Appl. Phys. **79**, 8091 (1996)
271. E. Dupont, M. Gao, Z.R. Wasilewski, H.C. Liu, Appl. Phys. Lett. **78**, 2067 (2001)
272. I. Gravé, A. Shakouri, N. Kruze, A. Yariv, Appl. Phys. Lett. **60**, 2362 (1992)
273. K. Kheng, M. Ramsteiner, H. Schneider, J.D. Ralston, F. Fuchs, P. Koidl, Appl. Phys. Lett. **61**, 666 (1992)
274. Y.H. Wang, S.S. Li, P. Ho, Appl. Phys. Lett. **62**, 93 (1993)
275. V. Berger, N. Vodjdani, P. Bois, S. Delaitre, Appl. Phys. Lett. **61**, 1898 (1990)
276. S.R. Parihar, S.A. Lyon, M. Santos, M. Shayegan, Appl. Phys. Lett. **55**, 2417 (1989)
277. E. Martinet, F. Luc, E. Rosencher, P. Bois, S. Delaitre, Appl. Phys. Lett. **60**, 895 (1992)
278. E. Martinet, E. Rosencher, F. Luc, P. Bois, E. Costard, S. Delaitre, Appl. Phys. Lett. **61**, 246 (1992)

279. N. Cohen, A. Zussman, G. Sarusi, Infrared Phys. Technol. **42**, 391 (2001)
280. H.C. Liu, C.Y. Song, E. Dupont, P. Poole, P.W. Wilson, B.J. Robinson, D.A. Thompson, Electron. Lett. **35**, 2055 (1999)
281. S.D. Gunapala, B.F. Levine, D. Ritter, R. Hamm, M.B. Panish, Appl. Phys. Lett. **60**, 636 (1992)
282. H.C. Liu, C.Y. Song, A. Shen, M. Gao, Z.R. Wasilewski, M. Buchanan, Appl. Phys. Lett. **77**, 2437 (2000)
283. H.C. Liu, C.Y. Song, A. Shen, M. Gao, E. Dupont, P.J. Poole, Z.R. Wasilewski, M. Buchanan, P.W. Wilson, B.J. Robinson, D.A. Thompson, Y. Ohno, H. Ohno, Infrared Phys. Technol. **42**, 163 (2000)
284. H. Schneider, C. Schönbein, G. Bihlmann, Appl. Phys. Lett. **68**, 1832 (1996)
285. G. Bender, E.C. Larkins, H. Schneider, J.D. Ralston, P. Koidl, Appl. Phys. Lett. **63**, 2920 (1993)
286. M.B. Reine, P.W. Norton, R. Starr, M.H. Weiler, M. Kestigian, B.L. Musicant, P. Mitra, T. Schimert, F.C. Case, I.B. Bhat, H. Ehsani, V. Rao, J. Electron. Mater. **24**, 669 (1995)
287. N. Cohen, G. Mizrahi, G. Sarusi, A. Sa'ar, Infrared Phys. Technol. **47**, 43 (2005)
288. H.C. Liu, J. Li, Z.R. Wasilewski, M. Buchanan, Electron. Lett. **31**, 832 (1995)
289. V. Ryzhii, M. Ershov, M. Ryzhii, I. Khmyrova, Jpn. J. Appl. Phys. **34**, L38 (1995)
290. H.C. Liu, L.B. Allard, M. Buchanan, Z.R. Wasilewski, Electron. Lett. **33**, 379 (1997)
291. V. Ryzhii, H.C. Liu, I. Khmyrova, M. Ryzhii, IEEE J. Quant. Electron. **33**, 1527 (1997)
292. M. Ershov, H.C. Liu, L.M. Schmitt, J. Appl. Phys. **82**, 1446 (1997)
293. L.B. Allard, H.C. Liu, M. Buchanan, Z.R. Wasilewski, Appl. Phys. Lett. **70**, 2784 (1997)
294. E. Dupont, H.C. Liu, M. Buchanan, Z.R. Wasilewski, D. St-Germain, P. Chevrette, Appl. Phys. Lett. **75**, 563 (1999)
295. E. Dupont, M. Byloos, M. Gao, M. Buchanan, C.-Y. Song, Z.R. Wasilewski, H.C. Liu, IEEE Photonics Technol. Lett. **14**, 182 (2002)
296. I. Schnitzer, E. Yablonovitch, C. Caneau, T.J. Gmitter, Appl. Phys. Lett. **62**, 131 (1993)
297. E. Dupont, H.C. Liu, M. Buchanan, S. Chiu, M. Gao, Appl. Phys. Lett. **76**, 4 (2000)
298. E. Dupont, X. Zhu, S. Chiu, S. Moisa, M. Buchanan, M. Gao, H.C. Liu, P.B. Corkum, Semicond. Sci. Technol. **15**, L15 (2000)
299. D. Ban, H. Luo, H.C. Liu, Z.R. Wasilewski, Y. Paltiel, A. Raizman, A. Sher, Appl. Phys. Lett. **86**, 201103 (2005)
300. H.C. Liu, M. Gao, M. Buchanan, Z.R. Wasilewski, P. Poole, SPIE **3491**, 214 (1998)
301. D. Ban, H. Luo, H.C. Liu, A.J. SpringThorpe, R. Glew, Z.R. Wasilewski, M. Buchanan, Electron. Lett. **39**, 1145 (2003)
302. D. Ban, H. Luo, H.C. Liu, Z.R. Wasilewski, M. Buchanan, IEEE Photonics Technol. Lett. **17**, 1477 (2005)
303. K.W. Berryman, S.A. Lyon, M. Segev, Appl. Phys. Lett. **70**, 1861 (1997)
304. J. Phillips, K. Kamath, P. Bhattacharya, Appl. Phys. Lett. **72**, 2020 (1998)
305. D. Pan, E. Towe, S. Kennerly, Appl. Phys. Lett. **73**, 1937 (1998)

306. S.J. Xu, S.J. Chua, T. Mei, X.C. Wang, X.H. Zhang, G. Karunasiri, W.J. Fan, C.H. Wang, J. Jiang, S. Wang, X.G. Xie, Appl. Phys. Lett. **73**, 3153 (1998)
307. L. Chu, A. Zrenner, G. Böhm, G. Abstreiter, Appl. Phys. Lett. **75**, 3599 (1999)
308. N. Horiguchi, T. Futatsugi, Y. Nakata, N. Yokoyama, T. Mankad, P.M. Petroff, Jpn. J. Appl. Phys. **38**, 2559 (1999)
309. L. Chu, A. Zrenner, G. Böhm, G. Abstreiter, Appl. Phys. Lett. **76**, 1944 (2000)
310. A.D. Stiff, S. Krishna, P. Bhattacharya, S. Kennerly, Appl. Phys. Lett. **79**, 421 (2001)
311. S. Sauvage, P. Boucaud, T. Brunhes, V. Immer, E. Finkman, J.-M. Gérard, Appl. Phys. Lett. **78**, 2327 (2001)
312. H.C. Liu, M. Gao, J. McCaffrey, Z.R. Wasilewski, S. Fafard, Appl. Phys. Lett. **78**, 79 (2001)
313. S. Raghavan, P. Potella, S. Stintz, B. Fuchs, S. Krishna, C. Morath, D.A. Cardimona, S.W. Kennerly, Appl. Phys. Lett. **81**, 1369 (2002)
314. B. Aslan, H.C. Liu, M. Korkusinski, S.-J. Cheng, P. Hawrylak, Appl. Phys. Lett. **82**, 630 (2003)
315. Z. Ye, J.C. Campbell, Z. Chen, E.-T. Kim, A. Madhukar, Appl. Phys. Lett. **83**, 1234 (2003)
316. H.C. Liu, B. Aslan, M. Korkusinski, S.-J. Cheng, P. Hawrylak, Infrared Phys. Technol. **44**, 503 (2003)
317. X.H. Su, S. Chakrabarti, A.D. Stiff-Roberts, J. Singh, P. Bhattacharya, Electron. Lett. **40**, 1082 (2004)
318. A. Weber, O. Gauthier-Lafaye, F.H. Julien, J. Brault, M. Gendry, Y. Désieres, T. Benyattou, Appl. Phys. Lett. **74**, 413 (1999)
319. V. Ryzhii, Semicond. Sci. Technol. **11**, 759 (1996)
320. V. Ryzhii, V. Pipa, I. Khmyrova, V. Mitin, M. Willander, Jpn. J. Appl. Phys. **39**, L1283 (2000)
321. U. Bockelmann, G. Bastard, Phys. Rev. B **42**, 8947 (1990)
322. J. Urayama, T.B. Norris, J. Singh, P. Bhattacharya, Phys. Rev. Lett. **86**, 4930 (2001)
323. V. Ryzhii, I. Khmyrova, V. Mitin, M. Stroscio, M. Willander, Appl. Phys. Lett. **78**, 3522 (2001)
324. S.-W. Lee, K. Hirakawa, Y. Shimada, Physica E **7**, 499 (2000)
325. J.-W. Kim, J.-E. Oh, S.-C. Hong, C.-H. Park, T.-K. Yoo, IEEE Electron Dev. Lett. **21**, 329 (2000)
326. L. Chu, A. Zrenner, M. Bichler, G. Abstreiter, Appl. Phys. Lett. **79**, 2249 (2001)
327. H.C. Liu, T.C.L.G. Sollner: High-frequency resonant-tunneling devices. In: *High-Speed Heterostructure Devices. Semiconductors and Semimetals*, vol. 41, ch. 6, ed. by R.A. Kiehl, T.C.L.G. Sollner (Academic, San Diego 1994) pp. 359–419
328. K.M.S.V. Bandara, B.F. Levine, R.E. Leibenguth, M.T. Asom, J. Appl. Phys. **74**, 1826 (1993)
329. K.K.M.N. Gurusinghe, K. Premaratne, T.G. Andersson, S.V. Bandara, J. Appl. Phys. **87**, 8575 (2000)
330. H.C. Liu, A.G. Steele, M. Buchanan, Z.R. Wasilewski, IEEE Electron Dev. Lett. **13**, 363 (1992)

331. H.C. Liu, J. Li, G.C. Aers, C.R. Leavens, M. Buchanan, Z.R. Wasilewski, Phys. Rev. B **51**, 5116 (1995)
332. A. Carbone, P. Mazzetti, Appl. Phys. Lett. **70**, 28 (1997)
333. M. Ershov, A.N. Korotkov, Appl. Phys. Lett. **71**, 1667 (1997)
334. H.C. Liu, M. Buchanan, G.C. Aers, Z.R. Wasilewski, Semicond. Sci. Technol. **6**, C124 (1991)
335. Y. Shimada, K. Hirakawa, Jpn. J. Appl. Phys. **37**, 1421 (1998)
336. V. Ryzhii, IEEE Trans. Electron Dev. **45**, 1797 (1998)
337. S. Bandara, S. Gunapala, S. Rafol, D. Ting, J. Liu, J. Mumolo, T. Trinh, Infrared Phys. Technol. **42**, 237 (2001)
338. V. Ryzhii, Jpn. J. Appl. Phys. **40**, 3137 (2001)
339. P.H. Wilson, H.C. Liu, Superlatt. Microstruct. **13**, 309 (1993)
340. D. Mandelik, M. Schniederman, V. Umansky, I. Bar-Joseph, Appl. Phys. Lett. **78**, 472 (2001)
341. V. Ryzhii, Jpn. J. Appl. Phys. **36**, L415 (1997)
342. K. Nonaka, *Semiconductor Device*, Jpn. Patent 4-364072 (1992)
343. H.C. Liu, A.G. Steele, M. Buchanan, Z.R. Wasilewski, *Intersubband Transitions in Quantum Wells*, ed. by E. Rosencher, B. Vinter, B.F. Levine (Plenum, New York, 1992), pp. 57–63
344. M. Schulz, L. Caldwell, Infrared Phys. Technol. **36**, 763 (1995)
345. S.D. Gunapala, S.V. Bandara, J.K. Liu, C.J. Hill, S.B. Rafol, J.M. Mumolo, J.T. Trinh, M.Z. Tidrow, P.D. LeVan, Semicond. Sci. Technol. **20**, 473 (2005)
346. W. Wittenstein, SPIE **3436**, 413 (1998)
347. M. Walther, F. Fuchs, H. Schneider, J. Schmitz, W. Pletschen, J. Braunstein, J. Ziegler, W. Cabanski, P. Koidl, G. Weimann, SPIE Proc. **3436**, 348 (1998)
348. M. Walther, J. Fleissner, H. Schneider, C. Schönbein, W. Pletschen, E. Diwo, K. Schwarz, J. Braunstein, P. Koidl, J. Ziegler, W. Cabanski, Inst. Phys. Conf. Ser. **166**, 427 (2000)
349. R. Breiter, W. Cabanski, J. Ziegler, M. Walther, H. Schneider, SPIE Proc. **4020**, 257 (2000)
350. W. Cabanski, R. Breiter, R. Koch, W. Gross, K.-H. Mauk, W. Rode, J. Ziegler, H. Schneider, M. Walther, R. Oelmaier, Infrared Phys. Technol. **43**, 257 (2002)
351. W. Cabanski, R. Breiter, R. Koch, K.-H. Mauk, W. Rode, J. Ziegler, H. Schneider, M. Walther, R. Oelmaier, SPIE Proc. **4369**, 547 (2001)
352. S.D. Gunapala, S.V. Bandara, A. Singh, J.K. Liu, S.B. Rafol, E.M. Luong, J.M. Mumolo, N.Q. Tran, J.D. Vincent, C.A. Shott, J. Long, P.D. LeVan, SPIE **3698**, 687 (1999)
353. A. Goldberg, T. Fischer, S. Kennerly, S. Wang, M. Sundaram, P. Uppal, M. Winn, G. Milne, M. Stevens, SPIE **4028**, 276 (2000)
354. H. Schneider, T. Maier, J. Fleissner, M. Walther, P. Koidl, G. Weimann, W. Cabanski, M. Finck, P. Menger, W. Rode, J. Ziegler, Electron. Lett. **40**, 831 (2004)
355. H. Schneider, T. Maier, J. Fleissner, M. Walther, P. Koidl, G. Weimann, W. Cabanski, M. Finck, P. Menger, W. Rode, J. Ziegler, SPIE **5726**, 35 (2005)
356. E.F. Costard, P.F. Bois, X. Marcadet, A. Nedelcu, SPIE **5406**, 646 (2004)
357. M.A. Fauci, R. Breiter, W. Cabanski, W. Fick, R. Koch, J. Ziegler, S.D. Gunapala, Infrared Phys. Technol. **42**, 337 (2001)

358. W.A. Cabanski, R. Breiter, W. Rode, J. Ziegler, H. Schneider, M. Walther, M.A. Fauci, in *Infrared Detectors and Focal Plane Arrays VII*, ed. by E.L. Dereniak, R.E. Sampson. SPIE Proceedings Series, vol. 4721 (SPIE, Bellingham, 2002), p. 165
359. J. Sandsten, P. Weibring, H. Edner, S. Svanberg, Opt. Exp. **6**, 92 (2000)
360. C. Garbe, U. Schimpf, B. Jaehne, SPIE **4710**, 171 (2002)
361. R. Brendel, M. Bail, B. Bodmann, J. Kensch, M. Schulz, Appl. Phys. Lett. **80**, 437 (2002)
362. P. Pohl, R. Brendel, Appl. Phys. Lett. **87**, 87 (2005)
363. S.V. Bandara, S.D. Gunapala, J.K. Liu, S.B. Rafol, C.J. Hill, D.Z. Ting, J.M. Mumolo, T.Q. Trinh, Infrared Phys. Technol. **47**, 15 (2005)
364. D.W. Beekman, J. Van Anda, Infrared Phys. Technol. **42**, 323 (2001)
365. M. Serna, Infrared Phys. Technol. **44**, 457 (2003)
366. C.J. Chen, K.K. Choi, L. Rokhinson, W.H. Chang, D.C. Tsui, Appl. Phys. Lett. **74**, 862 (1999)
367. E.M. Costard, P.F. Bois, X. Mercadet, E. Herniou, Infrared Phys. Technol. **42**, 291 (2001)
368. H.C. Liu, R. Dudek, T. Oogarah, P.D. Grant, Z.R. Wasilewski, H. Schneider, S. Steinkogler, M. Walther, P. Koidl, IEEE Circ. Dev. **19**, 9 (2003)
369. H.C. Liu, G.E. Jenkins, E.R. Brown, K.A. McIntosh, K.B. Nichols, M.J. Manfra, IEEE Electron Dev. Lett. **16**, 253 (1995)
370. S. Steinkogler, H. Schneider, R. Rehm, M. Walther, P. Koidl, Appl. Phys. Lett. **81**, 3401 (2002)
371. S. Steinkogler, H. Schneider, M. Walther, P. Koidl, Appl. Phys. Lett. **82**, 3925 (2003)
372. E.R. Brown, K.A. McIntosh, U.S. Patent 5,304,805 (19 Apr 1994)
373. R.K. Richards, D.P. Hutchinson, C.A. Bennett, M.L. Simpson, H.C. Liu, M. Buchanan, in *Long Wavelength Infrared Detectors and Arrays: Physics and Applications VI*, ed. by S.S. Li, H.C. Liu, M.Z. Tidrow, S.D. Gunapala. ECS Proceedings, vol. 98-21 (1998), p. 95
374. T. Kondoh, Y. Miura, S. Lee, R.K. Richards, D.P. Hutchinson, C.A. Bennett, Rev. Sci. Instrum. **74**, 1642 (2003)
375. R.K. Richards, D.P. Hutchinson, C.A. Bennett, T. Kondoh, Y. Miura, S. Lee, Rev. Sci. Instrum. **74**, 1646 (2003)
376. G. Sonnabend, D. Wirtz, F. Schmülling, R. Schieder, Appl. Opt. **41**, 2978 (2002)
377. W.C. Danchi, C.H. Townes, W. Fitelson, D.D.S. Hale, J.D. Monnier, S. Tevosjan, J. Weiner, SPIE **4838**, 33 (2003)
378. S. Ehret, H. Schneider, C. Schönbein, G. Bihlmann, J. Fleissner, Appl. Phys. Lett. **69**, 931 (1996)
379. M. Ershov, S. Satou, Y. Ikebe, J. Appl. Phys. **86**, 6442 (1999)
380. R. Rehm, H. Schneider, M. Walther, P. Koidl, Appl. Phys. Lett. **80**, 862 (2002)
381. S. Steinkogler, H. Schneider, R. Rehm, M. Walther, P. Koidl, P. Grant, R. Dudek, H.C. Liu, Infrared Phys. Technol. **44**, 355 (2003)
382. E.R. Brown, K.A. McIntosh, F.W. Smith, M.J. Manfra, Appl. Phys. Lett. **62**, 1513 (1993)
383. E.R. Brown, K.A. McIntosh, K.B. Nichols, F.W. Smith, M.J. Manfra: CO_2-laser heterodyne detection with GaAs/AlGaAs MQW structures. In: *Quantum Well Intersubband Transition Physics and Devices*, ed. by H.C. Liu, B.F. Levine, J.Y. Andersson (Kluwer, Dordrecht 1994) pp. 207–220

384. K.J. Siemsen, H.D. Riccius, Appl. Phys. A **35**, 177 (1984)
385. T. Koizumi, H. Tashiro, K. Nagasaka, S. Namba, Jpn. J. Appl. Phys. **23**, L809 (1984)
386. P.D. Grant, R. Dudek, L. Wolfson, M. Buchanan, H.C. Liu, Electron. Lett. **41**, 69 (2005)
387. H. Schneider, T. Maier, H.C. Liu, M. Walther, P. Koidl, Opt. Lett. **30**, 287 (2005)
388. T. Maier, H. Schneider, M. Walther, P. Koidl, H.C. Liu, Appl. Phys. Lett. **84**, 5162 (2004)
389. T. Hattori, Y. Kawashima, M. Daikoku, H. Inouye, H. Nakatsuka, Jpn. J. Appl. Phys. **39**, 4793 (2000)
390. W. Nessler, S. Ogawa, H. Nagano, H. Petek, J. Shimoyama, Y. Nakayama, K. Kisho, J. Electron Spectrosc. Relat. Phenom. **88**, 495 (1998)
391. R.A. Kaindl, K. Reimann, M. Woerner, T. Elsaesser, R. Hey, K. Ploog, Phys. Rev. B **24**, 161308 (2001)
392. T. Maier, H. Schneider, H.C. Liu, M. Walther, P. Koidl, Appl. Phys. Lett. **88**, 051117 (2006)
393. J.-C.M. Diels, J.J. Fontaine, I.C. McMichael, F. Simoni, Appl. Opt. **24**, 1270 (1985)
394. M. Woerner, R.A. Kaindl, F. Eickemeyer, K. Reimann, T. Elsaesser, A.M. Weiner, R. Hey, K.H. Ploog, Physica B **314**, 244 (2002)
395. T. Maier, H. Schneider, H.C. Liu, M. Walther, P. Koidl, Infrared Phys. Technol. **47**, 182 (2005)

Index

Springer Series in

OPTICAL SCIENCES

Volume 1

1 **Solid-State Laser Engineering**
By W. Koechner, 5th revised and updated ed. 1999, 472 figs., 55 tabs., XII, 746 pages

Published titles since volume 90

91 **Optical Super Resolution**
By Z. Zalevsky, D. Mendlovic, 2004, 164 figs., XVIII, 232 pages

92 **UV-Visible Reflection Spectroscopy of Liquids**
By J.A. Räty, K.-E. Peiponen, T. Asakura, 2004, 131 figs., XII, 219 pages

93 **Fundamentals of Semiconductor Lasers**
By T. Numai, 2004, 166 figs., XII, 264 pages

94 **Photonic Crystals**
Physics, Fabrication and Applications
By K. Inoue, K. Ohtaka (Eds.), 2004, 209 figs., XV, 320 pages

95 **Ultrafast Optics IV**
Selected Contributions to the 4th International Conference
on Ultrafast Optics, Vienna, Austria
By F. Krausz, G. Korn, P. Corkum, I.A. Walmsley (Eds.), 2004, 281 figs., XIV, 506 pages

96 **Progress in Nano-Electro Optics III**
Industrial Applications and Dynamics of the Nano-Optical System
By M. Ohtsu (Ed.), 2004, 186 figs., 8 tabs., XIV, 224 pages

97 **Microoptics**
From Technology to Applications
By J. Jahns, K.-H. Brenner, 2004, 303 figs., XI, 335 pages

98 **X-Ray Optics**
High-Energy-Resolution Applications
By Y. Shvyd'ko, 2004, 181 figs., XIV, 404 pages

99 **Mono-Cycle Photonics and Optical Scanning Tunneling Microscopy**
Route to Femtosecond Ångstrom Technology
By M. Yamashita, H. Shigekawa, R. Morita (Eds.) 2005, 241 figs., XX, 393 pages

100 **Quantum Interference and Coherence**
Theory and Experiments
By Z. Ficek and S. Swain, 2005, 178 figs., XV, 418 pages

101 **Polarization Optics in Telecommunications**
By J. Damask, 2005, 110 figs., XVI, 528 pages

102 **Lidar**
Range-Resolved Optical Remote Sensing of the Atmosphere
By C. Weitkamp (Ed.), 161 figs., XX, 416 pages

103 **Optical Fiber Fusion Splicing**
By A.D. Yablon, 2005, 137 figs., XIII, 306 pages

104 **Optoelectronics of Molecules and Polymers**
By A. Moliton, 2005, 229 figs., 592 pages

105 **Solid-State Random Lasers**
By M. Noginov, 2005, 131 figs., XII, 238 pages

106 **Coherent Sources of XUV Radiation**
Soft X-Ray Lasers and High-Order Harmonic Generation
By P. Jaeglé, 2006, 332 figs., XIII, 416 pages

107 **Optical Frequency-Modulated Continuous-Wave (FMCW) Interferometry**
By J. Zheng, 2005, 137 figs., XVIII, 254 pages

Springer Series in

OPTICAL SCIENCES

108 **Laser Resonators and Beam Propagation**
Fundamentals, Advanced Concepts and Applications
By N. Hodgson and H. Weber, 2005, 587 figs., XXVI, 794 pages

109 **Progress in Nano-Electro Optics IV**
Characterization of Nano-Optical Materials and Optical Near-Field Interactions
By M. Ohtsu (Ed.), 2005, 123 figs., XIV, 206 pages

110 **Kramers–Kronig Relations in Optical Materials Research**
By V. Lucarini, J.J. Saarinen, K.-E. Peiponen, E.M. Vartiainen, 2005, 37 figs., X, 162 pages

111 **Semiconductor Lasers**
Stability, Instability and Chaos
By J. Ohtsubo, 2005, 169 figs., XII, 438 pages

112 **Photovoltaic Solar Energy Generation**
By A. Goetzberger and V.U. Hoffmann, 2005, 139 figs., XII, 234 pages

113 **Photorefractive Materials and Their Applications 1**
Basic Effects
By P. Günter and J.P. Huignard, 2006, 169 figs., XIV, 421 pages

114 **Photorefractive Materials and Their Applications 2**
Materials
By P. Günter and J.P. Huignard, 2005, 100 figs., approx. XII, 300 pages

115 **Photorefractive Materials and Their Applications 3**
Applications
By P. Günter and J.P. Huignard, 2005, 100 figs., approx. XII, 300 pages

116 **Spatial Filtering Velocimetry**
Fundamentals and Applications
By Y. Aizu and T. Asakura, 2006, 112 figs., XII, 212 pages

117 **Progress in Nano-Electro-Optics V**
Nanophotonic Fabrications, Devices, Systems, and Their Theoretical Bases
By M. Ohtsu (Ed.), 2006, 122 figs., XIV, 188 pages

118 **Mid-infrared Semiconductor Optoelectronics**
By A. Krier (Ed.), 2006, 443 figs., XVIII, 751 pages

119 **Optical Interconnects**
The Silicon Approach
By L. Pavesi and G. Guillot (Eds.), 2006, 265 figs., XXII, 389 pages

120 **Relativistic Nonlinear Electrodynamics**
Interaction of Charged Particles with Strong and Super Strong Laser Fields
By H.K. Avetissian, 2006, 23 figs., XIII, 333 pages

121 **Thermal Processes Using Attosecond Laser Pulses**
When Time Matters
By M. Kozlowski and J. Marciak-Kozlowska, 2006, approx. 46 figs., XII, 232 pages

122 **Modeling and Analysis of Transient Processes in Open Resonant Structures**
New Methods and Techniques
By Y.K. Sirenko, N.P. Yashina, and S. Ström, 2007, approx. 110 figs., XIV, 346 pages

123 **Wavelength Filters in Fibre Optics**
By H. Venghaus (Ed.), 2007, approx. 210 figs., XXIV, 451 pages

124 **Light Scattering by Systems of Particles**
Null-Field Method with Discrete Sources: Theory and Programs
By A. Doicu, T. Wriedt, and Y.A. Eremin, 2006, 123 figs., XIII, 318 pages

125 **Electromagnetic and Optical Pulse Propagation 1**
Spectral Representations in Temporally Dispersive Media
By K.E. Oughstun, 2007, approx 74 figs., 452 pages

126 **Quantum Well Infrared Photodetectors**
Physics and Applications
By H. Schneider and H.C. Liu, 2007, 153 figs., XVI, 250 pages

MIX
Papier aus verantwortungsvollen Quellen
Paper from responsible sources
FSC® C105338

If you have any concerns about our products,
you can contact us on
ProductSafety@springernature.com

In case Publisher is established outside the EU,
the EU authorized representative is:
Springer Nature Customer Service Center GmbH
Europaplatz 3, 69115 Heidelberg, Germany

Printed by Libri Plureos GmbH
in Hamburg, Germany